T0145324

Lecture Notes in Control and Information Sciences

Volume 481

This series aims to report new developments in the fields of control and information sciences—quickly, informally and at a high level. The type of material considered for publication includes:

1. Preliminary drafts of monographs and advanced textbooks
2. Lectures on a new field, or presenting a new angle on a classical field
3. Research reports
4. Reports of meetings, provided they are

   (a) of exceptional interest and
   (b) devoted to a specific topic. The timeliness of subject material is very important.

Indexed by EI-Compendex, SCOPUS, Ulrich's, MathSciNet, Current Index to Statistics, Current Mathematical Publications, Mathematical Reviews, IngentaConnect, MetaPress and Springerlink.

More information about this series at http://www.springer.com/series/642

Eli Gershon · Uri Shaked

Advances in H_∞ Control Theory

Switched, Delayed, and Biological Systems

 Springer

Eli Gershon
Department of Electrical
and Electronics Engineering
Holon Institute of Technology
Holon, Israel

Uri Shaked
School of Electrical Engineering
Tel Aviv University
Tel Aviv, Israel

ISSN 0170-8643   ISSN 1610-7411   (electronic)
Lecture Notes in Control and Information Sciences
ISBN 978-3-030-16010-4   ISBN 978-3-030-16008-1   (eBook)
https://doi.org/10.1007/978-3-030-16008-1

Library of Congress Control Number: 2019935160

This Springer imprint is published by the registered company Springer Nature Switzerland AG
The registered company address is: Gewerbestrasse 11, 6330 Cham, Switzerland

Preface

This monograph may be viewed as a natural extension of our two previous monographes: the first is entitled "H_∞ Control and Estimation of State-multiplicative Linear Systems", Lecture Notes in Control and Information Sciences, LNCIS, Springer, vol. 318, and the second one is entitled "Advanced Topics in Control and Estimation of State-multiplicative Noisy Systems", Lecture Notes in Control and Information Sciences, LNCIS, Springer, vol. 439.

In the first book, the theory of stability, control, and estimation of state-multiplicative noisy systems has been developed by applying various state-of-the-art modern analytical techniques leading to the solutions of many problems that are typically encountered in the field of system control such as state feedback, static, and dynamical output-feedback control, filtering, and preview tracking. The second book extends the various problems that have been dealt with in the first book to their time-delay counterparts, in both the continuous and the discrete-time settings. It also contains, for the first time, the control and estimation theory of state-multiplicative nonlinear systems. In its last part, the book provides a glimpse at state-multiplicative, noisy, switched continuous-time systems.

The present monograph covers three topics that contain recent state-of-the-art results. The first part of the book includes an extended treatment of switched systems with dwell time, which aims mainly at deterministic, rather than stochastic, systems. It deals with the problems of control and estimation of continuous-time and discrete-time switched systems. In its second part, the book complements issues that were not covered in the second book concerning the control of retarded stochastic state-multiplicative noisy linear systems. These issues include dynamic and static output feedback and the use of predictor control for input-delayed systems. An extended treatment of robust control and estimation for delay-free stochastic systems is also brought there. The last part of the book brings a new approach to the control of biochemical pathways, applying both classical and modern control techniques. The theory that is brought in this book is demonstrated by the threonine synthesis and the glycolytic biochemical pathways.

Similarly to our first two books, the present monograph is addressed to engineers that are engaged in control systems research and development, to graduate students specializing in switching control, stochastic linear delay-free and retarded control and to applied mathematicians that are interested in system control by large.

The reader of this book should be familiar with the theory of switching control, stochastic linear systems, state-space methods, and optimal control theory. Some knowledge of stochastic processes, by large, would be an advantage, although the monograph provides a brief background of the subject. We note that some of the stochastic tools needed to master the subject material were given in our two previous books where principle concepts are introduced and explained. In the present book, these subjects are shortly repeated in the Appendix section.

The book consists of three parts and an Appendix section.

The first part of the book considers deterministic switched systems with dwell time and it includes the issues of robust stability and stabilization in Chaps. 2–4. The problem of robust estimation with dwell time is solved in Chap. 5 and in Chap. 6 the theory of robust stability and controller synthesis is brought for discrete-time switched systems. In Chap. 7, the problem of robust dynamic output-feedback control is formulated and solved. The first part also treats, in Chap. 8, the robust switched fault-tolerant control.

The second part of this book is focused mainly on non-switched stochastic systems and it consists of Chaps. 9–14. Starting with a short introduction of the subject material in Chap. 9, Chap. 10 complements the theory for non-delayed systems, which has been treated in our first book, by introducing the latest techniques that are applied to the solution of various robust control and estimation problems. In Chap. 11, the solution of the dynamic output feedback of delay-free discrete-time stochastic system is formulated and solved. Chapters 12–14 are a direct extension of the theory brought in our second book. Chapters 12 and 13 deal with retarded systems, they include the use of prediction in the synthesis of state-feedback controllers for input-delayed systems in Chap. 12, and in Chap. 13, the problem of zero-order control is formulated and solved. In Chap. 14, the problem of robust H_∞ control of discrete-time stochastic delay-free switched systems is addressed and solved.

The third part of the book consists of Chaps. 15–17. Followed by a short introduction of the control of biochemical processes brought in Chaps. 15 and 16 treats the H_∞ control of metabolic biochemical pathways which, to the best of our knowledge, has never been treated in the literature in that sense. This chapter includes a theoretical approach for evaluating the robustness of real-life biochemical pathways (demonstrated by the threonine synthesis pathway) via the H_∞ control framework. In Chap. 17, several other optimal measures, in addition to the H_∞ measure are considered which include an energy-to-energy measure, energy-to-peak, and the peak-to-peak norm measures. The numerical computations that are brought in Chaps. 16 and 17 are demonstrated by graphical simulations that clarify the results that are obtained by the application of the latter methods.

The Appendix section at the end of the book contains some basic features of stochastic control processes and some concepts of retarded systems. In Appendix A, a self-contained description of the σ-algebra and Ito calculus, which are needed for the study of state-dependent noisy systems, is brought. A brief description of the input–output method, which has become an essential technique in the study of retarded control systems is given in Appendix B.

Few words about the numbering scheme used in the book are in order. Each chapter is divided into sections. Thus, Sect. 2.5 refers to the fifth section within Chap. 2. In each chapter, theorems, lemmas, corollaries, examples, and figures are numbered consecutively within the chapter. In addition, each chapter includes the major needed references. These references are complemented by the References section at the end of each of the introductory chapters of the first three parts of the book (i.e., Chaps. 1, 9, and 15).

Acknowledgements

The material in this book is based on the research work performed by us together with our colleagues, Dr. Liron Allerhand and Prof. Emilia Fridman of Tel Aviv University. We thank our two colleagues for their contribution and for their insightful suggestions.

Tel Aviv, Israel
January 2019

Eli Gershon
Uri Shaked

Contents

Part I
Switched Linear Systems with Dwell Time

Chapter 1
Introduction

Abstract This chapter is the introduction to Part I of the book. It provides a thorough introduction to the area of switched systems, in general, and to switched systems with dwell time in particular. Switched systems are encountered in many fields of engineering (electrical, mechanical, aerospace, biological, and medical, to name a few). In many cases, switchings are also introduced, deliberately, to improve the behavior of the control systems. In this part of the book, we concentrate on switching with dwell time. These switchings are characterized by the fact that once a switching occurred there is a minimum time interval, called dwell time during which further switching cannot occur. In this part of the book, we consider the stability, the state-feedback control design and estimation of continuous- and discrete-time switched systems with dwell time in the presence of parameters uncertainties. The theory developed is applied to cases where deliberate switchings are introduced to the control design in order to improve the performance and to overcome an excessive parameter uncertainty. At the end of Part I of the book, we treat the output-feedback control of switched systems with dwell and also show how our results can be used in Fault-tolerant Control by assuming an infinitely large dwell time.

Abbreviation

AFTC	Active Fault-Tolerant Control
AWGN	Analog White Gaussian Noise
BRL	Bounded Real Lemma
BLS	Bilinear System
BLSS	Bilinear Stochastic System
DBLS	Deterministic Bilinear System
DII	Detection and Isolation Interval
DLMI	Difference LMI (discrete-time) or Differential LMI (continuous-time)
DRE	Difference Riccati Equation
FTC	Fault-Tolerant Control
GBM	Geometrical Brownian Motion
GS	Gain Scheduling
GSC	Gain Scheduling Control

© Springer Nature Switzerland AG 2019
E. Gershon and U. Shaked, *Advances in H∞ Control Theory*,
Lecture Notes in Control and Information Sciences 481,
https://doi.org/10.1007/978-3-030-16008-1_1

IDI   Instantaneous Detection and Isolation
LF   Lyapunov Function
LKF   Lyapunov Krasovskii Functional
LMI   Linear Matrix Inequality
LPD   Lyapunov Parameter Dependent
LTI   Linear Time Invariant
LTV   Linear Time Variant
OF   Output Feedback
P2P   Peak to Peak
PFTC   Passive Fault-Tolerant Control
SC   Supervisory Control
SDN   State-Dependent Noise
SF   State Feedback
SMC   Sliding Mode Control
SNR   Signal-to-Noise Ratio
SOF   Static Output Feedback
ZOH   Zero-Order Hold

1.1   Switched Linear Systems with Dwell Time

Switched systems is a class of systems whose dynamic behavior can change in a finite set of possible dynamics. The change in the systems dynamics is instantaneous. This change in the systems dynamics is referred to as switching, and the time instant in which it occurs is referred to as a switching instant. Each element in the set of possible dynamics is referred to as a subsystem. Uncertain switched systems are treated by considering uncertainty in the subsystems. Switched systems have received considerable attention in the past two decades [1], where the main interest was the stability and the stabilization of these systems.

Switched systems are encountered in many fields including: electronic systems, mechanical systems, aerospace systems, biological systems, and medical devices, and in cases where switching is deliberately introduced for control purposes.

Switching is very common in electronic systems: Switched mode power converters, DC motors, current snubbers, and pulse generators are just some examples [2].

In mechanical systems, switching is encountered when the system is designed to have discrete dynamics, such as gear transmissions in a car. Another cause for switching there is a rapid change in the working conditions, for example, a car switching from road to dirt driving, and vice versa.

In aerospace systems, the flight is often divided into several stages with a significant difference in the flight conditions [3], for example, a rocket engine running out of fuel causes a shift in the system dynamics.

In biological systems, switching is encountered when an organism, or a biological mechanism, switches from one strategy to another. On a large scale, this can include the fight-or-flight response to danger, which results in a significant shift in the organism behavior. On a smaller scale, these models can, for example, describe the instantaneous injection of insulin into the bloodstream in response to a high sugar intake, or the release of the spring-like mechanism in flea's legs.

Switching is also encountered in medical devices, for example, when a surgical robot end-effector is at the point of transition between one tissue to another.

Controller switching is often applied to improve performance and to adjust the control to the current operating conditions of the system. See for example [4].

It is common in the literature to assume that switching can occur arbitrarily fast [1]. The realistic case is, however, that the switching law is restricted in some manner. Various constraints can be applied to the switching law. An overview of possible constraints and their influence on the performance can be found in [5]. Of particular interest is the dwell-time constraint, which requires that after switching occurs there is a minimum time period, referred to as the dwell time, in which further switching cannot occur. For example, a car should not switch from dirt to road driving every 0.01 s, a surgical device should not penetrate a tissue arbitrarily fast, and an aircraft controller will not switch back and forth in less than 0.1 s.

The stability analysis of linear-switched systems with dwell time has been studied via various perspectives, see the survey paper [6] and the references therein. The latter paper also surveys results on average dwell time, which is a generalization of the dwell-time constraint. Average dwell-time constraints have also received considerable attention in the literature.

It has been shown in [7] that even the worst case (most destabilizing) switching law does obey some dwell-time constraint, at least for second-order systems. Switching without any dwell time can generate a dynamic which is a combination of the dynamics of the subsystems. This leads to some switched systems that are unstable for switching without dwell time and stable with a (short) dwell time [8]. Arbitrarily fast switching may also cause large state transients at the switching points. A dwell time may then be required for these transients to subside. This is probably one of the reasons why the area of switched systems with dwell time is becoming increasingly popular.

A strong sufficient condition for the stability of nominal switched systems with dwell time has been presented in [9]. This condition seems to be the best in terms of Lyapunov functions in quadratic form. This condition depends, however, on the exponent of the dynamic matrix, and it was, therefore, difficult to generalize to the uncertain case.

Controller switching with dwell time for a linear time-invariant system, whose dynamics belong to some discrete set has been studied in [10–13], where conditions for stabilization have been obtained. The latter results, called Supervisory Control (SC), have been extended to time-varying systems in [14]. In the SC, a given stabilizing controller is assigned to each of the possible dynamics of the system, referred to as subsystems. The switching between the controllers is done by assigning an observer to each of the subsystems, and the switching is made to the controller that fits the observer with the least error in some sense, under dwell time and hysteresis constraints. It was shown that SC can be used to overcome some of the limitations of traditional adaptive control [15]. Introductions to SC can be found in [1, 16].

In all these works, SC is dedicated to controller switching based on an estimator bank. Until now, SC has been able to guarantee input to state stability, but has not been extended to provide performance in the H_∞ sense, or in the H_2 sense. Another challenge that arises in SC is that the set of subsystems must be discrete.

Controller switching without dwell can be applied to enhance the performance of the control in the nonlinear case. It has been shown that the transient response of a system can be improved by controller switching [17, 18], and specifically by switching with dwell time [19]. Controller switching for wind turbines with highly nonlinear dynamics has been presented in [4], where the different controllers all stabilize the system, but are aimed at achieving different goals. The controllers are switched based on the tracking error and they are designed such that the stability of the overall system is preserved for any switching between them.

Sliding Mode Control (SMC) can also be considered as a form of switching. In SMC, the control signal switches between a maximal and a minimal value. It has been applied to a variety of applications and is considered to be a highly robust form of control, see for example [20–25] and the references therein. The main advantage of SMC is that it is highly robust to uncertainties in the system parameters. The main drawbacks of SMC are, however, that it requires infinitely fast switching, also known as chattering [25], that is inapplicable to systems with right half-plane zeros, and that it does not guarantee a bound on the L_2-gain of the system. Furthermore, the separation principle utilized in SMC is sensitive to measurement noise [25]. However, significant progress has been made in this respect with the introduction of practical relative degree [26].

Another field which is directly related to switched systems is Fault-Tolerant Control (FTC). FTC treats systems whose standard operation is interrupted by a malfunction in actuators, sensors, or other components of the system [27, 28]. The standard approach for modeling such a malfunction is to consider it to be an instantaneous shift in the systems dynamics. Faults can thus be treated as switching in the system dynamics, where it is assumed that once a fault has occurred, the system does not return to its original mode of operation.

FTC is achieved either by Passive Fault-Tolerant Control (PFTC) or by Active Fault-Tolerant Control (AFTC) [27–29]. In the former, a single controller is designed that is able to stabilize the system, and achieve the required performance, regardless of the fault. This controller does not require the knowledge of the fault. AFTC, on the other hand, reconfigures the controller by detecting and isolating the fault that occurred in the system. For AFTC, it is assumed that faults can be detected with sufficient accuracy, usually as they occur [30]. Fault detection and isolation have been treated by many ([28] and the references therein [31, 32]). A variety of control methods have been applied to FTC, for example, [33–36]. SMC is often applied for PFTC [34], and supervisory control has also recently been applied to AFTC [37].

A method that treats switched systems with dwell time and parameter uncertainty has recently been developed [38]. This method can handle large polytopic parameter uncertainties and, for the first time, it applies parameter-dependent Lyapunov Functions (LF) [39] to uncertain switched systems.

Switched systems are also considered in the context of switching law synthesis. Switching law synthesis is required in power systems such as switched convertors [2], in controller switching [1], and in several other applications [9]. Synthesis of state-dependent switching synthesis for linear-switched systems has been considered by many, see for example, the survey paper [40] and the references therein. State-dependent switching for an uncertain system that is based on a common LF has been presented in [41].

The above methods have also been extended to output-dependent switching [40, 42] when the state of the system is inaccessible. These switching laws are designed by adding additional states to the system (a filter), and using these states to determine the switching law. SMC can also be applied to generate output-dependent switching law, although it encounters the same drawback described above.

In this part of the book, the following linear-switched system is considered:

$$
\begin{aligned}
\dot{x} &= A_\sigma x + B_{1,\sigma} w + B_{2,\sigma} u, \quad x(0) = 0, \\
z &= C_{1,\sigma} x + D_{11,\sigma} w + D_{12,\sigma} u, \\
y &= C_{2,\sigma} x + D_{21,\sigma} w,
\end{aligned} \tag{1.1}
$$

where $x \in \mathcal{R}^n$ is the state, $z \in \mathcal{R}^r$ is the objective vector, $u \in \mathcal{R}^p$ is the control signal, $y \in \mathcal{R}^s$ is the measurement and $w \in \mathcal{R}^q$ is an exogenous disturbance in $\mathcal{L}_2$. This system is defined for all $t \geq 0$. The switching rule $\sigma(t) \in \{1 \ldots M\}$, for each $t \geq 0$, is such that $\Omega_{\sigma(t)} \in \{\Omega_1, \ldots, \Omega_M\}$, where

$$
\Omega_{\sigma(t)} = \begin{bmatrix} A_{\sigma(t)} & B_{1,\sigma(t)} \\ C_{1,\sigma(t)} & D_{11,\sigma(t)} \\ C_{2,\sigma(t)} & D_{21,\sigma(t)} \end{bmatrix}.
$$

The polytope Ω_i is defined by

$$
\Omega_i = \sum_{j=1}^{N} \eta_j(t) \Omega_i^{(j)}, \quad \sum_{j=1}^{N} \eta_j(t) = 1, \quad \eta_j(t) \geq 0 \tag{1.2}
$$

with vertices defined by

$$
\Omega_i^{(j)} = \begin{bmatrix} A_i^{(j)} & B_{1,i}^{(j)} \\ C_{1,i}^{(j)} & D_{11,i}^{(j)} \\ C_{2,i}^{(j)} & D_{21,i}^{(j)} \end{bmatrix}, \quad i = 1, \ldots M, \tag{1.3}
$$

where $A_i^{(j)}$, $B_{1,i}^{(j)}$, $C_{1,i}^{(j)}$, $C_{2,i}^{(j)}$, $D_{11,i}^{(j)}$, $D_{12,i}^{(j)}$ are constant matrices of the appropriate dimensions. The above model naturally imposes discontinuity in $\Omega_{\sigma(t)}$ since this matrix jumps instantaneously from Ω_i to Ω_q for some $i \neq q$ at the switching instances. We note that this means that each subsystem has a separate polytopic description, which is independent of the description of the other subsystems.

The dwell-time constraint implies that the switching law must be a piecewise constant, where each interval lasts at least T sec. The latter means that if the dwell time is T, and the switching instants are $\tau_1, \tau_2, \ldots$, then $\tau_{h+1} - \tau_h \geq T$, $\forall h \geq 1$.

In this part of the book, bounds on the performance of this system are derived. Some derivations in the book consider the switching law to be arbitrary, meaning that the performance is guaranteed for any switching law that obeys the dwell-time constraint. Other derivations consider the case where the switching law is chosen by the designer in order to guarantee some prescribed performance requirements. Two measures of performance will be presented. The first is the L_2-gain of the system, and the second is the variance of the output when w is a standard white-noise vector. For a comprehensive review of the meaning, importance and applications of L_2-gain, see [43, 44].

The main purpose of this part of the book is the development of robust control through the application of switching methodologies. Robust control, in the classic sense, assumes that the system's parameters are unknown, and the controller does not attempt to estimate them in any way. Instead, the controller is designed such that it guarantees some upper-bound on the performance of the system. In fact, one of the main reasons for applying feedback control is to lower the sensitivity of the system to parameter uncertainty [45]. This form of robustness is crucial even if the controller does attempt to estimate the parameters, since some uncertainty cannot be avoided. In some applications, the systems parameters can be easily measured with good accuracy, which allows the controller to be parameter dependent. This form of parameter-dependent control is referred to as Gain Scheduling (GS) [46]. GS is particularly useful in aircraft control, since the aircrafts dynamics is highly dependent on its Mach number as well as on its altitude, two variables that are easy to measure in real-time. GS is also useful for controlling motor vehicles such as cars and motorcycles, since their dynamics are highly dependent on their velocity, which is of course measured in real-time.

In order to improve the robust control design, switching is applied in this book in a form of gain scheduling to systems, whose parameters can be measured in real-time, or even partially measured. This is achieved by dividing the uncertainty region into several overlapping subregions. Once the uncertainty region has been divided into overlapping subregions, the system can be treated as a switched system where each of the subregions is a subsystem. It is assumed that the system's parameters cannot change arbitrarily fast, and therefore the overlap between neighboring regions guarantees some minimal dwell time between consecutive switching. It will be shown (in Sects. 4.2 and 4.6) that dividing the uncertainty region and treating the system as a switched system can lead to considerable improvement in performance, while requiring considerably less information about the system's parameters, compared to existing GS methodologies.

In the case where the system's parameters are not accessible, one must apply a form of SC in order to apply switching to robust control. While achieving SC for systems with continuous uncertainty regions is still an open problem, some progress in this direction is presented in this book. State and output switching laws are designed that constitute an essential part of SC. Filtering for switched systems is also considered, as filters are utilized in the design of the controller switching. Furthermore, in order to design SC, the designer must take into account the system models, controller design, filter design and controller switching laws synthesis. While this may be possible for some systems, it is unlikely that this problem can be fully considered for a high-order system with many uncertain parameters, due to the sheer size of the problem. It may, therefore, be beneficial to consider filter synthesis as a separate problem and incorporate these filters into the design at a later stage.

Switched systems are considered in the following 7 chapters.

Chapter 2, which is partially based on [38], considers the stability and the sta-bilizability of switched systems with dwell and with large parameter uncertainty. Sufficient conditions are given for the stability of these uncertain systems and to the stabilization of these systems by applying state feedback.

Chapter 3 applies the methodologies of the previous chapter to switching law synthesis, based on [47]. A state-dependent switching law, which is based on the local minimization of the Lyapunov function is used. A bound on the L_2-gain of this switching law, with a dwell-time constraint, is introduced for the first time.

Chapter 4 introduces a bounding sequence that leads to an efficient bound on the L_2-gain of switched systems, and it applies this bound to controller synthesis. This method, which is partially based on [47], is then used to design robust control for uncertain non-switched systems. It is shown that this method leads to improved performance with less information compared to standard GS.

Chapter 5 considers stochastic-switched systems with additive noise and is par-tially based on [48]. A bound on the variance of the objective vector is derived, and robust filters are found which are based on the latter bound. These filters are also compared to the Kalman filter and the advantages and disadvantages of these filters are discussed.

Chapter 6 generalizes the results of Chaps. 2 and 3 to the discrete-time case, based on [49]. It introduces a bound on the ℓ_2 of switched systems by using time-varying and switching dependent Lyapunov functions. Since modern control algo-rithms are implemented through digital controllers it was important to generalize the continuous-time results to the discrete-time case.

Chapter 7, which is partially based on [50], considers output-dependent switching. This kind of switching, which does not require the full knowledge of the states, is most important since, in practice, only the output of a system is accessible.

In Chap. 8, which is partially based on [51], fault-tolerant control is considered as a special case of switched system control. It is shown that by modeling faults as switching, the performance of switched systems can be significantly improved. Active FTC with delay in the detection and isolation of the faults is also considered, and a bound on the performance of the closed-loop is introduced for the first time. This is also important in the context of SC, since detecting a change in the system's parameters is never instantaneous.

Each of the abovementioned chapters has its own introduction and reference sections.

References

1. Liberzon, D.: Switching in Systems and Control. Birkhauser, Boston (2003)
2. Erickson, R.: Fundamentals of Power Electronics. Springer, New York (1997)
3. Shefer, M., Breakwell, J.V.: Estimation and Control with Cubic Nonlinearities. J. Opt. Theory Appl. **53**, 1–7 (1987)
4. Morse, A.S.: Beyond the linear limitations by combining switching and QFT: application to wind turbines pitch control systems. Int. J. Robust Nonlinear Control **19**, 40–58 (2009)
5. Hespanha, J.P.: Uniform stability of switched linear systems: extensions of LaSalle's invariance principle. IEEE Trans. Autom. Control **49**, 470–482 (2004)
6. Colaneri, P.: Dwell time analysis of deterministic and stochastic switched systems. Eur. J. Autom. Control **15**, 228–249 (2009)
7. Margaliot, M., Langholz, G.: Necessary and sufficient conditions for absolute stability: the case of second-order systems. IEEE Trans. Circuits Syst. Fundam. Theory Appl. **50**, 227–234 (2003)
8. Sun, Z., Ge, S.S.: Stability Theory of Switched Dynamical Systems. Springer, New York (2011)
9. Geromel, J., Colaneri, P.: Stability and stabilization of continuous-time switched linear systems. SIAM J. Control Optim. **45**(5), 1915–1930 (2006)
10. Morse, A.S.: Supervisory control of families of linear set-point controllers, part 1: exact matching. IEEE Trans. Autom. Control **41**, 1413–1431 (1996)
11. Morse, A.S.: Supervisory control of families of linear set-point controllers, part 2: robustness. IEEE Trans. Autom. Control **42**, 1500–1515 (1997)
12. Sun, Z., Ge, S.S.: Hysteresis-based switching algorithms for supervisory control of uncertain systems. Automatica **39**, 263–272 (2003)
13. De Persis, C., De Santis, R., Morse, A.S.: Supervisory control with state-dependent dwell-time logic and constraints. Automatica **40**(2), 269–275 (2004)
14. Liberzon, D., Vu, L.: Supervisory control of uncertain linear time-varying systems. IEEE Trans. Autom. Control **56**, 27–42 (2011)
15. Hespanha, J.P., Liberzon, D., Morse, A.S.: Overcoming the limitations of adaptive control by means of logic-based switching. Syst. Control Lett. **49**, 49–65 (2003)
16. Hespanha, J.: Tutorial on supervisory control. In: Proceedings of the 40th CDC01, Orlando, Florida (2001)
17. Feuer, A., Goodwin, G.C., Salgado, M.: Potential benefits of hybrid control for linear time invariant plants. Proc. ACC 2790–2794 (1997)
18. McClamroch, N.H., Kolmanovsky, I.: Performance benefits of hybrid control design for linear and nonlinear systems. Proc. IEEE **88**, 1083–1096 (2000)
19. Ishii, H., Francis, B.A.: Stabilizing a linear system by switching control with dwell time. IEEE Trans. Autom. Control **47**, 1962–1973 (2002)
20. Kunusch, C., Pulestonb, P.F., Mayoskyd, M.A., Fridman, L.: Experimental results applying second order sliding mode control to a PEM fuel cell based system. Control Eng. Pract. **21**(5), 719–726 (2013)
21. Giral, R., Martinez-Salamero, L., Leyva, R., Maixe, J.: Sliding-mode control of interleaved boost converters. IEEE Trans. Circuits Syst. I Fundam. Theory Appl. **47**(9), 1330–1339 (2000)
22. Sira-Ramreza, H., Luviano-Jurezb, A., Corts-Romero, J.: Robust input-output sliding mode control. IEEE Trans. Autom. Control **21**(5), 671–678 (2013)
23. Colbia-Vega, A., de Leo'n-Moralesa, J., Fridmanb, L., Salas-Penaa, O., Mata-Jime'neza, M.T.: Robust excitation control design using sliding-mode technique for multimachine power systems. Electr. Power Syst. Res. **78**, 1627–1634 (2008)
24. Imine, H., Fridman, L., Shraim, H., Djemai, M.: Sliding Mode Based Analysis and Identification of Vehicle Dynamics. Lecture Notes in Control and Information Sciences, vol. 414. Springer, Berlin (2011)
25. Shtessel, Y., Edwards, C., Friedman, L., Levant, A.: Sliding Mode Control and Observation. Birkhauser, Boston (2013)

26. Levant, A.: Ultimate robustness of homogeneous sliding modes. In: 2010 11th International Workshop on Proceedings of Variable Structure Systems (VSS), pp. 26–31 (2010)
27. Patton, R.J.: Fault-tolerant control: the 1997 situation. In: Proceedings of the 3rd IFAC Symposium on Fault Detection, Supervision and Safety for Technical Processes (1997)
28. Zhang, Y., Jiang, J.: Bibliographical review on reconfigurable fault-tolerant control systems. Annu. Rev. Control **32**(2), 229–252 (2008)
29. Kanev, S.: Robust fault tolerant control. Ph.D. thesis, University of Twente, The Netherlands (2004)
30. Zolghadri, A.: The challange of advanced model-based FDIR techniques for aerospace systems: the 2011 situations. Prog. Flight Dyn. Guid. Navig. Control Fault Detect. Avion. **6**(12), 231–248 (2013)
31. Chen, R.H., Speyer, J.L.: Robust multiple-fault detection filter. Int. J. Robust Nonlinear Control **12**(8), 675–696 (2002)
32. Chen, R.H., Mingori, D.L., Speyer, J.L.: Optimal stochastic fault detection filter. Automatica **39**(3), 377–390 (2003)
33. Blanke, M., Kinnaert, M., Lunze, J., Staroswiecki, M.: Diagnosis and Fault-Tolerant Control. Springer (2010)
34. Sami, M., Patton, R.J.: Fault tolerant adaptive sliding mode controller for wind turbine power maximisation. In: Proceedings of the 7th IFAC Symposium on Robust Control Design (ROCOND12), Aalborg, Denmark (2012)
35. Rosa, P.A.N., Casau, P., Silvestre, C., Tabatabaeipour, S., Stoustrup, J.: A set-valued approach to FDI and FTC: theory and implementation issues. In: Proceedings of the 8th IFAC Symposium on Fault Detection. Supervision and Safety of Technical Processes, Mexico City, Mexico (2012)
36. Zhang, Y., Jiang, J.: Integrated active fault-tolerant control using IMM approach. IEEE Trans. Aerosp. Elect. Sys. **37**(4), 1221–1235 (2001)
37. Yang, H., Jiang, B., Cocquempot, V.: Supervisory fault tolerant control design via switched system approach. In: Proceedings of the 1st Control and Fault-Tolerant Systems (SYSTOL10), Nice, France (2010)
38. Allerhand, L.I., Shaked, U.: Robust stability and stabilization of linear switched systems with dwell time. IEEE Trans. Autom. Control **56**, 381–386 (2011)
39. de Oliveira, M.C., Skelton, R.E.: em Stability Test for Constrained Linear Systems, Perspectives in Robust Control. In: Reza Moheimani, S.O. (ed.) Lecture Notes in Control and Information Sciences 268, Springer, London (2001)
40. Lin, H., Antsaklis, P.J.: Stability and stabilizability of switched linear systems: a survey of recent result. IEEE Trans. Autom. Control **54**(2), 308–322 (2009)
41. Zhai, G., Lin, H., Antsaklis, P.J.: Quadratic stabilizability of switched linear systems. Int. J. Control **77**(6), 598–605 (2004)
42. Geromel, J., Colaneri, P., Bolzern, P.: Dynamic output feedback control of switched linear systems. IEEE Trans. Autom. Control **53**(3), 720–733 (2008)
43. Scherer, C.: Theory of Robust Control. Lecture Note, Mechanical Engineering Systems and Control Group, Delft University of Technology, The Netherlands (2001)
44. Green, M., Limebeer, D.J.N.: Linear Robust Control. Prentice Hall (1995)
45. Horowitz, I.: Synthesis of Feedback Systems. Academic Press, New York (1963)
46. Apkarian, P., Gahinet, P.: A convex characterization of gain-scheduled H_∞ controllers. IEEE Trans. Autom. Control **40**(5), 853–864 (1995)
47. Allerhand, L.I., Shaked, U.: Robust state-dependent switching of linear systems with dwell time. IEEE Trans. Autom. Control **58**(4), 994–1001 (2012)
48. Allerhand, L.I., Shaked, U.: Robust filtering of linear systems via switching. Int. J. Control **86**(11), 2067–2074 (2013)
49. Allerhand, L.I., Shaked, U.: Robust control of linear systems via switching. IEEE Trans. Autom. Control **58**, 506–512 (2012)
50. Allerhand, L., Shaked, U.: Robust output-dependent switching of linear systems with dwell time. Technical Report 4/2018, Faculty of Engineering, Tel Aviv University
51. Allerhand, L., Shaked, U.: Robust switching-based fault tolerant control. IEEE Trans. Autom. Control **60**(8), 2272–2276 (2015)

Chapter 2
Robust Stability and Stabilization of Switched Systems with Dwell Time

Abstract Sufficient conditions for the stability of linear-switched systems with dwell-time and polytopic-type parameter uncertainties are presented. A Lyapunov function, in quadratic form, which is nonincreasing at the switching instants is assigned to each subsystem. During the dwell time, this function varies piecewise linearly in time after switching occurs and it becomes time invariant afterward. This function leads to asymptotic stability conditions for the nominal set of the subsystems that can be readily extended to the case where these subsystems suffer from polytopic-type parameter uncertainties. The method proposed is then applied to stabilization via state feedback, both for the nominal and the uncertain cases.

2.1 Introduction

We address the problem of stability of linear-switched systems with dwell-time and polytopic-type parameter uncertainties. The best stability result attained so far, that is both efficiently computable and yields a sufficient result, is the one of [1] (it can also be found in [2]). This result is based on a Lyapunov function of quadratic structure. It is obtained for the case without uncertainty (namely for a nominal system), and it is derived using Linear Matrix Inequalities (LMIs). The LMIs that are used in [1] depend on the exponent of the dynamics matrix of the subsystem, making it difficult to generalize to the uncertain case.

In [3], a piecewise linear in time Lyapunov function, in quadratic form, is used. We apply in this section such a function to switched systems with dwell time. The Lyapunov function applied is nonincreasing at the switching instants and is assigned to each subsystem. During the dwell time, this function varies piecewise linearly in time, after switching occurs, and it becomes time invariant afterward. Our choice of the Lyapunov function allows derivation of sufficient conditions for robust stability of the switched system in terms of LMIs. These conditions are less conservative than those obtained by bounding the matrix exponent of [1] by a scalar matrix.

The stability and stabilization problems are formulated in Sect. 2.2, where the previous result of [1] is described. A possible conservative way to extend the result of [1] to the uncertain case is also given there. In Sect. 2.3, the solution to the stability

© Springer Nature Switzerland AG 2019
E. Gershon and U. Shaked, *Advances in H∞ Control Theory*,
Lecture Notes in Control and Information Sciences 481,
https://doi.org/10.1007/978-3-030-16008-1_2

problem is presented, first for the nominal case and then to the case with uncertainty. The issue of stabilization is addressed in Sect. 2.4, where the cases of known and unknown switching laws are both treated via a Lyapunov function of a quadratic form. The results are extended there also to vertex dependent Lyapunov functions.

Notation: Throughout the chapter, the superscript $'$ stands for matrix transposition, $\mathcal{R}^n$ denotes the n-dimensional Euclidean space, and $\mathcal{R}^{n \times m}$ is the set of all $n \times m$ real matrices. For a symmetric $P \in \mathcal{R}^{n \times n}$, $P > 0$ means that it is positive definite. Matrices with eigenvalues in the open left half of the complex plane are referred to as stability matrices. Switching law of a switched system is denoted by $\sigma(t)$. In Sect. 2.5 few generalisations to the theory are presented and in Sect. 2.6 three demonstrative examples are brought.

2.2   Problem Formulation

We consider the following linear-switched system:

$$\dot{x}(t) = A_{\sigma(t)}x(t), \quad x(0) = x_0, \tag{2.1}$$

which is defined for all $t \geq 0$, where $x \in \mathcal{R}^n$ is the state. The switching rule $\sigma(t)$, for each $t \geq 0$, is such that $A_{\sigma(t)} \in \{A_1, \ldots, A_M\}$, where $A_i \in \mathcal{R}^{n \times n}, i = 1, \ldots M$, is a stability matrix which is assumed to reside within the following polytope:

$$\Omega_i = \{A_i | A_i = \sum_{j=1}^{N} \eta^j(t)A_i^{(j)}, \quad \sum_{j=1}^{N} \eta^j(t) = 1, \quad \eta^j(t) \geq 0\}. \tag{2.2}$$

The above model naturally imposes discontinuity in $A_{\sigma(t)}$ since this matrix jumps instantaneously from A_{i_1} to A_{i_2} for some $i_1 \neq i_2$ at the switching times.
In the stabilization setting, we consider the following linear-switched system:

$$\dot{x}(t) = A_{\sigma(t)}x(t) + B_{\sigma(t)}u(t), \quad x(0) = x_0, \tag{2.3}$$

where x and σ are defined above, $u \in \mathcal{R}^l$ is the control signal, and A_i and B_i are assumed to reside in the polytope:

$$\bar{\Omega}_i = \{[A_i, B_i] | A_i = \sum_{j=1}^{N} \eta^j(t)A_i^{(j)}, B_i = \sum_{j=1}^{N} \eta^j(t)B_i^{(j)},$$

$$\sum_{j=1}^{N} \eta^j(t) = 1, \quad \eta^j(t) \geq 0\}. \tag{2.4}$$

Note that A_i are no longer required to be stability matrices.

The focus of this chapter is on the stability under a constraint on the rate of allowed commutations (dwell-time analysis), which means that if the dwell time is T, and the switching instants are $\tau_1, \tau_2, \ldots$, then $\tau_{h+1} - \tau_h \geq T$, $\forall h \geq 1$. In [1] a simple sufficient condition for the stability of the nominal system is introduced.

Lemma 2.1 ([1]) *Given that for some positive scalar T there exists a collection of symmetric matrices $P_1, P_2, \ldots P_M$ of compatible dimensions that satisfy the following:*

$$P_m > 0 \quad P_m A_m + A'_m P_m < 0, \quad e^{A'_m T} P_q e^{A_m T} - P_m < 0, \quad \forall m = 1, \ldots M, \qquad (2.5)$$
$$q \neq m = 1, \ldots M.$$

Then, the system is globally asymptotically stable for dwell time greater than or equal to T.

In terms of Lyapunov functions with quadratic structure, this result seems to be the best possible. However, the LMIs in (2.5) depend on $e^{A_m T}$, which is not convex in A_m. This method cannot, therefore, be generalized to the uncertain case in a simple manner.

A possible conservative way to generalize the latter results to the polytopic-type uncertain case is to apply a bound on the exponential as inferred from [4].

Lemma 2.2 *Assume that for some positive scalars $T, \lambda_1, \ldots \lambda_M$ there exists a collection of symmetric matrices $P_1, P_2, \ldots P_M$ of compatible dimensions such that*

$$P_m > 0, \quad P_m A_m^{(j)} + A_m^{(j)'} P_m + \lambda_m P_m < 0,$$
$$e^{-\lambda_q T} P_q - P_m < 0, \quad \forall m = 1, \ldots M, \quad q \neq m = 1, \ldots M \quad \forall j = 1, \ldots N. \qquad (2.6)$$

Then, the system is globally asymptotically stable for dwell time greater than or equal to T.

Proof Condition (2.5c) stems from the requirement that switching from the mth subsystem to the qth one, the value of the Lyapunov function $V = x'(t) P_{\sigma(t)} x(t)$ just after switching is less than its value T seconds after the mth subsystem became active. An alternative way to guarantee the decrease of V is to require that the value of V, T seconds after the switching, is less than the value it has just prior to the switching. The latter requirement is satisfied if $e^{A'_q T} P_q e^{A_q T} - P_m < 0$. It follows, however, from [4] that (2.6b) implies that $e^{A'_q T} P_q e^{A_q T} < e^{-\lambda_q T} P_q$. The conditions of (2.6a–c) thus assure the latter drop in the value of V and consequently the asymptotic stability of the switched system (2.1). ∎

2.3 Time-Switching Stability

We introduce an alternative Lyapunov function which, although is more conservative, in the nominal case, than the one of [1], it provides an efficient way to deal with uncertainties. In order to derive the Lyapunov function, the following result, inspired by [3], is applied.

Lemma 2.3 *Assume that for some time interval $t \in [t_0, t_f]$, where $\delta = t_f - t_0$, there exist two symmetric matrices P_1 and P_2 of compatible dimensions that satisfy the following:*

$$P_1, P_2 > 0, \quad \frac{P_2 - P_1}{\delta} + P_1 A + A' P_1 < 0, \quad \frac{P_2 - P_1}{\delta} + P_2 A + A' P_2 < 0. \quad (2.7)$$

Then, for the system $\dot{x} = Ax$ the Lyapunov function

$$V(t) = x'(t) P(t) x(t), \quad \text{with } P(t) = P_1 + (P_2 - P_1) \frac{t - t_0}{\delta}$$

is strictly decreasing over the time interval $t \in [t_0, t_f]$.

Proof Differentiating $V(t)$ we get: $\dot{V}(t) = x'(\dot{P} + PA + A'P)x = x'(\frac{P_2 - P_1}{\delta} + PA + A'P)x$. Note that $\frac{P_2 - P_1}{\delta} + PA + A'P = \lambda_1 [\frac{P_2 - P_1}{\delta} + P_1 A + A' P_1] + \lambda_2 [\frac{P_2 - P_1}{\delta} + P_2 A + A' P_2]$, where

$$\lambda_1 = 1 - \frac{t - t_0}{\delta}, \quad \lambda_2 = \frac{t - t_0}{\delta}. \quad (2.8)$$

It thus follows from (2.7) that $\dot{V}(t) < 0$ in the time interval specified above. ∎

While the above proof is for a nominal system, the extension to the polytopic uncertainty case is immediate, choosing the same P_1 and P_2 for all the vertices of the uncertainty polytope.

We next present sufficient conditions for the stability of a nominal linear-switched systems. These conditions are more conservative than those presented in [1]. They are given however in terms of LMIs which are affine in the systems matrices, and they can thus be easily extended to the polytopic uncertainty case.

Theorem 2.1 *The nominal system (2.1) is globally asymptotically stable for any switching law with dwell time greater than or equal to $0 < T$ if there exist: a collection of symmetric matrices $P_{i,k}$, $i = 1, \ldots M$, $k = 0, \ldots K$ of compatible dimensions, where K is an integer that may be chosen a priori, according to the allowed computational complexity, and a sequence $\{\delta_k > 0, \quad k = 1, \ldots K, \quad \sum_{k=1}^{K} \delta_k = T\}$ such that, for all $i = 1, \ldots M$ the following holds.*

$$P_{i,k} > 0, \quad \frac{P_{i,k+1} - P_{i,k}}{\delta_{k+1}} + P_{i,k} A_i + A_i' P_{i,k} < 0,$$
$$\frac{P_{i,k+1} - P_{i,k}}{\delta_{k+1}} + P_{i,k+1} A_i + A_i' P_{i,k+1} < 0, \quad k = 0, \ldots K - 1 \quad (2.9)$$

$$P_{i,K} A_i + A_i' P_{i,K} < 0, \quad P_{i,K} - P_{l,0} \geq 0, \quad \forall l = 1, \ldots i - 1, i + 1, \ldots M. \quad (2.10)$$

Proof Let $i_0 = \sigma(0)$, and let $\tau_1, \tau_2, \ldots$ be the switching instants, where $\tau_{h+1} - \tau_h \geq T$, $\forall h = 1, 2, \ldots$. Define $\tau_{h,k} = \tau_h + \sum_{j=1}^{k} \delta_j$ for $k \geq 1$, and $\tau_{h,0} = \tau_h$. Note that the dwell-time constraint implies $\tau_{h,K} \leq \tau_{h+1,0} = \tau_{h+1}$. Choose the Lyapunov function $V(t) = x'(t) P(t) x(t)$, where $P(t)$ is chosen to be

$$P(t) = \begin{cases} P_{i,k} + (P_{i,k+1} - P_{i,k})\frac{t - \tau_{h,k}}{\delta_{k+1}} & t \in [\tau_{h,k}, \tau_{h,k+1}) \\ P_{i,K} & t \in [\tau_{h,K}, \tau_{h+1,0}), \quad h = 1, 2, \ldots \\ P_{i_0,K} & t \in [0, \tau_1) \end{cases} \quad (2.11)$$

and where i is the index of the subsystem that is active at time t.

Assume that at some switching instant τ_h the system switches from the ith subsystem to the lth subsystem. At the switching instant τ_h, we have

$$V(\tau_h^-) = x(\tau_h)'P_{i,K}x(\tau_h), \quad V(\tau_h) = x(\tau_h)'P_{l,0}x(\tau_h).$$

Therefore, for the Lyapunov function to be nonincreasing at the switching instants for any $x(\tau_h)$ we demand $P_{i,K} \geq P_{l,0}$ which is the condition (2.18b). After the dwell time, and before the next switching instant, we have that $V(t) = x(t)'P_{i,K}x(t)$, where $\dot{x}(t) = A_i x(t)$. Therefore, $\dot{V}(t) = x(t)'[P_{i,K}A_i + A_i'P_{i,K}]x(t)$ and (2.10a) guarantees that this expression is negative for any $x(t) \neq 0$. During the dwell time, we consider the time intervals $t \in [\tau_{h,k}, \tau_{h,k+1})$ where $\delta_{k+1} = \tau_{h,k+1} - \tau_{h,k}$. The matrix $P(t)$ changes then linearly from $P_{i,k}$ to $P_{i,k+1}$, and conditions (2.9a–c) guarantee that $V(t)$ is decreasing during this time interval, according to Lemma 2.3. ∎

Note that the LMIs in Theorem 2.1 are affine in the system matrices. Therefore, if the subsystems entail polytopic uncertainty, Theorem 2.1 should provide a solution to the uncertain system if the conditions hold at the vertices of all the subsystems. Robust stability in the case of uncertain subsystems with polytopic uncertainties is thus given by the following.

Corollary 2.1 *Consider the system (2.1a, b) and (2.2). Assume that for some dwell time $T > 0$ there exist a collection of symmetric matrices $P_{i,k}, i = 1, \ldots M, k = 0, \ldots K$ of compatible dimensions, where K is a pre-chosen integer, and a sequence $\{\delta_k > 0, \quad k = 1, \ldots K, \quad \sum_{k=1}^{K} \delta_k = T\}$ such that, for all $i = 1, \ldots M$, and $j = 1, \ldots N$ the following holds:*

$$P_{i,k} > 0, \quad \frac{P_{i,k+1} - P_{i,k}}{\delta_{k+1}} + P_{i,k}A_i^{(j)} + A_i^{(j)'}P_{i,k} < 0,$$

$$\frac{P_{i,k+1} - P_{i,k}}{\delta_{k+1}} + P_{i,k+1}A_i^{(j)} + A_i^{(j)'}P_{i,k+1} < 0, \quad k = 0, \ldots K - 1 \quad (2.12)$$

$$P_{i,K}A_i^{(j)} + A_i^{(j)'}P_{i,K} < 0, \quad P_{i,K} - P_{l,0} \geq 0, \quad \forall l = 1, \ldots i - 1, i + 1, \ldots M. \quad (2.13)$$

Then, the system (2.1) is globally asymptotically stable for any switching law with dwell time greater than or equal to T.

Remark 2.1 The result of Theorem 2.1 is more conservative than the one obtained in [1]. It is, however, applicable to the uncertain case (as in Corollary 2.1) and to stabilization via state feedback (see next section). Its conservatism decreases, however, the larger K becomes. It is shown in Example 2.1 below that for large enough K the result of Theorem 2.1 recovers the minimum dwell time achieved by the method of [1].

Remark 2.2 We note that if Theorem 2.1 (or Corollary 2.1) can be used to prove the stability for some dwell time $T > 0$, then it is possible to use it to prove stability for and dwell time $\bar{T} \geq T$. This is achieved by applying the same Lyapunov function used for T.

2.4 Stabilization

In this section, we introduce a way to generalize the results of the previous section in order to stabilize the system (2.3). To help achieve that goal, we introduce the following result.

Corollary 2.2 *Assume that for some dwell time $T > 0$ there exist a collection of symmetric matrices $Q_{i,k}$, $i = 1, \ldots M$, $k = 0, \ldots K$ of compatible dimensions, where K is a pre-chosen integer, and a sequence $\{\delta_k > 0, \quad k = 1, \ldots K, \quad \sum_{k=1}^{K} \delta_k = T\}$ such that, for all $i = 1, \ldots M$, and $j = 1, \ldots N$ the following holds:*

$$Q_{i,k} > 0, \quad -\tfrac{Q_{i,k+1}-Q_{i,k}}{\delta_{k+1}} + A_i^{(j)} Q_{i,k} + Q_{i,k} A_i^{(j)'} < 0,$$
$$-\tfrac{Q_{i,k+1}-Q_{i,k}}{\delta_{k+1}} + A_i^{(j)} Q_{i,k+1} + Q_{i,k+1} A_i^{(j)'} < 0, \quad k = 0, \ldots K - 1 \tag{2.14}$$

$$A_i^{(j)} Q_{i,K} + Q_{i,K} A_i^{(j)'} < 0, \quad -Q_{i,K} + Q_{l,0} \geq 0, \quad \forall l = 1, \ldots i - 1, i + 1, \ldots M. \tag{2.15}$$

Then, the system (2.1) is globally asymptotically stable for any switching law with dwell time greater than or equal to T.

Proof Let $Q_{i,k} = P_{i,k}^{-1}$. Multiplying the inequality $\dot{P} + PA + A'P < 0$ by $Q = P^{-1}$, from both sides, we obtain $-\dot{Q} + AQ + QA' < 0$. Taking $Q(t)$ to be piecewise linear, as $P(t)$ has been taken above, we obtain that (2.12a–c) and (2.13a) are equivalent to (2.14) and (2.15a), respectively. Since $Q_{i,k} = P_{i,k}^{-1}$, (2.13b) is equivalent to $P_{i,K} \geq P_{l,1}$, (2.15b) readily follows and Corollary 2.2 is thus equivalent to Corollary 2.1. ∎

In this chapter, we restrict our discussion to controllers of the form

$$u(t) = G(t)x(t). \tag{2.16}$$

Remark 2.3 The system (2.3) with the controller (2.16) is equivalent to the one of (2.1), where $A_{\sigma(t)}$ is replaced by $A_{\sigma(t)} + B_{\sigma(t)}G(t)$. Therefore, if $G(t)$ is taken to be constant, denoting $A_{cl,i}^{(j)} = A_i^{(j)} + B_i^{(j)}G$, and replacing, in Corollary 2.2, $A_i^{(j)}$ with $A_{cl,i}^{(j)}$, a set of Bilinear Matrix Inequalities (BMIs) is obtained which provides a sufficient condition for the stabilization of the system (2.3). These conditions are obtained via BMIs because the same state-feedback gain should be applied to all the subsystems.

Remark 2.4 BMIs can be either solved directly, by using a suitable solver such as PENBMI [5], or by applying local iterations, where at each step an LMI is solved.

In the latter case, *SDPT*3, [6], which is included in the CVX toolbox for convex optimization [7, 8], may be used if the inequalities in Theorem 2.2 are of high order. Both PENBMI and *SDPT*3 are conveniently accessed via the YALMIP Toolbox [9].

If it is possible to use a time-varying controller, and it is known which subsystem is active in real-time, it is possible to obtain a controller in terms of LMIs. As mentioned above, $A_i^{(j)}$ is replaced by $A_i^{(j)} + B_i^{(j)} G(t)$. Defining $Y(t) = G(t)Q(t)$, and taking $Y(t)$ to be piecewise linear, where it is linear over the same intervals as $Q(t)$, then the following sufficient conditions are obtained for robust stabilization in the case of uncertain subsystems with polytopic uncertainties and a controller $u(t) = G(t)x(t)$.

Theorem 2.2 *Assume that it is known online which subsystem is active at any time instant and that for some dwell time $T > 0$ there exist: a collection of matrices $Y_{i,k}, Q_{i,k} = Q'_{i,k}, i = 1, \ldots M, k = 0, \ldots K$ of compatible dimensions, where K is a pre-chosen integer, and a sequence $\{\delta_k > 0, \quad k = 1, \ldots K, \quad \sum_{k=1}^{K} \delta_k = T\}$ such that, for all $i = 1, \ldots M$, and $j = 1, \ldots N$ the following holds.*

$$Q_{i,k} > 0, \quad -\frac{Q_{i,k+1}-Q_{i,k}}{\delta_{k+1}} + A_i^{(j)} Q_{i,k} + B_i^{(j)} Y_{i,k} + Q_{i,k}A_i^{(j)'} + Y'_{i,k}B_i^{(j)'} < 0,$$
$$-\frac{Q_{i,k+1}-Q_{i,k}}{\delta_{k+1}} + A_i^{(j)} Q_{i,k+1} + B_i^{(j)} Y_{i,k+1} + Q_{i,k+1}A_i^{(j)'} + Y'_{i,k+1}B_i^{(j)'} < 0, \qquad (2.17)$$
$$k = 0, \ldots K - 1$$

$$A_i^{(j)} Q_{i,K} + B_i^{(j)} Y_{i,K} + Q_{i,K}A_i^{(j)'} + Y'_{i,K}B_i^{(j)'} < 0,$$
$$-Q_{i,K} + Q_{l,0} \geq 0, \quad \forall l = 1, \ldots i-1, i+1, \ldots M. \qquad (2.18)$$

Then, the system (2.1) is globally asymptotically stabilized by the controller $u(t) = G(t)u(t)$ for any switching law with dwell time greater than or equal to T, where

$$G(t) = (\lambda_1(t)Y_{i,k} + \lambda_2(t)Y_{i,k+1})[\lambda_1(t)Q_{i,k} + \lambda_2(t)Q_{i,k+1}]^{-1},$$

and where λ_i, $i = 1, 2$ are defined in (2.8) with $t_o = \tau_{h,k}$ and $\delta = \delta_{k+1}$.

Note that for this controller $G(t)$ changes during the dwell time after switching, and then remains constant until the next switching occurs. However, $G(t)$ does not change piecewise linearly in time, since $G(t) = Y(t)Q^{-1}(t)$. Also, $G(t)$ is discontinuous at the switching instants.

Remark 2.5 Enforcing P to be the same for all the subsystems, and therefore constant, and taking G_i to be constant between the switching instants, the result of Theorem 2.2 yields a standard Gain-Scheduled (GS) controller [10]. Dissimilar to the gain-scheduled controller, the time-varying controller of Theorem 2.2 does not attempt to guarantee the stability for any convex combination of the subsystems and it is thus less conservative.

2.5   Generalizations

In this section, several generalizations are presented. We begin with improving the robust stability conditions of Corollary 2.1. Extension of these results to robust stabilization can be readily obtained.

2.5.1   Parameter-Dependent Lyapunov Function

The condition of Corollary 2.1 applies the same Lyapunov function to all the points in the uncertainty polytopes $\Omega_i, i = 1, \ldots M$. The conservatism that is entailed in such a selection of the Lyapunov function can be reduced by using a vertex dependent Lyapunov function. Applying the method of [11] we obtain the following.

Corollary 2.3 *Assume that for some dwell time $T > 0$ there exist a collection of matrices $P_{i,k}^{(j)} = P_{i,k}^{(j)'}, Z_{i,k}, H_{i,k}, i = 1, \ldots M, k = 0, \ldots K, j = 1, \ldots N$ of compatible dimensions, where K is a pre-chosen integer, and a sequence $\{\delta_k > 0, \quad k = 1, \ldots K, \quad \sum_{k=1}^{K} \delta_k = T\}$ such that, for all $i = 1, \ldots M$, and $j = 1, \ldots N$, the following holds.*

$$P_{i,k}^{(j)} > 0, \quad \begin{bmatrix} \frac{P_{i,k+1}^{(j)} - P_{i,k}^{(j)}}{\delta_{k+1}} + Z_{i,k}A_i^{(j)} + A_i^{(j)'}Z_{i,k}' & * \\ -P_{i,k}^{(j)} + Z_{i,k}' - H_{i,k}'A_i^{(j)} & -H_{i,k}' - H_{i,k} \end{bmatrix} < 0$$

$$\begin{bmatrix} \frac{P_{i,k+1}^{(j)} - P_{i,k}^{(j)}}{\delta_{k+1}} + Z_{i,k+1}A_i^{(j)} + A_i^{(j)'}Z_{i,k+1}' & * \\ -P_{i,k+1}^{(j)} + Z_{i,k+1}' - H_{i,k+1}'A_i^{(j)} & -H_{i,k+1}' - H_{i,k+1} \end{bmatrix} < 0 \quad (2.19)$$

$$k = 0, \ldots K - 1$$

$$\begin{bmatrix} Z_{i,K}A_i^{(j)} + A_i^{(j)'}Z_{i,K}' & * \\ -P_{i,K}^{(j)} + Z_{i,K}' - H_{i,K}'A_i^{(j)} & -H_{i,K}' - H_{i,K} \end{bmatrix} < 0$$

$$and \qquad\qquad\qquad\qquad\qquad\qquad\qquad\qquad\qquad\qquad\qquad\qquad\qquad (2.20)$$

$$P_{i,K}^{(j)} - P_{l,0}^{(\bar{j})} \geq 0, \quad \forall\, l = 1, \ldots i - 1, i + 1, \ldots M, \bar{j} = 1, \ldots N.$$

Then, the system of (2.1) and (2.2) is globally asymptotically stable for any switching law with dwell time greater than or equal to T.

Proof Multiplying (2.19b, c) and (2.20a) by $J = \begin{bmatrix} I & -A_i^{(j)'} \\ 0 & I \end{bmatrix}$, from the left, and by J', from the right, the conditions of Corollary 2.1 are readily obtained.

2.5.2   Exponential Stability

Corollary 2.4 *Assume that for some dwell time $T > 0$ and for a given scalar $\lambda > 0$ there exist: a collection of symmetric matrices $P_{i,k}, i = 1, \ldots M, k = 0, \ldots K$ of* *compatible dimensions, where K is a pre-chosen integer, and a sequence $\{\delta_k > 0, \quad k = 1, \ldots K, \quad \sum_{k=1}^{K} \delta_k = T\}$ such that, for all $i = 1, \ldots M$, and $j = 1, \ldots N$, the following holds.*

$$P_{i,k} > 0, \quad \frac{P_{i,k+1}-P_{i,k}}{\delta_{k+1}} + P_{i,k}A_i^{(j)} + A_i^{(j)'}P_{i,k} + 2\lambda P_{i,k} < 0,$$
$$\frac{P_{i,k+1}-P_{i,k}}{\delta_{k+1}} + P_{i,k+1}A_i^{(j)} + A_i^{(j)'}P_{i,k+1} + 2\lambda P_{i,k+1} < 0, \quad k = 0, \ldots K-1 \tag{2.21}$$

$$P_{i,K}A_i^{(j)} + A_i^{(j)'}P_{i,K} + 2\lambda P_{i,K} < 0, \quad P_{i,K} - P_{l,0} \geq 0, \quad \forall\, l = 1, \ldots i-1, i+1, \ldots M. \tag{2.22}$$

Then, the system (2.1) is globally exponentially stable, with a decay rate $\geq \lambda$, for any switching law with dwell time greater than or equal to T.

Proof Choosing a Lyapunov function as in (2.11), we have from (2.21b, c) and (2.22a), that $P(t)A(t) + A^T(t)P(t) + 2\lambda P(t) < 0$ holds for all time instants, except for the set of switching instants, which constitute a zero measure set. Since (2.22b) guarantees that the Lyapunov function is nonincreasing at the switching instants, this shows that the conditions detailed in [4] hold, and therefore the system (2.1) is exponentially stable with a decay rate $\geq \lambda$. ∎

2.5.3 State Jumps

Assume that at the switching instant, τ_h, the systems switches form the ith subsystem to the qth subsystem and that the state vector is transformed according to the rule

$$x(\tau_h) = S_q^a S_i^d x(\tau_h^-), \tag{2.23}$$

where S_i^d is the state transformation applied when leaving subsystem i. The superscript d denotes "deactivation". The matrix S_q^a is the state transformation applied when switching to the qth subsystem, and the superscript a denotes "activation". This formulation suggests two-state switches, one for leaving a subsystem, and one for entering a new subsystem.

Corollary 2.5 *Assume that for some dwell time $T > 0$ there exist an integer K, a collection of symmetric matrices $P_{i,k}$, $i = 1, \ldots M$, $k = 0, \ldots K$ of compatible dimensions, and a sequence $\{\delta_k > 0, \quad k = 1, \ldots K, \quad \sum_{k=1}^{K} \delta_k = T\}$ such that, for all $i = 1, \ldots M$, and $j = 1, \ldots N$ the following holds.*

$$P_{i,k} > 0, \quad \frac{P_{i,k+1}-P_{i,k}}{\delta_{k+1}} + P_{i,k}A_i^{(j)} + A_i^{(j)'}P_{i,k} < 0,$$
$$\frac{P_{i,k+1}-P_{i,k}}{\delta_{k+1}} + P_{i,k+1}A_i^{(j)} + A_i^{(j)'}P_{i,k+1} < 0, \quad k = 0, \ldots K-1 \tag{2.24}$$

$$P_{i,K}A_i^{(j)} + A_i^{(j)'}P_{i,K} < 0, \quad \begin{bmatrix} -P_{i,K} & * \\ P_{q,0}S_q^a S_i^d & -P_{q,0} \end{bmatrix} \leq 0, \quad \forall\, q = 1, \ldots i-1, i+1, \ldots M. \tag{2.25}$$

Then, the system (2.1) with the state switching (2.23), is globally asymptotically stable, for any switching law with dwell time greater than or equal to T.

Proof Choosing the Lyapunov function to be as in (2.11), the latter conditions guarantee that the Lyapunov function is decreasing between the switching instants as in Corollary 2.1. In order to show that the Lyapunov function is nonincreasing at the switching instants, the following conditions must hold:

$$V(\tau_h^-) = x'(\tau_h^-)P_{i,K}x(\tau_h^-) \geq x'(\tau_h)P_{q,0}x(\tau_h) = V(\tau_h).$$

Substituting (2.23) into this inequality, and requiring it to hold for any $x(\tau_h^-)$ one obtains

$$P_{i,K} \geq S_i^{d'} S_q^{a'} P_{q,0} S_q^a S_i^d.$$

Using Schur complements [4], this condition leads to (2.25b). ∎

2.6 Examples

In the following, we apply the theory developed above where δ_k is taken to be $\delta_k = \frac{T}{K}$.

Example 2.1 (Stability) We consider the system taken from [2] with

$$A_1 = \begin{bmatrix} -1 & 0 & 1 \\ -1 & -1 & 0 \\ 0 & 1 & -1 \end{bmatrix}, \quad A_2 = \begin{bmatrix} -1 & 0 & 6 \\ -1 & -1 & -5 \\ 0 & 1 & -1 \end{bmatrix}.$$

Note that this system is not quadratically stable. Table 2.1 below shows the minimal dwell time for which the system is proved to be stable, calculated by Lemmas 2.1 and 2.2 and Theorem 2.1.

The table clearly demonstrates, on one hand, the superiority of the method of [1] in the case where A_1 and A_2 are perfectly known, and the fact that the result of Theorem 2.1 tends to the one of [1] when K is increased, on the other hand.

Example 2.2 (Robust stability) We add uncertainty to the subsystems of Example 2.1 and apply Corollary 2.1 and Lemma 2.2 where in the latter maximum λ_m is calculated, separately, over Ω_i for each subsystem.

$$A_1 = \begin{bmatrix} -1 & 0 & 1 \\ -1 & -1 & 0 \\ 0 & 1 & -1 \end{bmatrix} + \alpha \begin{bmatrix} 0 & 1 & 0 \\ 0 & 0 & 0 \\ 0 & 0 & 0 \end{bmatrix} A_2 = \begin{bmatrix} -1 & 0 & 6 \\ -1 & -1 & -5 \\ 0 & 1 & -1 \end{bmatrix} + \alpha \begin{bmatrix} 0 & 1 & 0 \\ 0 & 0 & 0 \\ 0 & 0 & 0 \end{bmatrix}, \quad \alpha \in [0, \bar{\alpha}].$$

Table 2.1 Minimum dwell time for Example 2.1

Method	Lemma 2.1 [1]	Lemma 2.2	$K = 1$	$K = 2$	$K = 3$	$K = 10$	$K = 20$	$K = 95$
Min T	0.4	1.02	0.87	0.75	0.71	0.48	0.43	0.4

Table 2.2 Min. dwell time for Example 2.2

$\bar{\alpha}$	Method					
	Lemma 2.2	$K = 1$	$K = 2$	$K = 3$	$K = 10$	$K = 20$
0.1	1.03	0.87 (0.87)	0.75 (0.76)	0.71 (0.71)	0.48 (0.49)	0.43 (0.44)
0.2	1.04	0.87 (0.88)	0.75 (0.76)	0.71 (0.72)	0.48 (0.5)	0.43 (0.45)
0.5	1.07	0.87 (0.91)	0.75 (0.79)	0.71 (0.75)	0.48 (0.56)	0.43 (0.49)
1	1.14	0.87 (0.98)	0.75 (0.87)	0.71 (0.81)	0.48 (0.74)	0.43 (0.7)

Table 2.2 shows the minimal dwell time for which the system is stable, found by the methods of Lemma 2.2 and Corollary 2.4. The corresponding results found by Corollary 2.1 are given there is brackets.

Example 2.3 (Stabilization) Two switching systems are brought here. The first represents a situation where the switching signal is measured in real-time, and it is impossible to stabilize the system with a constant state-feedback gain using the theory presented above.

The second system describes a case where the switching signal is unknown, and therefore a constant G is applied regardless of the switching. If the switching signal were known in real-time, it would be stabilizable even for instantaneous switching (zero dwell time).

First we consider the system (2.3), with:

$$A_1 = 10^{-4} \begin{bmatrix} 2 & 2 & 0 \\ 0 & 5 & 0 \\ 0 & 0 & -1 \end{bmatrix}, \quad A_2 = 10^{-4} \begin{bmatrix} 4 & 2 & 0 \\ 0 & 9 & 0 \\ 0 & 0 & -1 \end{bmatrix}, \quad \text{and } B_1 = B_2 = \begin{bmatrix} 1 \\ 2 \\ 0 \end{bmatrix}.$$

We note that both subsystems are stabilizable, but none is controllable. The subsystems are not quadratically stabilizable, and the maximal decay rate attainable for each of the subsystems is 10^{-4}. Assuming that the switching is known online, it is possible to stabilize this system, using Theorem 2.2, for a dwell time of $T = 0.26\,$s with $K = 1$, and for $T = 0.1$, for $K = 2$. Assuming the switching signal cannot be measured, it is found that it is impossible to find a constant feedback law that stabilizes the system by using Corollary 2.2 for the closed-loop.

The second switching system that we consider is one where:

$$A_1 = \begin{bmatrix} 0 & 2 \\ 1 & 1 \end{bmatrix}, \quad A_2 = \begin{bmatrix} 1 & 8 \\ 0 & -1 \end{bmatrix}, \quad B_1 = \begin{bmatrix} -1 \\ -1 \end{bmatrix}, \quad \text{and } B_2 = \begin{bmatrix} 1 \\ 0 \end{bmatrix}.$$

Also here, both subsystems are stabilizable, but none is controllable. Assuming that the switching is known on line, using Theorem 2.2 it is possible to stabilize this system for a dwell time as small as desired. Assuming, on the other hand, that the switching signal cannot be measured, and using a constant feedback law, it is found, using Corollary 2.2 for the closed-loop, as explained in Remark 2.3, that it is possible to stabilize the system for a dwell time $T = 0.01\,$s, using the gain matrix $G = [-1201\ 2403]$.

2.7 Conclusions

A new method for the stability analysis of a linear-switched system using a switching dependent Lyapunov function has been introduced which enables treatment of polytopic-type parameter uncertainties. While, for the nominal case, the method presented yields results that are more conservative than those of [1], it is easily extended to the uncertain case, and it can achieve results that are very close to the results in [1], in the nominal case, if the number of divisions of the dwell time interval, K, is taken large enough. In order to demonstrate the strength of the method that is introduced in this chapter, the result of [1] is extended to the uncertain case by bounding the matrix exponent there by a scalar matrix. Examples 2.1 and 2.2 then demonstrates the superiority of the new method.

The introduced method handles polytopic uncertainties; it can, however, be easily extended to include norm-bounded uncertainties as well. The results presented in this chapter can be easily generalized to cases of polytopic-type uncertainty where different subsystems have a different number of vertices. Other immediate extensions are to systems with a different dwell time for each subsystem, and systems whose switching is limited, in the sense that each subsystem can only be switched to a given subset of subsystems. This situation is encountered in modeling large uncertainties.

Throughout this chapter, we chose to make the Lyapunov function change during the dwell time after the switching occurred. Alternatively, it is possible to change the Lyapunov function during the dwell time prior to the switching.

References

1. Geromel, J., Colaneri, P.: Stability and stabilization of continuous-time switched linear systems. SIAM J. Control Optim. **45**(5), 1915–1930 (2006)
2. Colaneri, P.: Dwell time analysis of deterministic and stochastic switched systems. Eur. J. Autom. Control **15**, 228–249 (2009)
3. Boyarski, S., Shaked, U.: Time-convexity and time-gain-scheduling in finite-horizon robust H_∞-control. In: Proceedings of the 48th CDC09, Shanghai, China (2009)
4. Boyd, S., El Ghaoui, L., Feron, E., Balakrishnan, V.: Linear Matrix Inequality in Systems and Control Theory. SIAM Frontier Series (1994)
5. Kocvara, M., Stingl, M.: PENBMI User's Guide, (Version 2) (2005). www.penopt.com
6. Toh, K-C., Todd, M.J., Tutuncu, R.H..: A MATLAB software for semidefinite-quadratic-linear programming (2009). http://www.math.nus.edu.sg/~mattohkc/sdpt3.html
7. Grant, M., Boyd, S.: CVX, Matlab software for disciplined convex programming (web page and software). (2009). http://stanford.edu/~boyd/cvx
8. M. Grant and S. Boyd, Graph Implementations for Nonsmooth Convex Programs, Recent Advances in Learning and Control (a tribute to M. Vidyasagar), V. Blondel, S. Boyd, and H. Kimura, editors, pages 95-110. Lecture Notes in Control and Information Sciences, Springer (2008). http://stanford.edu/~boyd/graph_dcp.html
9. YALMIP: A toolbox for modeling and optimization in MATLAB. In: Lfberg, J. Proceedings of the CACSD Conference, Taipei, Taiwan (2004). http://control.ee.ethz.ch/~joloef/yalmip.php

10. Apkarian, P., Gahinet, P.: A convex characterization of gain-scheduled H_∞ conterollers. IEEE Trans. Automat. Control **40**(5), 853–864 (1995)
11. de Oliveira, M.C., Skelton, R.E.: Stability test for constrained linear systems. In: Reza Moheimani, S.O. (ed.) Perspectives in Robust Control. Lecture Notes in Control and Information Sciences, vol. 268. Springer, London (2001)

Chapter 3
Robust State-Dependent Switching of Linear Systems with Dwell Time

Abstract A state-dependent switching law that obeys a dwell-time constraint and guarantees the stability of a switched linear system is designed. Sufficient conditions are obtained for the stability of the switched systems when the switching law is applied in presence of polytopic-type parameter uncertainty. A Lyapunov function, in quadratic form, is assigned to each subsystem such that it is nonincreasing at the switching instants. During the dwell time, this function varies piecewise linearly in time. After the dwell, the system switches if the switching results in a decrease in the value of the LF. The proposed method is applicable also to robust stabilization via state feedback. It is further extended to guarantee a bound on the $\mathcal{L}_2$-gain of the switching system and it is also used in deriving state-feedback control law that robustly achieves a prescribed $\mathcal{L}_2$-gain bound.

3.1 Introduction

We treat in this chapter state-dependent switching under dwell-time constraints. An overview of different constraints that may be imposed on the switching law may be found in [1]. Using the results of Chap. 2, we investigate the bound on the achievable $\mathcal{L}_2$-gain of this system, and find state-feedback controllers that minimize these bounds. For review of stability of switched systems with dwell time and arbitrary switching, see also [2, 3]. Another approach for stabilizing a class of linear-switched systems by synthesis of limit cycles is found in [4]. For a review of the current methods for stabilizing a switched system by designing the switching law, without a dwell-time constraint, see the recent book [3] and the references therein. In the later reference, the effect of dwell time on the stability of switched system is also investigated.

It has been shown in [5] that the most destabilizing switching law for a switched system with stable subsystem applies, in many cases, a dwell time. Using the same analysis, it is found that the most stabilizing switching law for many switched systems with unstable subsystems obeys some dwell time. Adding a dwell-time constraint to a suboptimal switching law may thus achieve better results.

© Springer Nature Switzerland AG 2019
E. Gershon and U. Shaked, *Advances in H_∞ Control Theory*,
Lecture Notes in Control and Information Sciences 481,
https://doi.org/10.1007/978-3-030-16008-1_3

We present in this chapter a state-dependent switching law that obeys a dwell-time constraint with a prescribed dwell time. In practice, a dwell-time constraint is mandatory for every switched system, because there is always a limit to the rate in which the physical system can perform switching. The dwell time may be determined according to the capacity of the system's components, or by the designers desire to reduce discontinuities in the system's operation. For example, if the plant is a passenger aircraft, the designer may demand a large dwell-time constraint to refrain from making the ride bumpy for the passengers.

We first formulate the problem in Sect. 3.1. In Sect. 3.2, we extend the results of [6] to obtain stabilization results for nominal switched systems with dwell time and in Sect. 3.3 a method is introduced that can handle polytopic-type parameter uncertainty in these systems. Guaranteed $\mathcal{L}_2$-gain bounds are obtained in Sect. 3.4. Robust state-feedback synthesis is introduced in Sect. 3.5, where both the feedback gains and the switching law are derived that stabilize the uncertain system and minimize the bound on the $\mathcal{L}_2$-gain of the resulting closed-loop system. Three illustrative examples are given in Sect. 3.6.

3.2 Problem Formulation

We consider the following linear-switched system:

$$\begin{aligned}
\dot{x}(t) &= A_{\sigma(t)}x(t) + B_{\sigma(t)}w(t) + B_{2,\,\sigma(t)}u(t), \quad x(0) = 0, \\
z(t) &= C_{\sigma(t)}x(t) + D_{\sigma(t)}w(t) + D_{12,\,\sigma(t)}u(t),
\end{aligned} \tag{3.1}$$

where $x \in \mathcal{R}^n$ is the state, $z \in \mathcal{R}^r$ is the objective vector, $u \in \mathcal{R}^s$ is the control signal and $w \in \mathcal{R}^q$ is an exogenous disturbance in $\mathcal{L}_2$. This system is defined for all $t \geq 0$. The switching rule $\sigma(t) \in \{1 \ldots M\}$, for each $t \geq 0$, is such that $\Omega_{\sigma(t)} \in \{\Omega_1, \ldots, \Omega_M\}$, where

$$\Omega_{\sigma(t)} = \begin{bmatrix} A_{\sigma(t)} & B_{\sigma(t)} & B_{2,\sigma(t)} \\ C_{\sigma(t)} & D_{\sigma(t)} & D_{12,\sigma(t)} \end{bmatrix}.$$

The polytope Ω_i is defined by

$$\Omega_i = \sum_{j=1}^{N} \eta^j(t)\Omega_i^{(j)}, \quad \sum_{j=1}^{N} \eta^j(t) = 1, \quad \eta^j(t) \geq 0 \tag{3.2}$$

with vertices defined by

$$\Omega_i^{(j)} = \begin{bmatrix} A_i^{(j)} & B_i^{(j)} & B_{2,i}^{(j)} \\ C_i^{(j)} & D_i^{(j)} & D_{12,\,i}^{(j)} \end{bmatrix}, \quad i = 1, \ldots M, \tag{3.3}$$

where $A_i^{(j)}$, $B_i^{(j)}$, $B_{2,i}^{(j)}$, $C_i^{(j)}$, $D_i^{(j)}$, $D_{12,i}^{(j)}$ are constant matrices of the appropriate dimensions. The above model naturally imposes discontinuity in $\Omega_{\sigma(t)}$ since this matrix jumps instantaneously from Ω_{i_1} to Ω_{i_2} for some $i_1 \neq i_2$ at the switching instances.

Note that this means that each subsystem has a separate polytopic description, which is independent of the description of the other subsystems.

The present chapter is focused on stabilizing the system via state-feedback control and on achieving a prescribed bound on the $\mathcal{L}_2$-gain of the system under a dwell-time constraint. The latter means that if the dwell time is T, and the switching instants are $\tau_1, \tau_2, \ldots$, then $\tau_{h+1} - \tau_h \geq T$, $\forall h \geq 1$.

The measure of performance in this work is the bound on the $\mathcal{L}_2$-gain of the system, from the disturbance w to the objective vector z.

3.3 Nominal Time-Switching Stability

Theorem 3.1 *Assume there exist in $\mathcal{R}^{n \times n}$ a collection of positive definite matrices $P_i, i = 1, \ldots M$ and a set of scalars $\lambda_{i,l} \geq 0$, $i = 1 \ldots M$, $l \in L_i$, such that*

$$P_i A_i + A_i' P_i + \Sigma_{l \in L_i} \lambda_{i,l} (e^{A_l' T} P_l e^{A_l T} - P_i) < 0, \quad \forall i = 1, \ldots, M. \quad (3.4)$$

Assume also that the nominal system (3.1) switched at $t = \tau_h$ to $\sigma(\tau_h) = i$. Then, the following law for the next switching at $t = \tau_{h+1}$

$$\sigma(t) = i \ \forall t \in [\tau_h, \tau_h + T)$$
$$\sigma(t) = i \ \forall t > \tau_h + T, \ if \ x(t)' P_i x(t) \leq x(t)' e^{A_l' T} P_l e^{A_l T} x(t), \quad \forall l \in L_i \quad (3.5)$$
$$\sigma(\tau_{h+1}) = argmin_l \ x(\tau_{h+1})' e^{A_l' T} P_l e^{A_l T} x(\tau_{h+1}), \quad otherwise,$$

globally asymptotically stabilizes the system (3.1) with dwell-time constraint $T > 0$.

Proof Let $i_0 = \sigma(0)$, and let $\tau_1, \tau_2, \ldots$ be the switching instants, where $\tau_{h+1} - \tau_h \geq T$, $\forall h = 1, 2, \ldots$. Choosing the Lyapunov Function (LF) $V(t) = x'(t) P_{\sigma(t)} x(t)$ we have, between $\tau_h + T$ and τ_{h+1}, that

$$\dot{V}(t) = x'(t)(P_i A_i + A_i' P_i) x(t)$$

along the system trajectory $x(t)$, and from the definition of the switching law (3.5) we have that
$$x'(t)(e^{A_l' T} P_l e^{A_l T} - P_i) x(t) \geq 0, \quad \forall l \in L_i.$$

Hence:

$$\dot{V}(t) \le x'(t)[P_i A_i + A'_i P_i + \Sigma_{l \in L_i} \lambda_{i,l}(e^{A'_i T} P_l e^{A_l T} - P_i)]x(t)$$

and it follows from (3.4) that

$$P_i A_i + A'_i P_i + \Sigma_{l \in L_i} \lambda_{i,l}(e^{A'_i T} P_l e^{A_l T} - P_i) < 0 \quad \text{and thus} \quad \dot{V}(t) < 0.$$

At the switching instant $t = \tau_{h+1}$, $V(\tau_{h+1}^-) = x'(\tau_{h+1})P_i x(\tau_{h+1})$, and $V(\tau_{h+1} + T) = x'(\tau_{h+1})e^{A'_i T} P_l e^{A_l T} x(\tau_{h+1})$. Thus, $V(\tau_{h+1} + T) - V(\tau_{h+1}^-) = x'(\tau_{h+1})(e^{A'_i T} P_l e^{A_l T} - P_i)x(\tau_{h+1})$. From the definition of the switching law, we have $x'(\tau_{h+1})(e^{A'_i T} P_l e^{A_l T} - P_i)x(\tau_{h+1}) < 0$ and therefore $V(\tau_{h+1} + T) - V(\tau_{h+1}^-) < 0$. ∎

We note that the switching strategy of (3.5) is not claimed to produce an optimal result in any sense. It only provides a possible solution for the stabilization problem. In the case where at least one of the subsystems is stable, stability is achieved by merely switching to this system.

The result of Theorem 3.1 cannot be easily applied to the uncertain case where the parameters of A_i are only known to reside in Ω_i, $i = 1, \ldots, M$. We introduce below a LF-based approach to deal with the uncertainty. The results obtained will converge to those of [6] in the limit where the dwell time tends to zero.

3.4 Robust Switching Stability

We first derive the following result, inspired by [7].

Lemma 3.1 *Assume that for some time interval $t \in [t_0, t_f]$, where $\delta = t_f - t_0$ there exist two symmetric matrices P_1 and P_2 of compatible dimensions that satisfy the following:*

$$P_1, P_2 > 0, \quad \frac{P_2 - P_1}{\delta} + P_1 A + A' P_1 < 0, \quad \frac{P_2 - P_1}{\delta} + P_2 A + A' P_2 < 0. \quad (3.6)$$

Then, for the system $\dot{x} = Ax$, the LF $V(t) = x'(t)P(t)x(t)$, with $P(t) = P_1 + (P_2 - P_1)\frac{t - t_0}{\delta}$, is strictly decreasing over the time interval $t \in [t_0, t_f]$.

Proof The derivative of $V(t)$ is: $\dot{V}(t) = x'(\dot{P} + PA + A'P)x = x'(\frac{P_2 - P_1}{\delta} + PA + A'P)x$. Since $\frac{P_2 - P_1}{\delta} + PA + A'P = \lambda_1[\frac{P_2 - P_1}{\delta} + P_1 A + A'P_1] + \lambda_2[\frac{P_2 - P_1}{\delta} + P_2 A + A'P_2]$, where $\lambda_1 = 1 - \frac{t - t_0}{\delta}$ and $\lambda_2 = \frac{t - t_0}{\delta}$, we obtain from (3.6) that $\dot{V}(t) < 0$ in the time interval specified above. ∎

Note that while the latter result has been obtained for a nominal system, the extension to the polytopic uncertainty case is readily achieved by choosing the same P_1 and P_2 for all the vertices of the uncertainty polytope.

Based on Lemma 3.1 we obtain the following, for a prescribed integer K that is chosen according to the allowed computational complexity. A larger K leads to less conservatism in the conditions.

Theorem 3.2 *Given that the nominal system (3.1) switched at $t = \tau_h$ to $\sigma(\tau_h) = i$ and assume there exist a collection of positive-definite matrices $P_{i,k}$, $i = 1, \ldots M$, $k = 0, \ldots K$ in $\mathcal{R}^{n \times n}$, and a set of scalars $\lambda_{i,l} \geq 0$, $i = 1 \ldots M$, $l \in L_i$, such that*

$$\frac{P_{i,k+1} - P_{i,k}}{T/K} + P_{i,k} A_i + A_i' P_{i,k} < 0,$$
$$\frac{P_{i,k+1} - P_{i,k}}{T/K} + P_{i,k+1} A_i + A_i' P_{i,k+1} < 0, \quad k = 0, \ldots K - 1 \quad (3.7)$$
$$P_{i,K} A_i + A_i' P_{i,K} + \Sigma_{l \in L_i} \lambda_{i,l}(P_{l,0} - P_{i,K}) < 0.$$

Then, the following law for the next switching at $t = \tau_{h+1}$:

$$\sigma(t) = i \ \ \forall t \in [\tau_h, \tau_h + T)$$
$$\sigma(t) = i \ \ \forall t > \tau_h + T, \ if \ x(t)' P_{i,K} x(t) \leq x(t)' P_{l,0} x(t), \ \forall l \in L_i \quad (3.8)$$
$$\sigma(\tau_{h+1}) = argmin_l \ x(\tau_{h+1})' P_{l,0} x(\tau_{h+1}), \ otherwise,$$

globally asymptotically stabilizes the system (3.1) with a constraint of dwell time $T > 0$.

Proof Let $i_0 = \sigma(0)$, and let $\tau_1, \tau_2, \ldots$ be the switching instants, where $\tau_{h+1} - \tau_h \geq T$, $\forall h = 1, 2, \ldots$. Define $\tau_{h,k} = \tau_h + kT/K$ for $k \geq 1$, and $\tau_{h,0} = \tau_h$. Note that the dwell time constraint implies $\tau_h + T \leq \tau_{h+1,0} = \tau_{h+1}$. We consider the LF: $V(t) = x'(t)P(t)x(t)$, where $P(t)$ is chosen to be:

$$P(t) = \begin{cases} P_{i,k} + (P_{i,k+1} - P_{i,k})\frac{t - \tau_{h,k}}{T/K} & t \in [\tau_{h,k}, \tau_{h,k+1}) \\ P_{i,K} & t \in [\tau_{h,K}, \tau_{h+1,0}), \quad h = 1, 2, \ldots, \quad (3.9) \\ P_{i_0,K} & t \in [0, \tau_1) \end{cases}$$

and where i is the index of the subsystem that is active at time t.

After the dwell time, and before the next switching instant, namely, for $t \in [\tau_h + T, \tau_{h+1})$, we have that $V(t) = x(t)' P_{i,K} x(t)$, where $\dot{x}(t) = A_i x(t)$, and where, by the definition of the switching law, $x(t)' P_{i,K} x(t) \leq x(t)' P_{l,0} x(t)$, $\forall l \in L_i$, along the system trajectory $x(t)$. Therefore,

$$\dot{V}(t) = x(t)'[P_{i,K} A_i + A_i' P_{i,K}]x(t) \leq x(t)'[P_{i,K} A_i + A_i' P_{i,K}$$

$$+ \Sigma_{l \in L_i} \lambda_{i,l}(P_{l,0} - P_{i,K})]x(t).$$

Before the next switching occurs, we have, by (3.7c), that for any $x(t) \neq 0$ and especially for the trajectory of the system, this expression is negative. During the dwell time, we consider the time intervals $t \in [\tau_{h,k}, \tau_{h,k+1})$ where $\tau_{h,k+1} = \tau_{h,k} + T/K$. The matrix $P(t)$ changes then linearly from $P_{i,k}$ to $P_{i,k+1}$, and, according to Lemma 3.1, conditions (3.7a, b) then guarantee that $V(t)$ is decreasing during this time interval. At the switching instant, we have from (3.8b) that there is at least one subsystem that switching to would decrease the value of the LF. From (3.8c), we find that the system switches to the subsystem that yields the minimum value for the LF, and therefore the LF is nonincreasing at the switching instant.   ∎

The method for solving the inequalities in Theorem 3.2, and in the results that will be derived below, deserves an additional explanation. The main difficulty in their solution is the non-convex nature of these inequalities which prevents application of standard LMI solution packages. The fact that $\lambda_{i,l} \geq 0$, $i = 1 \ldots M$, $l \in L_i$ are scalars may allow in the future efficient computational tools of solution. In the present chapter we adopt a simpler approach. We restrict ourselves to equal $\lambda_{i,l}$ and thus remain with only one tuning parameter. We note that using a line-search for this tuning parameter, the resulting inequalities can be solved by computationally efficient methods that are available in the literature for solving LMIs.

Lemma 3.2 *For a dwell time $T \to 0$, the conditions of Theorem 3.2 recover those of [6].*

Proof We denote $\epsilon_k = k\sqrt{T/K}$, and consider $P_{i,k} = P_{i,0} - \epsilon_k I$, $i = 1, \ldots M$, $k = 0, \ldots K$ which implies that $\frac{P_{i,k+1} - P_{i,k}}{T/K} = -\frac{1}{\epsilon_1} I$. Conditions (3.7a, b) then readily follow. Substituting $P_{i,K} = P_{i,0} - \epsilon_K I$ into (3.7c) we obtain

$$P_{i,0} A_i + A_i' P_{i,0} + \Sigma_{l \in L_i} \lambda_{i,l} (P_{l,0} - P_{i,0} + \epsilon_K I) - \epsilon_K (A_i + A_i') < 0,$$

which, due to the fact that $\epsilon_K \to 0$, leads to the condition obtained in [6].   ∎

3.5 L_2-gain

Given the system (3.1), we seek a sufficient condition for the asymptotic stability of the above system and for the following performance criterion to hold for a prescribed positive scalar γ and the initial condition $x(0) = 0$.

$$J = \int_0^\infty (z'z - \gamma^2 w'w)dt \leq 0, \ \forall \ w \in \mathcal{L}_2.$$

Denoting:

$$\bar{J} = lim_{t \to \infty}[V(t) + \int_0^t (z'z - \gamma^2 w'w)ds]$$

we have that $J \leq \bar{J}$, since $V(t) \geq 0 \ \forall t$. Taking into account that $V(t)$ is differentiable for all t, except for the switching instances, and that $x(0) = 0$, we have

$$lim_{t \to \infty} V(t) = \sum_{h=0}^\infty \int_{\tau_h}^{\tau_{h+1}} \dot{V}(t)dt + \sum_{h=1}^\infty (V(\tau_h) - V(\tau_h^-)),$$

where $\tau_0 = 0$. From the switching law (3.8) we find that $V(t)$ is nonincreasing at the switching instances, and thus $V(\tau_h) - V(\tau_h^-) \le 0 \ \forall h > 0$, and

$$lim_{t \to \infty} V(t) \le \sum_{h=0}^{\infty} \int_{\tau_h}^{\tau_{h+1}} \dot{V}(t)dt. \tag{3.10}$$

Denoting

$$\tilde{J} = \sum_{h=0}^{\infty} \int_{\tau_h}^{\tau_{h+1}} \dot{V}(t)dt + \int_0^{\infty} (z'z - \gamma^2 w'w)ds$$

$$= \sum_{h=0}^{\infty} \int_{\tau_h}^{\tau_{h+1}} (\dot{V}(s) + z'z - \gamma^2 w'w)ds$$

it is found from (3.10) that: $J \le \bar{J} \le \tilde{J}$. Consequentially, if $\tilde{J} \le 0$, the L_2-gain of the system will be less than or equal to γ, since at the switching instances the LF does not increase.

Theorem 3.3 (L_2-gain) *Assume that for a prescribed scalar $\gamma > 0$ there exist a collection of positive-definite matrices $P_{i,k}, i = 1, \ldots M, k = 0, \ldots K$ in $\mathcal{R}^{n \times n}$ and a set of scalars $\lambda_{i,l} \ge 0$, $i = 1 \ldots M$, $l \in L_i$, such that*

$$\begin{bmatrix} \Delta_{i,k} + P_{i,k}A_i + A_i'P_{i,k} & P_{i,k}B_i & C_i' \\ * & -\gamma^2 I & D_i' \\ * & * & -I \end{bmatrix} < 0,$$

$$\begin{bmatrix} \Delta_{i,k} + P_{i,k+1}A_i + A_i'P_{i,k+1} & P_{i,k+1}B_i & C_i' \\ * & -\gamma^2 I & D_i' \\ * & * & -I \end{bmatrix} < 0, \quad k = 0, \ldots K-1, \tag{3.11}$$

and

$$\begin{bmatrix} P_{i,K}A_i + A_i'P_{i,K} + \bar{\Gamma}_i & P_{i,K}B_i & C_i' \\ * & -\gamma^2 I & D_i' \\ * & * & -I \end{bmatrix} < 0,$$

where: $\bar{\Gamma}_i = \sum_{l \in L_i} \lambda_{i,l}(P_{l,0} - P_{i,K})$ and $\Delta_{i,k} = \frac{P_{i,k+1} - P_{i,k}}{T/K}$.

Then, the switching law (3.8) globally asymptotically stabilizes the system (3.1), with a constraint of dwell time $T > 0$, and it guarantees an L_2-gain bound of γ.

Proof Let $P(t)$ be as in (3.9). Condition (3.11a) guarantees that the LF is positive for any $x \ne 0$. From the standard derivation of the Bounded Real Lemma (BRL) for linear systems [8], we have that the following conditions are equivalent:

$$\dot{V} + z'z - \gamma^2 w'w < 0,$$

$$\begin{bmatrix} \dot{P}(t) + P(t)A + A'P(t) & P(t)B & C' \\ * & -\gamma^2 I & D^T \\ * & * & -I \end{bmatrix} < 0, \tag{3.12}$$

where the matrices A, B, C, D are as in (3.1). Following the same method that has been used in the proof of Theorem 3.1, conditions (3.11a, b) guarantee that (3.12b) is satisfied during the dwell time, and condition (3.11c) guarantees that (3.12b) holds after the dwell time. The switching law (3.8) guarantees that the LF is nonincreasing at the switching instants. We, therefore, have that $\tilde{J}$ is negative and thus the $\mathcal{L}_2$-gain of the system (3.1) is less than γ.   ∎

Note that in Theorem 3.3, γ can be taken as a decision variable and it is possible to minimize it by solving the LMIs. Therefore, it is possible to minimize the bound on the $\mathcal{L}_2$-gain using our method. This applies to all the following results in this chapter.

The conditions of Theorem 3.3 are unsuitable for design by state feedback since they lead to non-convex conditions. A dual result that is applicable to this design is derived next.

Corollary 3.1 (*Dual L_2-gain*) *Assume that for a prescribed positive scalar γ there exist a collection of positive definite matrices $Q_{i,k}, i = 1, \ldots M, k = 0, \ldots K$ in $\mathcal{R}^{n \times n}$, where K is a prechosen integer, and a set of scalars $\lambda_{i,l} \geq 0, \ i = 1 \ldots M, \ l \in L_i$, such that*

$$
\begin{bmatrix}
\Delta_{i,k}^d + A_i Q_{i,k} + Q_{i,k} A_i' & * & * \\
B_i' & -\gamma^2 I & * \\
C_i Q_{i,k} & D_i & -I
\end{bmatrix} < 0,
$$

$$
\begin{bmatrix}
\Delta_{i,k}^d + A_i Q_{i,k+1} + Q_{i,k+1} A_i' & * & * \\
B_i' & -\gamma^2 I & * \\
C_i Q_{i,k+1} & D_i & -I
\end{bmatrix} < 0, \quad k = 0, \ldots K-1,
$$

and

$$
\begin{bmatrix}
A_i Q_{i,K} + Q_{i,K} A_i' + \Gamma_i & * & * & * & * \\
B_i' & -\gamma^2 I & * & * & * \\
C_i Q_{i,K} & D_i & -I & * & * \\
Q_{i,K} & 0 & 0 & \frac{-1}{\lambda_{i,l_1^{(i)}}} Q_{l_1^{(i)},0} & * \\
\vdots & & \vdots & \ddots & \ddots & \vdots \\
Q_{i,K} & & & \cdots & & \frac{-1}{\lambda_{i,l_{M-1}^{(i)}}} Q_{l_{M-1}^{(i)},0}
\end{bmatrix} < 0,
$$

(3.13)

$\forall l = 1, \ldots i-1, i+1, \ldots M,$

where:

$$
\Gamma_i = -Q_{i,K} \Sigma_{l \in L_i} \lambda_{i,l} \quad and \quad \Delta_{i,k}^d = -\frac{Q_{i,k+1} - Q_{i,k}}{T/K}.
$$

Then, the switching law (3.8), where we substitute $P_{i,K} = Q_{i,K}^{-1}$ and $P_{l,0} = Q_{l,0}^{-1}$, globally asymptotically stabilizes the system (3.1), with the constraint of dwell time $T > 0$, and it guarantees an $\mathcal{L}_2$-gain bound of γ for this system.

Proof Let $Q(t) = P^{-1}(t)$. We depart from the linearity in time assumption on $P(t)$ and assign this linearity to $Q(t)$, namely,

$$Q(t) = \begin{cases} Q_{i,k} + (Q_{i,k+1} - Q_{i,k})\frac{t - T_{h,k}}{T/K} & t \in [\tau_{h,k}, \tau_{h,k+1}) \\ Q_{i,K} & t \in [\tau_{h,K}, \tau_{h+1,0}) \\ Q_{i_0,K} & t \in [0, \tau_1) \end{cases}, \quad h = 1, 2, \ldots.$$

We also consider next the dual condition of (3.12b). From [8] we have that the following conditions are equivalent:

$$\dot{V} + z'z - \gamma^2 w'w < 0,$$

$$\begin{bmatrix} \dot{P}(t) + P(t)A + A'P(t) & P(t)B & C' \\ * & -\gamma^2 I & D' \\ * & * & -I \end{bmatrix} < 0,$$

and

$$\begin{bmatrix} -\dot{Q}(t) + AQ(t) + Q(t)A' & B & C'Q(t) \\ * & -\gamma^2 I & D' \\ * & * & -I \end{bmatrix} < 0,$$

where $Q(t) = P(t)^{-1}$.

Since $Q(t)$ is now piecewise linear in time, we obtain conditions (3.13a, b), by applying the same method that has been used in the proof of Theorem 3.3.

In order to derive (3.13c), we multiply (3.11c) by $\begin{bmatrix} Q_{i,K} & 0 & 0 \\ 0 & I & 0 \\ 0 & 0 & I \end{bmatrix}$, from both sides.

Since after the dwell time both P and Q are constant, until the next switching occurs, the latter multiplication leads to the following inequality

$$\begin{bmatrix} A_i Q_{i,K} + Q_{i,K} A'_i + \Xi & B_i & Q_{i,K} C'_i \\ * & -\gamma^2 I & D'_i \\ * & * & -I \end{bmatrix} < 0,$$

where $\Xi = Q_{i,K}[\Sigma_{l \in L_i} \lambda_{i,l}(Q_{l,0}^{-1} - Q_{i,K}^{-1})]Q_{i,K} = Q_{i,K}[\Sigma_{l \in L_i} \lambda_{i,l} Q_{l,0}^{-1}]Q_{i,K} - Q_{i,K}$ $\Sigma_{l \in L_i} \lambda_{i,l}$. By applying Schurs complements to the first term in the right side of the latter equality, (3.13c) is obtained. ∎

Remark 3.1 The above conditions are all affine in the systems matrices, and they can, therefore, be readily applied to the uncertain case simply by solving these conditions for all the vertices of the uncertainty polytope, namely by replacing A_i, B_i, $B_{2,i}$, C_i, D_i, $D_{12,i}$ with $A_i^{(j)}$, $B_i^{(j)}$, $B_{2,i}^{(j)}$, $C_i^{(j)}$, $D_i^{(j)}$, $D_{12,i}^{(j)}$, respectively, and by solving for all $j = 1 \ldots N$.

Remark 3.2 Corollary 3.1 provides also an alternative sufficient condition for stability by taking $\gamma \to \infty$. The application of this condition in the uncertain case may lead to results that differ from those obtained by Theorem 3.2.

3.6   State Feedback

The synthesis of state-feedback controllers is achieved by replacing $A_i^{(j)}$ and $C_i^{(j)}$ by $A_i^{(j)} + B_{2,i}^{(j)} G_{i,k}$ and $C_i^{(j)} + D_{12,i}^{(j)} G_{i,k}$, respectively. These controllers are time varying even between the switching instances. This demand for time-varying controllers will be removed in the sequel. By applying the latter substitution to (3.13), and denoting $Y_{i,k} = G_{i,k} Q_{i,k}$ the following result is obtained for a prechosen integer K.

Corollary 3.2 (Quadratic state feedback) *Assume that for a prescribed scalar $\gamma > 0$ there exist: a collection of matrices $Q_{i,k} > 0 \in \mathcal{R}^{n \times n}$, and $Y_{i,k} \in \mathcal{R}^{s \times n}$, $i = 1, \ldots M, k = 0, \ldots K$, and a set of scalars $\lambda_{i,l} \geq 0$, $i = 1 \ldots M$, $l \in L_i$, such that*

$$\begin{bmatrix} \Delta_{i,k} + \Phi_{i,h} & * & * \\ B_i^{(j)'} & -\gamma^2 I & * \\ C_i^{(j)} Q_{i,h} + D_{12,i}^{(j)} Y_{i,h} & D_i^{(j)} & -I \end{bmatrix} < 0, \quad k = 0, \ldots K-1, \quad h = k, k+1,$$

and

$$\begin{bmatrix} \Phi_{i,K} + \Gamma_i & * & * & * & * \\ B_i^{(j)'} & -\gamma^2 I & * & * & * \\ C_i^{(j)} Q_{i,K} + D_{12,i}^{(j)} Y_{i,K} D_i^{(j)} & -I & * & * & * \\ Q_{i,K} & 0 & 0 \frac{-1}{\lambda_{i,l_1}^{(i)}} Q_{l_1^{(i)},0} & * \\ \vdots & \vdots & \ddots & \ddots & \vdots \\ Q_{i,K} & & & \cdots & \frac{-1}{\lambda_{i,l_{M-1}^{(i)}}} Q_{l_{M-1}^{(i)},0} \end{bmatrix} < 0, \quad (3.14)$$

where

$$\Phi_{i,k} = A_i^{(j)} Q_{i,k} + Q_{i,k} A_i^{(j)'} + B_{2,i}^{(j)} Y_{i,k} + Y_{i,k}' B_{2,i}^{(j)'},$$

$$\Gamma_i = -Q_{i,K} \Sigma_{l \in L_i} \lambda_{i,l}, \text{ and } \Delta_{i,k} = -\frac{Q_{i,k+1} - Q_{i,k}}{T/K}.$$

Then, the switching law (3.8), where we substitute $P_{i,K} = Q_{i,K}^{-1}$ and $P_{l,0} = Q_{l,0}^{-1}$, and the feedback law $u(t) = G(t) x(t)$, where $G(t) =$

$$\begin{cases} [Y_{i,k} + \frac{t-\tau_{h,k}}{T/K}(Y_{i,k+1} - Y_{i,k})][Q_{i,k} + \frac{t-\tau_{h,k}}{T/K}(Q_{i,k+1} - Q_{i,k})]^{-1} & t \in [\tau_{h,k}, \tau_{h,k+1}) \\ Y_{i,K} Q_{i,K}^{-1} & t \in [\tau_{h,K}, \tau_{h+1,0}) \\ Y_{i_0,K} Q_{i_0,K}^{-1} & t \in [0, \tau_1) \end{cases}$$

(3.15)

globally asymptotically stabilize the system (3.1) with the uncertainty (3.2),(3.3) and with the constraint of dwell time $T > 0$. The latter control law also guarantees an $\mathcal{L}_2$-gain bound of γ for the switching system.

Proof In order to preserve the convexity for time values in the interval $[\tau_{h,k}, \tau_{h,k+1})$, we choose $Y_{(t)} = Y_{i,k} + \frac{t-\tau_{h,k}}{T/K}(Y_{i,k+1} - Y_{i,k})$. The result then follows from the fact that $G_i(t) = Y_i(t) Q_i^{-1}(t)$. ∎

The state-feedback law of Corollary 3.2 may not be easy to implement. In order to obtain a simpler feedback gain, we introduce an alternative method to guarantee the $\mathcal{L}_2$-gain of the system, recalling that $Q(t)$ must be time-varying. This method will allow us to obtain simpler controllers, by applying the method of [9].

Corollary 3.3 (L_2-gain) *Assume that for prescribed scalar $\gamma > 0$ and integer K there exist a collection of matrices $Q_{i,k} > 0$, $Z_{i,k}$, $H_{i,k} \in \mathcal{R}^{n \times n}$, $i = 1, \ldots M, k = 0, \ldots K$, and a set of scalars $\lambda_{i,l} \geq 0$, $i = 1 \ldots M$, $l \in L_i$, such that*

$$
\begin{bmatrix}
\Delta_{i,k}^{(j)} + A_i^{(j)} Z_{i,h} + Z_{i,h}' A_i^{(j)'} & * & * & * \\
B_i^{(j)'} & -\gamma^2 I & * & * \\
C_i^{(j)} Z_{i,h} & D_i^{(j)} & -I & * \\
Z_{i,h} - Q_{i,h} - H_{i,h}' A_i^{(j)'} & 0 & -H_{i,h}' C_i^{(j)'} - \hat{H}_{i,h}
\end{bmatrix} < 0, \quad k = 0, \ldots K-1, \quad h = k, k+1,
$$

and

$$
\begin{bmatrix}
\Psi_{i,j} & * & * & & * & * \\
B_i^{(j)'} & -\gamma^2 I & * & & * & * & * \\
C_i^{(j)} Z_{i,K} D_i^{(j)} & -I & & * & * & * \\
\psi_{i,j} & 0 & -H_{i,K}' C_i^{(j)'} & -\hat{H}_{i,K} & * & * \\
Q_{i,K} & 0 & 0 & \frac{-1}{\lambda_{i,l_1^{(i)}}} Q_{l_1^{(i)},0} & * \\
\vdots & \vdots & \ddots & & \ddots & \vdots \\
Q_{i,K} & & & & \cdots \frac{-1}{\lambda_{i,l_{M-1}^{(i)}}} Q_{l_{M-1}^{(i)},0}
\end{bmatrix} < 0, \tag{3.16}
$$

where

$$
\Gamma_i = -Q_{i,K} \Sigma_{l \in L_i} \lambda_{i,l}, \quad \Delta_{i,k}^{(j)} = -\frac{Q_{i,k+1} - Q_{i,k}}{T/K}, \quad \hat{H}_{i,h} = H_{i,h} + H_{i,h}'
$$

$$
\Psi_{i,j} = A_i^{(j)} Z_{i,K} + Z_{i,K}' A_i^{(j)'} + \Gamma_i, \quad \psi_{i,j} = Z_{i,K} - Q_{i,K} - H_{i,K}' A_i^{(j)'}.
$$

Then, the switching law (3.8), where we substitute $P_{i,K} = Q_{i,K}^{-1}$ and $P_{l,0} = Q_{l,0}^{-1}$, globally asymptotically stabilizes the system (3.1) with the uncertainty (3.2),(3.3), and with a constraint of dwell time $T > 0$. It also guarantees a bound γ for the $\mathcal{L}_2$-gain of the system.

Proof We denote

$$
R = \begin{bmatrix}
I & 0 & 0 & -A_i^{(j)} \\
0 & I & 0 & 0 \\
0 & 0 & I & -C_i^{(j)} \\
0 & 0 & 0 & I
\end{bmatrix}
\quad \text{and} \quad
\bar{R} = \begin{bmatrix}
I & 0 & 0 & -A_i^{(j)} & 0 & \cdots \\
0 & I & 0 & 0 & 0 & \cdots \\
0 & 0 & I & -C_i^{(j)} & 0 & \cdots \\
0 & 0 & 0 & I & 0 & \cdots \\
0 & 0 & 0 & 0 & I & \cdots \\
\vdots & \vdots & \vdots & \ddots & \ddots & \ddots
\end{bmatrix}.
$$

The matrix $\bar{R}$ is a square matrix with $(M+1)m + r + q$ rows. Multiplying (3.16a) by R, from the left, and by R', from the right, and multiplying (3.16b) by $\bar{R}$, from the left, and by $\bar{R}'$, from the right, the conditions of Corollary 3.1 are readily obtained.   ∎

We note that the method of [9] is usually used to reduce the conservatism entailed in applying a single-quadratic LF by using a parameter-dependent LF. In our case, since the switching law depends on the LF, one cannot apply a parameter-dependent LF without having a perfect knowledge of the parameters. Still, the introduction of the slack variables $Z_{i,h}$, and $H_{i,h}$ allows one to obtain simpler controllers (e.g., constant feedback gains between switching).

Replacing $A_i^{(j)}$ and $C_i^{(j)}$ by $A_i^{(j)} + B_{2,i}^{(j)} G_{i,k}$ and $C_i^{(j)} + D_{12,i}^{(j)} G_{i,k}$, respectively, denoting $Y_{i,k} = G_{i,k} Z_{i,k}$, and demanding $H_{i,k} = \eta Z_{i,k}$ the following result is obtained, for a prechosen integer K and prescribed $\gamma > 0$.

Corollary 3.4 (State feedback) *Assume there exist: a collection of matrices* $Q_{i,k} > 0$, $Z_{i,k}$, $H_{i,k} \in \mathcal{R}^{n \times n}$, $Y_{i,k} \in \mathcal{R}^{s \times n}$, $i = 1, \dots M, k = 0, \dots K$, *and a set of scalars* $\lambda_{i,l} \geq 0$, $i = 1 \dots M$, $l \in L_i$, *such that, for a tuning scalar parameter* η, *the following holds.*

$$\begin{bmatrix} \Delta_{i,k,j} + \Phi_{i,h,j} & * & * & * \\ B_i^{(j)'} & -\gamma^2 I & * & * \\ \Upsilon_{i,h,j} & D_i^{(j)} & -I & * \\ \psi_{i,h,j} & 0 & \bar{\Upsilon}_{i,h,j} - \hat{H}_{i,h} \end{bmatrix} < 0, \quad k = 0, \dots K-1, \quad h = k, k+1,$$

and

$$\begin{bmatrix} \Psi_{i,j} & * & * & * & * & & \\ B_i' & -\gamma^2 I & * & * & * & * & * \\ \Upsilon_{i,K} & D_i & -I & * & * & * & \\ \psi_{i,K,j} & 0 & \bar{\Upsilon}_{i,K} - \hat{H}_{i,K} & * & * & \\ Q_{i,K} & 0 & 0 & 0 & \frac{-1}{\lambda_{i,l_1^{(i)}}} Q_{l_1^{(i)},0} & * \\ \vdots & \vdots & \vdots & \ddots & & \ddots & \vdots \\ Q_{i,K} & & & \cdots & & \frac{-1}{\lambda_{i,l_{M-1}^{(i)}}} Q_{l_{M-1}^{(i)},0} \end{bmatrix} < 0, \qquad (3.17)$$

where

$$H_{i,k} = \eta Z_{i,k}, \quad \Phi_{i,h,j} = A_i^{(j)} Z_{i,h} + Z_{i,h}' A_i^{(j)'} + B_{2,i}^{(j)} Y_{i,h} + Y_{i,h}' B_{2,i}^{(j)'},$$

$$\Gamma_i = -Q_{i,K} \Sigma_{l \in L_i} \lambda_{i,l}, \quad \Delta_{i,k,j} = -\frac{Q_{i,k+1}^{(j)} - Q_{i,k}^{(j)}}{T/K},$$

$$\Psi_{i,j} = A_i^{(j)} Z_{i,K} + Z_{i,K}' A_i^{(j)'} + B_{2,i}^{(j)} Y_{i,K} + Y_{i,K}' B_{2,i}^{(j)'} + \Gamma_i,$$

$$\Upsilon_{i,h,j} = C_i^{(j)} Z_{i,h} + D_{12,i}^{(j)} Y_{i,h},$$

$$\psi_{i,h,j} = Z_{i,h} - Q_{i,h} - H_{i,h}' A_i^{(j)'} - \eta Y_{i,h}' B_{2,i}^{(j)'},$$

$$\bar{\Upsilon}_{i,h,j} = -H_{i,h}' C_i^{(j)'} - \eta Y_{i,h}' D_{12,i}^{(j)'}, \quad \hat{H}_{i,h} = H_{i,h} + H_{i,h}'.$$

Then, the switching law (3.8), where we substitute $P_{i,K} = Q_{i,K}^{-1}$ *and* $P_{l,0} = Q_{l,0}^{-1}$, *and the feedback law* $u(t) = G(t)x(t)$, *where* $G(t)$ *is given as in (3.15) with* $Z_{i,k}$ *replacing* $Q_{i,k}$, *globally asymptotically stabilizes the system (3.1) with the uncertainty*

(3.2),(3.3), and with a constraint of dwell time $T > 0$. It also guarantees a bound γ for the $\mathcal{L}_2$-gain of the system.

The matrices $Q_i(t)$ must be time-varying; one can require, however, $Z_i(t)$ and $Y_i(t)$ to be constant between the different switching instances, namely $Z_{i,k_1} = Z_{i,k_2}$, $Y_{i,k_1} = Y_{i,k_2}$ $\forall k_1, k_2 = 0 \ldots K$, thus obtaining a controller that is constant between the switching instances. Alternatively, one can obtain a controller that changes piecewise linearly by requiring only $Z_i(t)$ to be constant between the different switching instances, namely $Z_{i,k_1} = Z_{i,k_2}$, $\forall k_1, k_2 = 0 \ldots K$.

3.7 Examples

Example 3.1 *(L_2-gain)* We consider the nominal system of [6] where

$$A_1 = \begin{bmatrix} 0 & 1 \\ 2 & -9 \end{bmatrix} \quad \text{and} \quad A_2 = \begin{bmatrix} 0 & 1 \\ -2 & 8 \end{bmatrix}.$$

These subsystems are both unstable. Using Theorem 3.2, this system is found to be stabilized by the switching law (3.8), with a dwell time of 0.002s and $K = 1$, taking $\lambda_{1,2} = \lambda_{2,1} = 300$.

Considering the nominal system (3.1) with

$$A_1 = \begin{bmatrix} 0 & 1 \\ 2 & -9 \end{bmatrix}, \quad A_2 = \begin{bmatrix} 0 & 1 \\ -2 & 8 \end{bmatrix}, \quad B_1 = B_2 = \begin{bmatrix} 0 \\ 0.1 \end{bmatrix}, \quad C_1 = C_2 = \begin{bmatrix} 0.1 & 0 \end{bmatrix},$$
$$D_1 = D_2 = 0,$$

the system is found by Theorem 3.3 to be stabilized by the switching law (3.8), with a dwell time of $0.001\,\text{s}$ and $K = 1$ with an $\mathcal{L}_2$-gain bound of 2.65, using $\lambda_{1,2} = \lambda_{2,1} = 360$. Similarly, taking $K = 5$ we obtain $\gamma = 1.94$, and for $K = 10$ we obtain $\gamma = 1.87$.

Example 3.2 *(Robust state feedback)* We consider the system (3.1) with

$$A_1 = \begin{bmatrix} 0 & 1 \\ 2 + \delta & -9 \end{bmatrix}, \quad A_2 = \begin{bmatrix} 0 & 1 \\ -2 + \delta & 8 \end{bmatrix}, \quad B_1 = B_2 = \begin{bmatrix} 0 \\ 1 \end{bmatrix}, \quad B_{2,1} = B_{2,2} = \begin{bmatrix} 1 \\ 1 \end{bmatrix}$$
$$C_1 = C_2 = \begin{bmatrix} 1 & 0 \\ 0 & 0 \end{bmatrix}, \quad D_1 = D_2 = \begin{bmatrix} 0 \\ 0 \end{bmatrix} \quad D_{12,1} = D_{12,2} = \begin{bmatrix} 0 \\ 5 \end{bmatrix},$$

where $|\delta| \le 50$ is an uncertain scalar parameter. Seeking state-feedback gains that are constant between the switching instances, this system is found, using Corollary 3.4, to be stabilized by the switching law (3.8), for a dwell time of $0.001\,\text{s}$ and $K = 1$, with an $\mathcal{L}_2$-gain bound of 0.13, by taking $\lambda_{1,2} = \lambda_{2,1} = 5$ and $\eta = 0.005$. The corresponding feedback gains are $G_1 = [-21.9226 \ -0.7509]$ and $G_2 = [-27.1678 \ -20.5190]$. We note that allowing the feedback gains to be time-varying during the dwell time does not improve the minimal bound on γ. The $\mathcal{L}_2$-gain bound that can be guaranteed by a time-varying controller using Corollary 3.2 is 0.35.

Example 3.3 (*L_2-gain*) We consider a Boost Converter which is a switched power converter whose purpose is to amplify DC voltage with high efficiency [10]. The state vector of the system is $x = [I_L \ V_C \ 1]'$, where I_L is the inductors current, V_C is the capacitors voltage, and the third state is approximately one during the system's activity. The objective vector is the output voltage, measured on the load. The disturbance in this description is an output current disturbance. The first subsystem describes the system dynamics when the inductor is connected to the load, and the second is when the inductor is connected to the ground. We have [10]:

$$A_1 = \begin{bmatrix} -\frac{r_L}{L} & -\frac{1}{L} & \frac{V_{in}}{L} \\ \frac{1}{C} & -\frac{1}{C}\frac{1}{r_o} & 0 \\ 0 & 0 & -\epsilon \end{bmatrix}, \quad A_2 = \begin{bmatrix} -\frac{r_L}{L} & 0 & \frac{V_{in}}{L} \\ 0 & -\frac{1}{C}\frac{1}{r_o} & 0 \\ 0 & 0 & -\epsilon \end{bmatrix},$$

$$B_1 = B_2 = \begin{bmatrix} 0 \\ \frac{1}{C} \\ 0 \end{bmatrix}, \quad C_1 = C_2 = \begin{bmatrix} 0 \ 1 \ -V_{des} \end{bmatrix},$$

$$D_1 = D_2 = 0,$$

with: $V_{in} = 100V$, $r_L = 2\Omega$, $L = 10^{-4}H$, $r_o \in [50 - 100]\Omega$, $C = 4.7 \cdot 10^{-4}F$, $V_{des} = 200V$, $\epsilon = 10^{-4}$. In the later, L is the inductor inductance, r_L is the inductor serial resistance, C is the capacitor capacitance, V_{in} is the input voltage, V_{des} is the desired output voltage, and r_o is the load's resistance.

Using Theorem 3.3, this system is shown to achieve an L_2-gain bound of 5.41 by the switching law (3.8), for a given dwell time of $5 \cdot 10^{-4}$ s. This is found for $K = 1$, with $\lambda_{1,2} = \lambda_{2,1} = 2680$. We note that taking a smaller dwell time, or eliminating it altogether, a higher bound on the L_2-gain is obtained.

3.8 Conclusions

A new method for the analysis and synthesis of state-dependent switching law under a dwell time constraint is introduced, using a switching-dependent LF of quadratic structure. This method enables the treatment of systems with polytopic-type parameter uncertainties, and it can also be easily extended to treat norm-bounded parameter uncertainties. When the dwell time tends to zero, our results converge to those of [6]. The new method is extended to verify bounds on the L_2-gain of the switching system and to derive a state-feedback control law that robustly stabilizes the system and guarantees a prescribed L_2-gain bound for the closed-loop system.

Corollary 3.1 is introduced in order to provide a simple platform for solving the state-feedback control problem. Application of its inequalities to the derivation of the bound on the L_2-gain of the system may lead to results that are different from those obtained by Theorem 3.2. In the latter derivation, the results of Corollary 3.1 may be quite conservative when dealing with unstable switched systems.

Deliberately introducing a dwell-time constraint in the switching law can, in some cases, lead to a reduced bound on the $\mathcal{L}_2$-gain of the system, e.g., Example 3.3. This is in line with the result of [5], and it is probably due to the fact that the switching law is not optimal.

The results in this chapter have been focused on switching between subsystems. Their extension to switching controllers is the subject of the next chapter.

In the example section, it has been demonstrated that, in some cases, addition of a dwell time constraint improves the system's performance. This may happen because the switching law that is presented in this chapter is suboptimal, and thus adding or changing the dwell time may, in some cases, lead to a smaller $\mathcal{L}_2$-gain.

References

1. Hespanha, J.P.: Uniform stability of switched linear systems: extensions of LaSalle's invariance principle. IEEE Trans. Autom. Control **49**, 470–482 (2004)
2. Colaneri, P.: Dwell time analysis of deterministic and stochastic switched systems. Eur. J. Autom. Control **15**, 228–249 (2009)
3. Sun, Z., Ge, S.S.: Stability Theory of Switched Dynamical Systems. Springer, New York (2011)
4. Boniolo, I., Colaneri, P., Bolzern, P., Corless, M., Shorten, R.: On the design and synthesis of limit cycles using switching linear systems. Int. J. Control **83**, 915–927 (2010)
5. Margaliot, M., Langholz, G.: Necessary and sufficient conditions for absolute stability: the case of second-order systems. IEEE Trans. Circuits Syst. Fundam. Theory Appl. **50**, 227–234 (2003)
6. Geromel, J., Colaneri, P.: Stability and stabilization of continuous-time switched linear systems. SIAM J. Control Optim. **45**(5), 1915–1930 (2006)
7. Boyarski, S., Shaked, U.: Time-convexity and time-gain-scheduling in finite-horizon robust H_∞-control. In: Proceedings of the 48th CDC09, Shanghai, China (2009)
8. Boyd, S., El Ghaoui, L., Feron, E., Balakrishnan, V.: Linear Matrix Inequality in Systems and Control Theory. SIAM Frontier Series (1994)
9. de Oliveira, M.C., Skelton, R.E.: Stability test for constrained linear systems. In: Reza Moheimani, S.O. (ed.) Perspectives in Robust Control. Lecture Notes in Control and Information Sciences, vol. 268. Springer, London (2001)
10. Erickson, R.: Fundamentals of Power Electronics. Springer, New York (1997)

Chapter 4
Robust Control of Linear Systems via Switching

Abstract An alternative approach to robust control is presented where the uncertainty polytope is divided into overlapping smaller regions and where each of these regions is assigned to a separate subsystem. Assuming that there is an online information on which of the regions the parameters of the system move to, the method of the previous chapters for H_∞ design of switched system with dwell time is applied. In order to handle the switching between the subsystems, a Lyapunov Function (LF) in a quadratic form, which is nonincreasing at the switching instances, is assigned to each subsystem. This function is used to determine the stability and to find a bound on the L_2-gain of the switched system. The obtained results are then used to solve the corresponding robust H_∞ state-feedback and static output-feedback control problems. Two practical examples are given which show that, by deliberately introducing switching, the obtained designs are less conservative than those achieved by the available standard techniques.

4.1 Introduction

One of the main reasons for applying feedback control is to lower the sensitivity of the system to parameter uncertainty [1]. Various methods have been suggested in the frequency domain to tackle uncertainty [1–4] via appropriately shaping the system loop transmission. It was the seminal work of [5] that enabled H_∞-analysis and design of systems in the state space. It was soon realized that the H_∞-control problem can be solved using Linear Matrix Inequalities (LMIs) [6, 7]. In many control problems, these LMIs were found to be affine in the system parameters, which allowed a solution to these problems, in the case where the uncertainty is of the polytopic type, by simultaneously solving the LMIs for all the vertices of the uncertainty polytope [8, 9]. These LMIs have also been used to solve the robust H_∞ control problem using Gain Scheduling (GS), whereby measuring the system parameters online a convex combination of the vertex controllers is continuously formed [10].

© Springer Nature Switzerland AG 2019

E. Gershon and U. Shaked, *Advances in H∞ Control Theory*,
Lecture Notes in Control and Information Sciences 481,
https://doi.org/10.1007/978-3-030-16008-1_4

The above solutions via LMIs are based on the convexity of these inequalities, which requires one or more decision variables there to be constant over the entire uncertainty polytope. This requirement leads to conservative solutions, especially in problems with large uncertainty. Attempts were made in the past to reduce this conservatism by applying some transformations on the resulting LMIs [11], but the fact remained that some matrices in the resulting LMIs had to be constant over the uncertainty polytope. This fact has lead to an overdesign that may even fail to produce a solution in cases with large uncertainties.

The method that was developed in Chap. 3 can handle large polytopic parameter uncertainties and, for the first time, it applies parameter-dependent LF to uncertain switched systems. We transform in the present chapter the problem of robust control of system with large polytopic parameter uncertainty to one of controlling switched systems with dwell time. Dividing the uncertainty polytope to overlapping sub-polytopes, and grossly measuring these parameters to determine which sub-polytope these parameters reside in, a separate subsystem is assigned to each sub-polytope and the problem becomes one of controlling a switched system.

In [13], a piecewise linear in time LF in a quadratic form has been used to reduce the robust H_∞ design conservatism, and in Chap. 3 this function has been used to solve the problems of stability and stabilization of switched linear systems with dwell time. We apply in the present chapter the same LF to find and minimize the L_2-gain of uncertain switched linear systems with dwell time. The LF applied is nonincreasing at the switching instants and is assigned separately to each subsystem. During the dwell time, this function varies piecewise linearly in time and it becomes time invariant afterward. Our choice of the LF allows derivation of sufficient conditions for robust stability, and finding an upper-bound on the L_2-gain of the uncertain switched system, all in terms of LMIs.

Our purpose is to apply switching to robust control of uncertain linear systems with grossly measured parameters. In the example section we demonstrate the fact that, by applying switching in this setting, improved results can be obtained, in comparison to existing robust control design methods and to the Gain-Scheduling approach that requires accurate information about the system's parameters.

The system considered in the chapter is described in Sect. 4.2, where the transformation to a switched system is introduced. Applying the results obtained in Chap. 3 for the robust stability of switched systems, a criterion which is similar to the Bounded Real Lemma (BRL) [9], is developed in Sect. 4.3 which allows a derivation of an upper-bound to the L_2-gain of the uncertain transformed system. The resulting criterion is used in Sect. 4.4 to solve the robust state-feedback control problem in cases where the switching signal is measured online. In Sect. 4.5, we address the problem of robust H_∞ static output-feedback control. A sufficient condition for the existing of switching constant output-feedback gains that stabilize the uncertain system and guarantee a prescribed bound on the H_∞-norm of the closed-loop system is derived. In Sect. 4.6, the theory developed is applied to the control of the longitudinal short period mode of a F4E fighter aircraft with additional canards. Both state- and output-feedback switching gains are obtained there that, in spite of large parameter uncertainty, stabilize the plane and guarantee a given bound to its L_2-gain . A comparison to results that are obtained using the standard robust H_∞ and GS designs is given there.

4.2 Problem Formulation

We consider the following linear system:

$$
\begin{aligned}
\dot{x}(t) &= Ax(t) + B_1 w(t) + B_2 u(t), \quad x(0) = 0, \\
z(t) &= C_1 x(t) + D_{11} w(t) + D_{12} u(t), \\
y(t) &= C_2 x(t) + D_{21} w(t),
\end{aligned}
\tag{4.1}
$$

where $x \in \mathcal{R}^n$ is the state, $w \in \mathcal{R}^q$ is an exogenous disturbance in $\mathcal{L}_2$, $z \in \mathcal{R}^r$ is the objective vector, $u \in \mathcal{R}^p$ is the control signal, and $y \in \mathcal{R}^s$ is the measurement. The systems parameters are uncertain and they are assumed to reside in the following polytope:

$$
\Omega = \begin{bmatrix} A & B_1 & B_2 \\ C_1 & D_{11} & D_{12} \\ 0 & D_{21} & 0 \end{bmatrix} = \sum_{j=1}^{N} \eta_j(t)\Omega^{(j)}, \quad \sum_{j=1}^{N} \eta_j(t) = 1, \ \eta_j(t) \geq 0, \text{ and}
\tag{4.2}
$$

$$
\Omega^{(j)} = \begin{bmatrix} A^{(j)} & B_1^{(j)} & B_2^{(j)} \\ C_1^{(j)} & D_{11}^{(j)} & D_{12}^{(j)} \\ 0 & D_{21}^{(j)} & 0 \end{bmatrix}, \quad j = 1, \dots N.
$$

We assume that C_2 is known. In the case where the latter matrix encounters uncertainty, a standard routine which is used in GS design [14] can be applied. Applying this technique, the parameter dependence of C_2 is inserted into A.

The conventional way of treating this uncertainty in stability analysis and in feedback design is to consider the whole polytope as one entity. The present chapter suggests an alternative approach. We divide the polytope Ω into overlapping sub-polytopes: $\Omega = \bigcup_{i=1,\dots M} \Omega_i$ where

$$
\Omega_i = \begin{bmatrix} A_i & B_{1,i} & B_{2,i} \\ C_{1,i} & D_{11,i} & D_{12,i} \\ 0 & D_{21,i} & 0 \end{bmatrix} = \sum_{j=1}^{N} \eta_i^j(t)\Omega_i^{(j)}, \quad \sum_{j=1}^{N} \eta_i^j(t) = 1, \ \eta_i^j(t) \geq 0, \text{ and}
\tag{4.3}
$$

$$
\Omega_i^{(j)} = \begin{bmatrix} A_i^{(j)} & B_{1,i}^{(j)} & B_{2,i}^{(j)} \\ C_{1,i}^{(j)} & D_{11,i}^{(j)} & D_{12,i}^{(j)} \\ 0 & D_{21,i}^{(j)} & 0 \end{bmatrix}.
$$

Fig. 4.1 The polytope Ω, divided into 2 overlapping regions

We consider then the whole system as a switched system, where each of these sub-polytopes is considered to be the uncertainty polytope of a subsystem of the original system. We assume that the rate of change of the parameters is bounded, which along with the overlapping between the sub-polytopes implies a dwell-time constraint on the switching between the subsystems. We also assume that the system's parameters are grossly measured in real-time, so that it is known in which region the system parameters reside at any given time. Dissimilar to the GS method [10], the knowledge of the exact location of the parameters inside Ω is not required.

We illustrate in Fig. 4.1 the split of Ω for a single uncertain parameter ($N = 2$). The overlap of the dashed and the dotted regions is a solid line. The dotted and the dashed regions would each correspond and define a subsystem of the switched system, while the size of the solid line and the bound on the rate of change of the single parameter determine the dwell time.

In the sequel, we consider the switched system that arises from the method described above. In the example section, we demonstrate how this method is applied to state-feedback control and static output-feedback designs and how the proposed approach allows us to obtain lower bounds on the L_2-gain of the system.

The switched version of (4.1a–d) is the following:

$$\begin{aligned}
\dot{x}(t) &= A_{\sigma(t)}x(t) + B_{1,\sigma(t)}w(t) + B_{2,\sigma(t)}u(t), \quad x(0) = 0, \\
z(t) &= C_{1,\sigma(t)}x(t) + D_{11,\sigma(t)}w(t) + D_{12,\sigma(t)}u(t), \\
y(t) &= C_2 x(t) + D_{21,\sigma(t)}w(t),
\end{aligned} \tag{4.4}$$

where the switching rule $\sigma(t)$, for each $t \geq 0$, is such that $\Omega_{\sigma(t)} \in \{\Omega_1, \ldots, \Omega_M\}$.

In this chapter, we assume a dwell-time constraint on the switching signal $\sigma(t)$, which means that if the dwell time is T, and the switching instances are $\tau_1, \tau_2, \ldots$, then $\tau_{s+1} - \tau_s \geq T, \quad \forall s \geq 1$.

4.3 L_2-gain

Given the system (4.4) where $A_i \in \mathcal{R}^{n \times n}, i = 1, \ldots M$ are stability matrices, and the overlapped sub-polytopes (4.3). This system emerges from (4.1a–d). We seek a sufficient condition for the asymptotic stability of the system and for the following performance criterion to hold for a prescribed scalar γ.

$$J = \int_0^\infty (z'z - \gamma^2 w'w)dt \leq 0, \ \forall \, w \in \mathcal{L}_2. \tag{4.5}$$

For this purpose, we consider the corresponding switched system (4.4). Let $i_0 = \sigma(0)$, and let $\tau_1, \tau_2, \ldots$ be the switching instances, where $\tau_{s+1} - \tau_s \geq T$ $\forall s = 1, 2, \ldots$. Define $\tau_{s,k} = \tau_s + k\frac{T}{K}$ for $k = 0, \ldots, K$. Note that the dwell-time constraint implies $\tau_{s,K} \leq \tau_{s+1,0} = \tau_{s+1}$. As in [12], we choose the LF $V(t)$:

$$V(t) = x'(t) P_{\sigma(t)}(t) x(t), \tag{4.6}$$

where $P_{\sigma(t)}(t)$ is chosen to be

$$P_{\sigma(t)}(t) = \begin{cases} P_{i,k} + (P_{i,k+1} - P_{i,k})\frac{t - \tau_{s,k}}{T/K} & t \in [\tau_{s,k}, \tau_{s,k+1}), \quad k = 0, 1, \ldots K - 1, \\ P_{i,K} & t \in [\tau_{s,K}, \tau_{s+1,0}) \\ P_{i_0,K} & t \in [0, \tau_1) \end{cases}$$

$$s = 1, 2, \ldots, \tag{4.7}$$

where i is the index of the subsystem that is active at time t and where K is an integer that is chosen a priori, according to the allowed computational complexity.

Remark 4.1 The Lyapunov function of (4.6) and Theorem 4.1 below are derived for a prescribed integer K that is chosen according to the allowed computational burden. A larger K leads to results that are less conservative at the price of larger computational complexity. It has been demonstrated in [12] that, letting K go to infinity, the best result that has been obtained so far (in [20]) for the stability of switched systems, without uncertainty but with dwell- time constraint and quadratic LF, is recovered.

We denote

$$\bar{J} = lim_{t \to \infty}\{V(t) + \int_0^\infty (z'z - \gamma^2 w'w)ds\}. \tag{4.8}$$

Since $V(t) \geq 0$ $\forall t$, we have that $J \leq \bar{J}$. Taking into account that $V(t)$ is differentiable for all t, except for the switching instances, and that $x(0) = 0$, we write

$$lim_{t \to \infty} V(t) = \sum_{s=0}^\infty \int_{\tau_s}^{\tau_{s+1}} \dot{V}(t)dt + \sum_{s=1}^\infty (V(\tau_s) - V(\tau_s^-)), \tag{4.9}$$

where $\tau_0 = 0$. If the conditions for nonincreasing $V(t)$ at the switching instances are satisfied, we find that $V(\tau_h) - V(\tau_h^-) \leq 0$, $\forall h > 0$, and then $lim_{t \to \infty} V(t) \leq \sum_{s=1}^\infty \int_{\tau_s}^{\tau_{s+1}} \dot{V}(t)dt$.

Denoting

$$\tilde{J} = \sum_{s=1}^\infty \int_{\tau_s}^{\tau_{s+1}} \dot{V}(t)dt + \int_0^\infty (z'z - \gamma^2 w'w)ds = \sum_{s=1}^\infty \int_{\tau_s}^{\tau_{s+1}} (\dot{V}(s) + z'z - \gamma^2 w'w)ds$$

$$\tag{4.10}$$

we thus obtain that $J \leq \bar{J} \leq \tilde{J}$. Consequently, if $\tilde{J} \leq 0$, and the above LF does not increase at the switching instants, the L_2-gain of the switched system will be less than or equal to γ. These conditions are included in the following theorem.

Theorem 4.1 *The L_2-gain of a nominal version of (4.4) (a system without uncertainty) is less than a prescribed $\gamma > 0$ for a dwell time of $\bar{T} \geq T$ if there exists a collection of matrices $P_{i,k} > 0$, $i = 1, \ldots M$, $k = 0, \ldots K$ of compatible dimensions, where K is a prescribed integer, such that, for all $i = 1, \ldots M$ the following holds:*

$$\begin{bmatrix} \frac{P_{i,k+1}-P_{i,k}}{T/K} + P_{i,h}A_i + A_i'P_{i,h} & P_{i,h}B_{1,i} & C_{1,i}' \\ * & -\gamma^2 I & D_{11,i}' \\ * & * & -I \end{bmatrix} < 0,$$

$$h = k, k+1, \quad k = 0, \ldots K-1,$$

$$\begin{bmatrix} P_{i,K}A_i + A_i'P_{i,K} & P_{i,K}B_{1,i} & C_{1,i}' \\ * & -\gamma^2 I & D_{11,i}' \\ * & * & -I \end{bmatrix} < 0, \quad and \quad P_{i,K} - P_{l,0} \geq 0, \quad \forall l \in L_i.$$

$$(4.11)$$

Proof Let $P(t)$ be as in (4.7). The positive definitness of $P_{i,k}$ guarantees that the LF is positive for any $x \neq 0$. From the standard derivation of the BRL for linear systems [9], we have that the following conditions are equivalent:

$$\dot{V} + z'z - \gamma^2 w'w < 0,$$

$$\begin{bmatrix} \dot{P}(t) + P(t)A + A'P(t) & P(t)B & C' \\ * & -\gamma^2 I & D' \\ * & * & -I \end{bmatrix} < 0,$$

$$\begin{bmatrix} -\dot{Q}(t) + AQ(t) + Q(t)A' & B & C'Q(t) \\ * & -\gamma^2 I & D' \\ * & * & -I \end{bmatrix} < 0, \qquad (4.12)$$

where $Q(t) = P^{-1}(t)$.

It follows from [12] that condition (4.11c) means that once a switching occurs, $P(t)$ switches in such a way that it does not increase at the switching instant of time. From the switching instant on, $P(t)$ becomes piecewise linear, where over the time interval $t \in [\tau_{s,k}, \tau_{s,k+1}]$, it changes linearly from $P_{i,k}$ to $P_{i,k+1}$, where i is the index of the active subsystem at $t \in [\tau_s, \tau_{s+1}]$.

Before the first switching occurs, $P(t)$ is constant and by (4.11b) it guarantees that $V(t)$ decreases while the system remains in the same subsystem. It follows then from (4.11a) that $V(t)$ is strictly decreasing during the dwell time, the LMIs (4.11b) guarantees that the LF is strictly decreasing for any $t \in [\tau_{s,K}, \tau_{s+1,0})$, and (4.11d) guarantees that the LF is nonincreasing at the switching instances. Since the switching points are distinct, the system is asymptotically stable according to Lasalle's invariance principle.

The first diagonal block of the LMIs (4.11a) and (4.11b) corresponds to $\dot{V}$ of (4.10). The second and the third columns and rows in these LMIs correspond to $w'w$ and $z'z$, respectively. The result of the theorem thus follows using the derivation of the standard BRL [9]. ∎

Corollary 4.1 *The L_2-gain of the system (4.1) with the uncertainty (4.2) and the corresponding sub-polytopes (4.3) is less than a positive γ for a dwell time of $\bar{T} \geq T$ if there exists a collection of matrices $P_{i,k} > 0$, $i = 1, \ldots M$, $k = 0, \ldots K$ of compatible dimensions, where K is a prescribed integer, such that, for all $i = 1, \ldots M$, and $j = 1, \ldots N$ the following holds:*

$$\begin{bmatrix} \frac{P_{i,k+1}-P_{i,k}}{T/K} + P_{i,h}A_i^{(j)} + A_i^{(j)'}P_{i,h} & P_{i,h}B_{1,i}^{(j)} & C_{1,i}^{(j)'} \\ * & -\gamma^2 I & D_{11,i}^{(j)'} \\ * & * & -I \end{bmatrix} < 0$$

$$h = k, k+1, \text{ and } k = 0, \ldots K-1,$$

$$\begin{bmatrix} P_{i,K}A_i^{(j)} + A_i^{(j)'}P_{i,K} & P_{i,K}B_{1,i}^{(j)} & C_{1,i}^{(j)'} \\ * & -\gamma^2 I & D_{11,i}^{(j)'} \\ * & * & -I \end{bmatrix} < 0, \text{ and } P_{i,K} - P_{l,0} \geq 0, \quad \forall l \in L_i.$$

$$(4.13)$$

Corollary 4.2 *The L_2-gain of the system (4.1) with the uncertainty (4.3) is less than a positive γ for a dwell time of $\bar{T} \geq T$ if there exists a collection of matrices $Q_{i,k} > 0$, $i = 1, \ldots M$, $k = 0, \ldots K$ of compatible dimensions, where K is a prescribed integer, such that, for all $i = 1, \ldots M$, and $j = 1, \ldots N$ the following holds:*

$$\begin{bmatrix} -\frac{Q_{i,k+1}-Q_{i,k}}{T/K} + Q_{i,h}A_i^{(j)'} + A_i^{(j)}Q_{i,h} & B_{1,i}^{(j)} & Q_{i,h}C_{1,i}^{(j)'} \\ * & -\gamma^2 I & D_{11,i}^{(j)'} \\ * & * & -I \end{bmatrix} < 0,$$

$$h = k, k+1, \text{ and } k = 0, \ldots K-1,$$

$$\begin{bmatrix} Q_{i,K}A_i^{(j)'} + A_i^{(j)}Q_{i,K} & B_{1,i}^{(j)} & Q_{i,K}C_{1,i}^{(j)'} \\ * & -\gamma^2 I & D_{11,i}^{(j)'} \\ * & * & -I \end{bmatrix} < 0,$$

and

$$Q_{i,K} - Q_{l,0} \leq 0, \quad \forall l \in L_i. \tag{4.14}$$

Proof Let $Q(t) = P^{-1}(t)$. We depart from the piecewise linearity in time assumption on $P(t)$, and assign this piecewise linearity to $Q(t)$, namely,

$$Q_{\sigma(t)}(t) = \begin{cases} Q_{i,k} + (Q_{i,k+1} - Q_{i,k})\frac{t-\tau_{s,k}}{T/K} & t \in [\tau_{s,k}, \tau_{s,k+1}) \\ Q_{i,K} & t \in [\tau_{s,K}, \tau_{s+1,0}), \quad s = 1, 2, \ldots. \\ Q_{i_0,K} & t \in [0, \tau_1) \end{cases}$$

$$(4.15)$$

Substituting $Q(t)$ into (4.12c), and taking into account that $Q(t)$ is now piecewise linear in time, we obtain conditions (4.14a) for the period of the dwell time. Since after the dwell time, $Q(t)$ is constant until the next switching occurs, the latter multiplication leads to the condition (4.14b).   ∎

Assigning a single LF to each subsystem entails conservatism in the case where these subsystems encounter uncertainty. In order to reduce the conservatism, we apply the method of [11], that leads to a parameter-dependent LF.

Lemma 4.1 *For a nominal linear system* $\dot{x} = Ax + Bw$, $z = Cx + Dw$ *the following conditions guarantee an* L_2-*gain bound on the system and are equivalent:*
(I) There exists $Q > 0$ *of the appropriate dimensions such that*

$$\begin{bmatrix} -\dot{Q} + AQ + QA' & B & C'Q \\ * & -\gamma^2 I & D' \\ * & * & -I \end{bmatrix} < 0.$$

(II)There exist $Q > 0$, R, H *of the appropriate dimensions such that*

$$\begin{bmatrix} -\dot{Q} + AR(t) + R'A' & B & R'C' & Q - R' + AH \\ * & -\gamma^2 I & D' & 0 \\ * & * & -I & CH(t) \\ * & * & * & -H - H' \end{bmatrix} < 0. \qquad (4.16)$$

Proof To prove that (II) leads to (I), multiply (4.16a) from the left by R_T and R'_T from the right and the condition of (4.16b) is obtained, where

$$R_T = \begin{bmatrix} I & 0 & 0 & A \\ 0 & I & 0 & 0 \\ 0 & 0 & I & C \end{bmatrix}.$$

To prove that (I) leads to (II), choose $R = Q$, and $H = \epsilon Q$ where $0 < \epsilon \to 0$.   ∎

Since in condition (4.16b), the matrix Q is not multiplied by any of the systems matrices, one can take it to be vertex dependent, while still preserving the convexity of these conditions.

Taking condition (4.16b), and letting $Q_{\sigma(t)}(t)$ be piecewise linear in time as in (4.15), the following conditions are obtained.

Corollary 4.3 *The* L_2-*gain of the system (4.1) is less than a prescribed positive scalar* γ *if there exists a collection of matrices* $R_{i,k}$, $H_{i,k}$, $Q_{i,k}^{(j)} > 0$, $i = 1, \ldots M$, $j = 1, \ldots, N$, $k = 0, \ldots K$ *of compatible dimensions, where* K *is a prescribed integer, such that, for all* $i = 1, \ldots M$, *and* $j = 1, \ldots N$, *the following holds:*

$$
\begin{bmatrix}
-\frac{Q^j_{i,k+1}-Q^j_{i,k}}{T/K}+\Psi_{i,h} & B^{(j)}_{1,i} & R'_{i,h}C^{(j)'}_{1,i} & Q^{(j)}_{i,h}-R^T_{i,h}+A^{(j)}_i H_{i,h} \\
* & -\gamma^2 I & D^{(j)'}_{11,i} & 0 \\
* & * & -I & C^{(j)}_{1,i}H_{i,h} \\
* & * & * & -H_{i,h}-H'_{i,h}
\end{bmatrix} < 0,
$$

$$
h = k, k+1, \text{ and } k = 0, \dots K-1,
$$

$$
\begin{bmatrix}
\Psi_{i,K} & B^{(j)}_{1,i} & R'_{i,K}C^{(j)'}_{1,i} & Q^{(j)}_{i,K}-R'_{i,K}+A^{(j)}_i H_{i,K} \\
* & -\gamma^2 I & D^{(j)'}_{11,i} & 0 \\
* & * & -I & C^{(j)}_{1,i}H_{i,K} \\
* & * & * & -H_{i,K}-H'_{i,K}
\end{bmatrix} < 0,
$$

$$
\text{and} \quad Q^j_{i,K}-Q^j_{l,0} \le 0, \quad \forall l = L_i,
$$

(4.17)

where $\Psi_{i,h} = R'_{i,h}A^{(j)'}_i+A^{(j)}_i R_{i,h}$.

4.4 State-Feedback Control

Given the system (4.1a–d) with the uncertainty (4.3), where $w \in \mathcal{L}_2$. A state-feedback controller is sought that stabilizes the system, and satisfies the performance criterion (4.5).

Theorem 4.2 *The L_2-gain of the system (4.1) with the uncertainty (4.3) and the state feedback $u(t) = G_{\sigma(t)}(t)x(t)$ is less than positive γ for a dwell time of $\bar{T} \ge T$ if there exists a collection of matrices $Q_{i,k} > 0$, Yi,k $i = 1, \dots M$, $k = 0, \dots K$ of compatible dimensions, where K is a prescribed integer, such that, for all $i = 1, \dots M$ and $j = 1, \dots N$, the following holds:*

$$
\begin{bmatrix}
\Upsilon_{1,1} & B^{(j)}_{1,i} & Q_{i,h}C^{(j)'}_{1,i}+Y'_{i,h}D^{(j)'}_{12,i} \\
* & -\gamma^2 I & D^{(j)'}_{11,i} \\
* & * & -I
\end{bmatrix} < 0,
$$

$$
h = k, k+1, \quad k = 0, \dots K-1
$$

$$
\begin{bmatrix}
\bar{\Upsilon}_{1,1} & B^{(j)}_{1,i} & Q_{i,K}C^{(j)'}_{1,i}+Y'_{i,K}D^{(j)'}_{12,i} \\
* & -\gamma^2 I & D^{(j)'}_{11,i} \\
* & * & -I
\end{bmatrix} < 0
$$

$$
Q_{i,K}-Q_{l,0} \le 0, \quad \forall l = 1, \dots i-1, i+1, \dots M,
$$

(4.18)

where

$$
\Upsilon_{1,1} = -\frac{Q_{i,k+1}-Q_{i,k}}{T/K}+Q_{i,h}A^{(j)'}_i+A^{(j)}_i Q_{i,h}+Y'_{i,h}B^{(j)'}_{2,i}+B^{(j)}_{2,i}Y_{i,h}
$$
$$
\bar{\Upsilon}_{1,1} = Q_{i,K}A^{(j)'}_i+A^{(j)}_i Q_{i,K}+Y'_{i,K}B^{(j)'}_{2,i}+B^{(j)}_{2,i}Y_{i,K}.
$$

The feedback gain matrix $G_{\sigma(t)}(t)$ is then given by

$$
G_{\sigma(t)}(t) = \begin{cases} G_{1,k} & t \in [\tau_{h,k}, \tau_{s,k+1}) \\ Y_{i,K} Q_{i,K}^{-1} & t \in [\tau_{s,K}, \tau_{s+1,0}) \\ Y_{i_0,K} Q_{i_0,K}^{-1} & t \in [0, \tau_1), \end{cases}
\tag{4.19}
$$

where

$$
G_{1,k} = [Y_{i,k} + \frac{t - \tau_{s,k}}{T/K}(Y_{i,k+1} - Y_{i,k})][Q_{i,k} + \frac{t - \tau_{s,k}}{T/K}(Q_{i,k+1} - Q_{i,k})]^{-1}.
$$

Proof Let $Y_{\sigma(t)}(t) \triangleq G_{\sigma(t)}(t)Q_{\sigma(t)}(t)$. In order to preserve the convexity in time of $Y_{\sigma(t)}(t)$, we demand that $Y_{\sigma(t)}(t)$ is piecewise linear in time as $Q_{\sigma(t)}(t)$ has been taken in (4.15). Replacing in Corollary 4.1 $A_i^{(j)}$ and $C_i^{(j)}$ by $A_i^{(j)} + B_{2,i}^{(j)} G_i(t)$ and $C_{1,i}^{(j)} + D_{12,i}^{(j)} G_i(t)$, respectively, the conditions of Theorem 4.2 readily follow. ∎

The state-feedback law of Theorem 4.2 may not be easy to implement. In order to obtain a simpler feedback gain, we introduce an alternative method to guarantee the L_2-gain of the system, recalling that $Q(t)$ must be time-varying. We apply the method of Corollary 4.3 and replace there $A_i^{(j)}$ and $C_i^{(j)}$ by $A_i^{(j)} + B_{2,i}^{(j)} G_i(t)$ and $C_{1,i}^{(j)} + D_{12,i}^{(j)} G_i(t)$, respectively. The following conditions are thus obtained.

Corollary 4.4 *The L_2-gain of the system (4.1) is less than a prescribed positive scalar γ if there exists a collection of matrices $R_{i,k}$, $Q_{i,k}^{(j)} > 0$, $i = 1, \ldots M$, $j = 1, \ldots, N$, $k = 0, \ldots K$ of compatible dimensions, and a scalar β, where K is a prescribed integer, such that, for all $i = 1, \ldots M$ and $j = 1, \ldots N$, the following holds:*

$$
\begin{bmatrix} -\frac{Q_{i,k+1}^j - Q_{i,k}^j}{T/K} + \Gamma_{i,h} & B_{1,i}^{(j)} & R_{i,h}' C_{1,i}^{(j)'} + Y_{i,h}' D_{12,i}^{(j)'} & \bar{Q}_{1,4} \\ * & -\gamma^2 I & D_{11,i}^{(j)'} & 0 \\ * & * & -I & \beta C_{1,i}^{(j)} R_{i,h} + \beta D_{12,i}^{(j)} Y_{i,h} \\ * & * & * & -\beta R_{i,h} - \beta R_{i,h}' \end{bmatrix} < 0,
$$

where $\bar{Q}_{1,4} = Q_{i,h}^{(j)} - R_{i,h}^T + \beta A_i^{(j)} R_{i,h} + \beta B_{2,i}^{(j)} Y_{i,h}$,

$h = k, k+1, \quad k = 0, \ldots K - 1$,

$$
\begin{bmatrix} \Psi_{i,K} & B_{1,i}^{(j)} & R_{i,K}' C_{1,i}^{(j)'} + Y_{i,K}' D_{12,i}^{(j)'} & Q_{i,K}^{(j)} - R_{i,K}' + \beta A_i^{(j)} R_{i,K} + \beta B_{2,i}^{(j)} Y_{i,K} \\ * & -\gamma^2 I & D_{11,i}^{(j)'} & 0 \\ * & * & -I & \beta C_{1,i}^{(j)} R_{i,K} + \beta D_{12,i}^{(j)} Y_{i,K} \\ * & * & * & -\beta R_{i,K} - \beta R_{i,K}' \end{bmatrix} < 0,
$$

and $Q_{i,K}^j - Q_{l,0}^j \le 0, \quad \forall l = L_i$,

(4.20)

where $\Gamma_{i,h} = R_{i,h}' A_i^{(j)'} + A_i^{(j)} R_{i,h} + Y_{i,h}' B_{2,i}^{(j)'} + B_{2,i}^{(j)} Y_{i,h}$.
The state-feedback gain is then given by

$$G_{\sigma(t)}(t) = \begin{cases} \bar{G}_{1,k} & t \in [\tau_{s,k}, \tau_{s,k+1}) \\ Y_{i,K} R_{i,K}^{-1} & t \in [\tau_{s,K}, \tau_{s+1,0}) \\ Y_{i_0,K} R_{i_0,K}^{-1} & t \in [0, \tau_1), \end{cases} \qquad (4.21)$$

where

$$\bar{G}_{1,k} = [Y_{i,k} + \frac{t - \tau_{s,k}}{T/K}(Y_{i,k+1} - Y_{i,k})][R_{i,k} + \frac{t - \tau_{s,k}}{T/K}(R_{i,k+1} - R_{i,k})]^{-1}.$$

Proof Let $Y_{\sigma(t)}(t) \overset{\Delta}{=} G_{\sigma(t)}(t) R_{\sigma(t)}(t)$ and $H_{\sigma(t)}(t) = \beta R_{\sigma(t)}(t)$. In order to preserve the convexity in time of $Y_{\sigma(t)}(t)$, we demand that $Y_{\sigma(t)}(t)$ is piecewise linear in time as $Q_{\sigma(t)}(t)$ has been. The conditions of Corollary 4.4 readily follow then. ∎

These state-feedback gains may not be easy to implement. However, in the present formulation, we can demand $R_{i,k}$ and $Y_{i,k}$ to be independent of k, which will make the state-feedback gain constant between the switching instances, while allowing $Q_{\sigma(t)}(t)$ to be time-varying during the dwell time.

4.5 Output-Feedback Control

This section deals with output-feedback control of systems with polytopic-type parameter uncertainties, a specific application of which is state estimation for systems with uncertain parameters. The problem of output-feedback control can be formulated as a static output-feedback control problem [6]. This may provide a solution to the output-feedback control problem using the solution , if exists, to the corresponding problem of static output feedback.

Consider the system (4.4) which is obtained from (4.1) by applying the polytope splitting of Sect. 4.2. Also consider the static output-feedback controller $u = G_{\sigma(t)}(t)y$. Denoting $\xi = col\{x, \ y\}$ we have

$$E\dot{\xi} = A_{cl,\sigma(t)}\xi + B_{cl,\sigma(t)}w, \quad z = C_{cl,\sigma(t)}\xi + D_{cl,\sigma(t)}w, \qquad (4.22)$$

where

$$A_{cl,\sigma(t)} = \begin{bmatrix} A_{\sigma(t)} & B_{2,\sigma(t)}G_{\sigma(t)} \\ C_2 & -I \end{bmatrix}, \quad B_{cl,\sigma(t)} = \begin{bmatrix} B_{1,\sigma(t)} \\ D_{21} \end{bmatrix},$$

$$C_{cl,\sigma(t)} = \begin{bmatrix} C_{1,\sigma(t)} & D_{12,\sigma(t)}G_{\sigma(t)} \end{bmatrix}, \quad D_{cl,\sigma(t)} = D_{11,\sigma(t)}, \text{ and } E = diag\{I, \ 0\}, \qquad (4.23)$$

and where $A_{cl,\sigma(t)}$, $B_{cl,\sigma(t)}$, $C_{cl,\sigma(t)}$, and $D_{cl,\sigma(t)}$ belong to the polytope that emerges from $\Omega_{\sigma(t)}$.

The L_2-gain conditions of Corollary 4.2 are not suitable for the descriptor system of (4.22). We, therefore, derive the corresponding conditions for (4.22) in which we shall substitute the closed-loop matrices of (4.23).

Consider the LF

$$V = x' P_{\sigma(t)}(t) x = \xi' \bar{P}' E \xi, \text{ where } \bar{P}_{\sigma(t)} = \begin{bmatrix} P_{\sigma(t)}(t) & 0 \\ -\alpha^{-1} C_2 P_{\sigma(t)}(t) & \alpha^{-1} \hat{P}_{\sigma(t)} \end{bmatrix}.$$

We readily find that

$$\bar{Q}_{\sigma(t)}(t) = \bar{P}_{\sigma(t)}^{-1}(t) = \begin{bmatrix} Q_{\sigma(t)}(t) & 0 \\ \hat{Q}_{\sigma(t)} C_2 & \alpha \hat{Q}_{\sigma(t)} \end{bmatrix}, \quad Q_{\sigma(t)}(t) = P_{\sigma(t)}^{-1}(t) \text{ and } \hat{Q}_{\sigma(t)} = \hat{P}_{\sigma(t)}^{-1}.$$

Differentiating V between the switching instances, we obtain

$$\dot{V} = \xi' \dot{\bar{P}}'_{\sigma(t)}(t) E \xi + 2\xi' \bar{P}'_{\sigma(t)}(t)(A_{cl,\sigma(t)}\xi + B_{cl,\sigma(t)}w),$$

and, similarly to (4.12c), we derive the following conditions for the L_2-gain of the switched system to be less than γ for all the points in the splitted polytopes:

$$\begin{bmatrix} -\dot{Q}_{\sigma(t)}(t)+\Im_{\sigma(t)}(t) & \Phi_{12} & B_{1,\sigma(t)} & \tilde{Q}_{1,4} \\ * & -\alpha(\hat{Q}_{\sigma(t)} + \hat{Q}'_{\sigma(t)}) & D_{21} & \alpha Y'_{\sigma(t)} D'_{12,\sigma(t)} \\ * & * & -\gamma^2 I & D'_{11,\sigma(t)} \\ * & * & * & -I \end{bmatrix} < 0, \quad (4.24)$$

where

$$\tilde{Q}_{1,4} = Q_{\sigma(t)}(t) C'_{1,\sigma(t)} + C'_2 Y'_{\sigma(t)} D'_{12,\sigma(t)},$$

and where $Y(t) = G_{\sigma(t)} \hat{Q}_{\sigma(t)}$, $\Im_{\sigma(t)}(t) = A_{\sigma(t)} Q_{\sigma(t)}(t) + Q'_{\sigma(t)}(t) A'_{\sigma(t)} + B_{2,\sigma(t)}$
$Y_{\sigma(t)} C_2 + C'_2 Y'_{\sigma(t)} B'_{2,\sigma(t)}$ and $\Phi_{12} = \alpha B_{2,\sigma(t)} Y_{\sigma(t)} + Q_{\sigma(t)}(t) C'_2 - C'_2 \hat{Q}'_{\sigma(t)}$.
Choosing $Q_{\sigma(t)}(t)$ to be piecewise linear in time, as $Q_{\sigma(t)}(t)$ has been in (4.15), the following result is obtained.

Theorem 4.3 *The L_2-gain of the system (4.1) with the uncertainty (4.3) and the output feedback $u(t) = G_{\sigma(t)}(t) y(t)$ is less than positive γ for a dwell time of $\bar{T} \geq T$ if, for a prescribed scalar α and an integer K, there exist: a collection of matrices $Q_{i,k} > 0 \in \mathcal{R}^{n \times n}$, $\hat{Q}_i \in \mathcal{R}^{s \times s}$, and $Y_i \in \mathcal{R}^{p \times s}$, $i = 1, \ldots M$, $j = 1, \ldots N$, that satisfy*

$$\begin{bmatrix} -\frac{Q_{i,k+1}-Q_{i,k}}{T/K} + \Xi_{i,h}^{(j)} & \bar{\Xi}_{i,h}^{(j)} & B_{1,i}^{(j)} & Q_{i,h} C_{1,i}^{(j)'} + C'_2 Y'_i D_{12,i}^{(j)'} \\ * & -\alpha(\hat{Q}_i + \hat{Q}'_i) & D_{21,i}^{(j)} & \alpha Y'_i D_{12,i}^{(j)'} \\ * & * & -\gamma^2 I & D'_{11,i} \\ * & * & * & -I \end{bmatrix} < 0$$

$$k = 0 \ldots K - 1, \; h = k, k+1,$$

$$\begin{bmatrix} \Xi_{i,K}^{(j)} & \bar{\Xi}_{i,K}^{(j)} & B_{1,i}^{(j)} & Q_{i,K} C_{1,i}^{(j)'} + C'_2 Y'_i D_{12,i}^{(j)'} \\ * & -\alpha(\hat{Q}_i + \hat{Q}'_i) & D_{21,i}^{(j)} & \alpha Y'_i D_{12,i}^{(j)'} \\ * & * & -\gamma^2 I & D'_{11,i} \\ * & * & * & -I \end{bmatrix} < 0,$$

and

$$Q_{i,K} - Q_{l,0} \leq 0, \quad \forall l \in L_i,$$

(4.25)

where

$$\Xi_{i,h}^{(j)} = A_i^{(j)} Q_{i,h} + Q_{i,h}' A_i^{(j)'} + B_{2,i}^{(j)} Y_i C_2 + C_2' Y_i' B_{2,i}^{(j)'},$$

and

$$\bar{\Xi}_{i,h}^{(j)} = \alpha B_{2,i}^{(j)} Y_i + Q_{i,h} C_2' - C_2' \hat{Q}_i'.$$

In this configuration, $G_{\sigma(t)}$ is given by: $G_{\sigma(t)} = Y_{\sigma(t)} \hat{Q}_{\sigma(t)}^{-1}$.

We note that the result of Theorem (4.3) for controllers that are constant between the switching instances is obtained without resorting to the more complicated conditions that are based on Corollary (4.3).

Remark 4.2 Corollary 4.4 and Theorem 4.3 apply a tuning parameter. Optimal value of this parameter can be found by performing a line-search, where at each point an LMI is solved. This method of searching for a single parameter is considered to be computationally efficient.

4.6 Examples

Example 4.1 (State feedback) We consider the problem of stabilizing and attenuating disturbances acting on the longitudinal short period mode of the F4E fighter aircraft with additional canards, which is taken from [15, 16].

The state-space model for this aircraft is given by

$$\frac{d}{dt}\begin{bmatrix} N_z \\ q \\ \delta_e \end{bmatrix} = \begin{bmatrix} a_{11} & a_{12} & a_{13} \\ a_{21} & a_{22} & a_{23} \\ 0 & 0 & -30 \end{bmatrix}\begin{bmatrix} N_z \\ q \\ \delta_e \end{bmatrix} + \begin{bmatrix} b_1 \\ 0 \\ 30 \end{bmatrix} u + I w \qquad (4.26)$$

with the objective and measurement vectors given by $z = \begin{bmatrix} N_z & q & u \end{bmatrix}^T$ and $y = \begin{bmatrix} N_z & q \end{bmatrix}^T$.

In this model, N_z is the *normal acceleration*, q is the *pitch rate*, and δ_e is the *elevator angle*.

The parameters of the model for 4 Operating Points (OPs) are described in the following.

O.P.	Mach	Altitude (ft)	a_{11}	a_{12}	a_{13}	a_{21}	a_{22}	a_{23}	b_1
1	0.5	5000	−0.9896	17.41	96.15	0.2648	−0.8512	−11.39	−97.78
2	0.9	35000	−0.6607	18.11	84.34	0.08201	−0.6587	−10.81	−272.2
3	0.85	5000	−1.702	50.72	263.5	0.2201	−1.418	−31.99	−85.09
4	1.5	35000	−0.5162	29.96	178.9	−0.6896	−1.225	−30.38	−175.6

and it is assumed that between the O.P.s the system parameters are a convex combination of the four parameter sets of the table. We seek a controller that satisfies (4.5)

Table 4.1 Min. values of γ for Example 4.1

Method	Quadratic Robust	GS	$K = 1$	$K = 10$
γ	3.85	3.39	2.9 (2.48)[2.07]	2.79 (2.32)[1.61]

for

$$C_1 = \begin{bmatrix} 1 & 0 & 0 \\ 0 & 1 & 0 \\ 0 & 0 & 0 \end{bmatrix} \quad \text{and} \quad D_{12} = \begin{bmatrix} 0 \\ 0 \\ 1 \end{bmatrix}.$$

Three controller design methods are presented: The first controller is a robust controller with a constant gain and quadratic LF [15]. The second is a Gain Scheduled controller (GS) with a quadratic LF [10] and the third is a switched controller, designed using either Corollary 4.4 or Theorem 4.2. We design the switched controller by splitting the parameters' polytope into four regions as described in the following figure.

Note that these regions overlap, which allows us to assume some dwell time between constitutive switching instances. We use the result of Corollary 4.4 to design a switched state-feedback controller, assuming that the switching signal is measured online, and a dwell time $T = 3$ s. The results are given in Table 4.1 above, where the minimum values of γ for two values of K are obtained for a constant G between the switching instances. The corresponding results that are obtained by Theorem 4.2 (a time-varying controller) are given there in simple brackets. The corresponding time-varying gains that are obtained by Corollary 4.4 (with $\beta = 0.04$) are given there in square brackets.

For $K = 1$, the constant state-feedback gains that are obtained via Corollary 4.4 are $G_1 = \begin{bmatrix} 0.175 & 1.30 & -0.294 \end{bmatrix}$, $G_2 = \begin{bmatrix} 0.244 & 1.37 & -1.01 \end{bmatrix}$, $G_3 = \begin{bmatrix} 0.327 & 3.42 & -0.871 \end{bmatrix}$, and $G_4 = \begin{bmatrix} 0.21 & 1.44 & -0.922 \end{bmatrix}$.

The results show that while the gain-scheduled state-feedback controller hardly improves on the standard quadratic state-feedback control, the switched controller yields results that are better than the one achieved by GS. This improvement is a result of the decreased size of the uncertainty polytopes that correspond to the individual subsystems.

Example 4.2 (Static Output-feedback Control) We consider the system of Example 4.1, where the state vector is measured via

$$C_2 = \begin{bmatrix} 1 & 0 & 0 \\ 0 & 1 & 0 \end{bmatrix}, \quad \text{and} \quad D_{21} = 10^{-2} I_2,$$

where the two measurement noise signals that affect the measurement y are independent of the three input disturbances already defined in Example 4.1, and where the controller is given by $u = G_{\sigma(t)} y$ with a constant $G_{\sigma(t)}$ between the switching instances.

Table 4.2 Values of γ for Example 4.2

Method	Quadratic Robust	[17]	GS	$K = 1$	$K = 10$	$K = 100$
γ	8.97	3.92	4.11	3.48	3.35	3.2

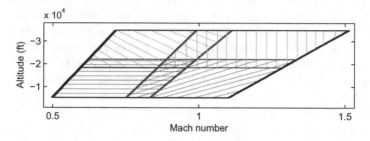

Fig. 4.2 The overlapping sub-polytopes

Once again, we compare the quadratic stabilizing static output-feedback result obtained, say in [17, 18], with the corresponding result that is obtained by the GS method and the solutions that are derived by applying Theorem 4.3 for $\alpha = 0.3$ for various values of K.

The results are summarized in Table 4.2.

For $K = 1$ the SOF gains are

$$G_1 = \begin{bmatrix} 0.135 & 1.139 \end{bmatrix}, \quad G_2 = \begin{bmatrix} 0.112 & 0.773 \end{bmatrix}, \quad G_3 = \begin{bmatrix} 0.127 & 1.373 \end{bmatrix}$$

and

$$G_4 = \begin{bmatrix} 0.104 & 0.854 \end{bmatrix}.$$

We note that there are many methods that were suggested in the literature for designing SOF controllers. We compared the results of Theorem (4.3) with the one in [17] that was similarly obtained using the descriptor representation and a single Q (Table 4.2).

4.7 Conclusions

A design method is introduced that can be used to reduce current existing conservatism in the design of linear non-switched systems with polytopic-type time-varying parameter uncertainties. By dividing the uncertainty polytope into overlapping sub-polytopes, and by determining, a priori, the dwell time to be a bound on the time required by the parameters of the system to cross the overlapping regions in the uncertainty polytope, the problem can be considered as one of finding a controller for a switching system with dwell time. Applying a special time-varying LF, which is nonincreasing at the switching instants, sufficient conditions for robust stability and the existence of a prescribed bound on the L_2-gain of the original uncertain system are derived in terms of LMIs.

The results in this chapter are based on the fact that the existing solutions for robust control of linear systems with large parameter uncertainty entails a significant overdesign that stems from the requirement that some decision variables in the LMI conditions should be vertex independent in order to preserve convexity. This requirement for convexity is also encountered when applying the GS method. The splitting that is applied in the present chapter of the uncertainty polytope into smaller sub-polytopes reduces the above overdesign and consequently, as demonstrated in the examples of Sect. 4.6, a better disturbance attenuation is achieved.

The question remains how to divide the uncertainty polytope and how many sub-polytopes should be applied. This is still an open question to be investigated in the future. In many practical cases, however, the division into sub-polytopes follows naturally from the operating points of the system. In the F4E example of Sect. 4.6 the four operating points dictate the split into four sub-polytopes there.

Two examples are given that demonstrate the applicability of the theory developed. They distinctly show the advantage of considering switching between the uncertainty sub-polytopes. It is shown there that the proposed method not only improves the result obtained by using the standard quadratic stable synthesis, but it is also favorably compared to the corresponding GS method of design where an exact online measurement of the parameters is required.

A solution to the robust output-feedback control problem is obtained by finding switching constant control gains that stabilize the uncertain system and achieve a prescribed bound on the disturbance attenuation level of the resulting switched system. These gains are obtained by applying an LF that does not depend on the parameters of the system. A parameter-dependent LF can also be applied by using the method of Corollary 4.3.

References

1. Horowitz, I.: Synthesis of Feedback Systems. Academic Press, New York (1963)
2. Zames, G.: Feedback and optimal sensitivity: model reference transformation, multiplicative seminorms, and approximate inverses. IEEE Trans. Autom. Control **26**, 301–320 (1981)
3. Doyle, J.C., Stein, G.: Multivariable feedback design: concepts for a classical/modern synthesis. IEEE Trans. Autom. Control **26**, 4–16 (1981)
4. Doyle, J., Francis, B.A., Tannenbaum, A.R.: Feedback Control Theory. Macmillan, New York (1992)
5. Doyle, J.C., Glover, K., Khargonekar, P., Francis, B.: State-space solution to standard H_2 and H_∞ control problems. IEEE Trans. Autom. Control **34**, 831–847 (1989)
6. Gahinet, P., Apkarian, P.: A linear matrix inequality approach to H_∞ control. Int. J. Robust Nonlinear Control **4**, 421–448 (1994)
7. Scherer, C.: Theory of Robust Control, Lecture Note, Mechanical Engineering Systems and Control Group, Delft University of Technology, The Netherlands (2001)

8. Scherer, C., Weiland, S.: Linear Matrix Inequalities in Control, Lecture Notes at Delft University of Technology and Eindhoven University of Technology (2005)
9. Boyd, S., El Ghaoui, L., Feron, E., Balakrishnan, V.: Linear Matrix Inequality in Systems and Control Theory. SIAM Frontier Series (1994)
10. Apkarian, P., Gahinet, P.: A convex characterization of gain-scheduled H_∞ controllers. IEEE Trans. Autom. Control 40(5), 853–864 (1995)
11. de Oliveira, M.C., Skelton, R.E.: Stability test for constrained linear systems. In: Reza Moheimani, S.O. (Ed.) Perspectives in Robust Control. Lecture Notes in Control and Information Sciences, vol. 268. Springer, London (2001)
12. Allerhand, L.I., Shaked, U.: Robust stability and stabilization of linear switched systems with dwell time. IEEE Trans. Autom. Control 56, 381–386 (2011)
13. Boyarski, S., Shaked, U.: Time-convexity and time-gain-scheduling in finite-horizon robust H_∞-control. In: Proceedings of the 48th CDC09. Shanghai, China (2009)
14. Gahinet, P., Nemirovski, A., laub, A.J., Chilali, M.: LMI Control Toolbox. The MathWorks Inc., Natick, Mass. USA (1995)
15. Petersen, I.R.: Quadratic stabilizability of uncertain linear systems: existence of a nonlinear stabilizing control does not imply existence of a stabilizing control. IEEE Trans. Autom. Control 30, 291–293 (1985)
16. Ackermann, J.: Longitudinal control of fighter aircraft F4E with additional canards. In: Sondergeld, K.P. (compiler) (Germany: A Collection of Plant Models and Design Specifications for Robust Control.) DFVLR, Oberpfaffenhoffen
17. Yaesh, I., Shaked, U.: Robust reduced-order output-feedback H_∞ control. In: Proceedings of the Symposium on Robust Control Design. Haifa, Israel (2009)
18. Shaked, U.: A LPV approach to robust H_2 and H_∞ static output-feedback design. Trans. Autom. Control 48, 866–872 (2003)
19. Peres, P.L.D., Geromel, J.C., Souza, S.R.: H_∞ control design by static output-feedback. In: Proceedings of the IFAC Symposium on Robust Control Design, pp. 243–248. Rio de Janeiro, Brazil (1994)
20. Geromel, J., Colaneri, P.: Stability and stabilization of continuous-time switched linear systems. SIAM J. Control Optim. 45(5), 1915–1930 (2006)

Chapter 5
Robust Estimation of Linear-Switched Systems with Dwell Time

Abstract In order to overcome the deficiencies of the standard approaches to robust filtering of systems with large parameter uncertainties, a new method of estimation for non-switched, exponentially stable, uncertain systems is presented. We divide the uncertainty region into several overlapping subregions where, for each of these subregions, a separate filter is assigned. The resulting system is then treated as a switched system that switches between the subregions. The theory developed is demonstrated by two simple examples. In the first, a filtering solution is obtained for a switched system with error variance that is very close to the best achievable result for a single subsystem. The second example demonstrates the benefit, from the estimation error point of view, of dividing the uncertainty polytope of an uncertain non-switched system into two overlapping sub-polytopes.

5.1 Introduction

Stochastic analysis of Linear Time-Varying (LTV) non-switched systems with white-noise inputs is well documented in the literature [1, 2]. For LTV systems, with parameters that are a priori known, an accurate analysis of the systems state covariance is achieved by solving a differential Lyapunov equation [1]. If the system parameters encounter uncertainty, only a bound on the covariance of the state vector can be obtained. For Linear Time-Invariant (LTI) systems, this bound can be found by solving a set of Linear Matrix Inequalities (LMIs) [3]. In the LTV case, it is necessary to solve Differential LMIs (DLMIs) in order to obtain a bound on the variance [4].

For switched systems with white input noise, and switching signal that is known a priori, the standard LTV analysis can be applied [5]. In the case where the switching signal is arbitrary and unknown, it may still be possible to find a bound on the covariance of the state, by treating the system as an uncertain non-switched system, where the subsystems constitute the vertices of the artificial uncertainty polytope.

For switched systems without dwell, a quadratic filter can then be used to derive a switching-independent filter. A quadratic filter is a single filter that applies the same Lyapunov Function (LF) for all the points in the uncertainty polytope. It is obtained by solving the filtering LMIs for all the systems vertices using the same values for

© Springer Nature Switzerland AG 2019
E. Gershon and U. Shaked, *Advances in H∞ Control Theory*,
Lecture Notes in Control and Information Sciences 481,
https://doi.org/10.1007/978-3-030-16008-1_5

the decision variables. For systems with dwell time, a quadratic filter can still be used. However, here, the quadratic filter is more conservative since it does not allow the bound on the covariance to be time dependent [6–8].

In the present chapter, we introduce an analysis that reduces the conservatism which is encountered in the analysis of switched systems without dwell. We derive an upper-bound on the covariance of the states of the switched system which depends on the switching signal. The obtained results are then applied to filtering problems in switched systems with driving and measurement white-noise signals, where general-type filters are considered. These filters assume that the switching signal is not given a priori but is available in real-time.

Filtering is considered in this chapter for systems with parameters that are perfectly known (nominal systems) and for systems with parameter uncertainties.

The theory that is developed in the present chapter for switched systems is applied also to overcome the deficiencies of the standard approaches to robust filtering of systems with large parameter uncertainties. A new method of estimation for non-switched, exponentially stable, uncertain systems is introduced. Similarly to the method used in Chap. 4, we divide the uncertainty region into several overlapping sub-regions where, for each of these subregions, a separate filter is assigned. The resulting system is then treated as a switched system that switches between the sub-regions.

The theory developed is demonstrated by two simple examples. In the first, a filtering solution is obtained for a switched system with error variance that is very close to the best achievable result for a single subsystem. The second example demonstrates the benefit, from the estimation error point of view, of dividing the uncertainty polytope of an uncertain non-switched system into two overlapping sub-polytopes.

5.2   Problem Formulation

Consider the following Linear Time-Varying (LTV) system:

$$\dot{x}(t) = A(t)x(t) + B(t)w(t),$$
$$z(t) = C_1(t)x(t),$$

(5.1)

where $x(t) \in \mathcal{R}^n$ is the state vector, $w(t) \in \mathcal{R}^q$ is a standard Aalog White Gaussian Noise(AWGN) signal, and $z(t) \in \mathcal{R}^r$ is the objective vector. For convenience, when possible, we shall omit in the sequel the dependence of the vectors and matrices on the time t. The initial value of the state vector is assumed to be normally distributed with a zero mean and covariance $\bar{Q}_0$. We also assume that the system is exponentially stable.

A switched system is a special case of the system (5.1), where

$$\dot{x} = A_{\sigma(t)}x + B_{\sigma(t)}w,$$
$$z = C_{1,\sigma(t)}x,$$

(5.2)

and where the switching rule $\sigma(t) \in \{1 \ldots \bar{\sigma}\}$, for all $t \geq 0$.

Each value of σ defines a subsystem. We assume that the parameters of these subsystems are uncertain and that they reside in the following polytopes:

$$\Omega_\sigma = \sum_{j=1}^{N} \eta_j(t)\Omega_\sigma^{(j)}, \quad \sum_{j=1}^{N} \eta_j(t) = 1, \ \eta_j(t) \geq 0,$$

where we denote

$$\Omega_i^{(j)} = \begin{bmatrix} A_i^{(j)} & B_i^{(j)} \\ C_i^{(j)} & 0 \end{bmatrix}, \quad i \in L_i.$$

The initial value of the state vector is assumed to be normally distributed with zero mean and covariance $\bar{Q}_0$. We also assume that the system is exponentially stable and that there is a dwell time constraint on the switching signal $\sigma(t)$, which means that if the dwell time is T, and the switching instances are $\tau_1, \tau_2, \ldots$, then $\tau_{s+1} - \tau_s \geq T, \quad \forall s \geq 1$.

Let $i_0 = \sigma(0)$. We define $\tau_{s,k} = \tau_s + k\frac{T}{K}$ for $k \geq 1$, and $\tau_{s,0} = \tau_s$. Note that the dwell time constraint implies $\tau_{s,K} \leq \tau_{s+1,0} = \tau_{s+1}$.

The first problem that we address in this chapter is to obtain a minimum upper-bound on the covariance of z over all the uncertainty polytopes Ω_σ, $\sigma = 1, \ldots, N$.

For this problem, we consider the following LTV system:

$$\begin{aligned} \dot{x}(t) &= A(t)x(t) + B_1(t)w(t), \\ z(t) &= C_1(t)x(t), \\ y(t) &= C_2(t)x(t) + D_{21}w(t), \end{aligned} \tag{5.3}$$

where z is the objective vector to be estimated, $y \in \mathcal{R}^s$ is the measurement, and x, and w are defined above.

The filter is given by the state-space representation with the matrices A_c, B_c, C_c, where y is the filter's input, and the filter's output is the estimated state.

As a special case of the system (5.3), we consider the following switched system:

$$\begin{aligned} \dot{x} &= A_\sigma x + B_{1,\sigma}w, \\ z &= C_{1,\sigma}x, \\ y &= C_{2,\sigma}x + D_{21,\sigma}w, \end{aligned} \tag{5.4}$$

where x, w, z, y, σ are defined above.

The systems parameters are uncertain and they are assumed to reside in the following polytopes:

$$\Omega_\sigma = \sum_{j=1}^{N}\eta_j(t)\Omega_\sigma^{(j)}, \quad \sum_{j=1}^{N}\eta_j(t) = 1, \ \eta_j(t) \geq 0, \ \text{and}$$

$$\Omega_i = \begin{bmatrix} A_i & B_{1,i} \\ C_{1,i} & 0 \\ C_{2,i} & D_{21,i} \end{bmatrix}, \quad \Omega_i^{(j)} = \begin{bmatrix} A_i^{(j)} & B_{1,i}^{(j)} \\ C_{1,i}^{(j)} & 0 \\ C_{2,i}^{(j)} & D_{21,i}^{(j)} \end{bmatrix}, \quad \begin{aligned} j &= 1, \ldots N, \\ i &= 1, \ldots \bar{\sigma}. \end{aligned} \tag{5.5}$$

Here, we seek a procedure for estimating z which, based on the measurement y, will provide a minimum bound on the estimation error covariance.

5.3 Covariance Analysis

Let $\bar{Q}$ denote the covariance of the state of the system (5.1). It is well known that $\bar{Q}$ satisfies the following Lyapunov differential equation.

$$-\dot{\bar{Q}} + A\bar{Q} + \bar{Q}A' + BB' = 0, \quad Q(0) = Q_0.$$

Since the system encounters parameter uncertainty, we seek an upper-bound Q on the covariance matrix $\bar{Q}$. From [3] we have that, since our system is exponentially stable any $Q > 0$ that satisfies

$$- \dot{Q} + AQ + QA' + BB' < 0 \qquad (5.6)$$

is an upper-bound on $\bar{Q}(t)$ for any positive initial value $Q(0)$. This will be the bound for all $t > 0$ when $\bar{Q}_0 \leq Q(0)$ and it will hold for large enough t, otherwise. We shall, therefore assume, throughout this chapter, that time t is such that the effect of the initial values of the state (and its covariance) have subsided to zero.

Using Schur complements [9], the condition (5.6) is equivalent to

$$Q > 0, \quad \begin{bmatrix} -\dot{Q} + AQ + QA' & B \\ * & -I \end{bmatrix} < 0. \qquad (5.7)$$

Denoting $P = Q^{-1}$, multiplying (5.6) by P from both sides and using Schur complements, the following equivalent condition is obtained:

$$P > 0, \quad \begin{bmatrix} \dot{P} + PA + A'P & PB \\ * & -I \end{bmatrix} < 0. \qquad (5.8)$$

A bound on the covariance of the objective vector $z = C_1x$ of (5.1), denoted by M, is obtained [3] by

$$\begin{bmatrix} M & C_1Q \\ * & Q \end{bmatrix} > 0 \quad \Leftrightarrow \quad \begin{bmatrix} M & C_1 \\ * & P \end{bmatrix} > 0. \qquad (5.9)$$

In the sequel, we pre-choose an integer K that determines the complexity of our computations. The larger K is, less conservative bounds will be achieved.

In the theorem below, we obtain an upper-bound on the covariance Q of the state vector of the switched system (5.2), and a scalar upper-bound γ^2 on the variance (the trace of the covariance) of the objective vector.

Theorem 5.1 *Consider the switched system (5.2). The variance of the objective vector z is bounded by a prescribed scalar γ^2, for a dwell time of $\bar{T} \geq T$, if there exists a collection of matrices $Q_{i,k} > 0, M_{i,k} > 0, i = 1, \ldots \bar{\sigma}, k = 0, \ldots K$ of compatible dimensions, such that, for all $i = 1, \ldots \bar{\sigma}$, and $j = 1, \ldots N$ the following holds:*

$$\begin{bmatrix} -\frac{Q_{i,k+1}-Q_{i,k}}{T/K} + Q_{i,h}A_i^{(j)'} + A_i^{(j)}Q_{i,h} & B_i^{(j)} \\ * & -I \end{bmatrix} < 0 \quad h = k, k+1,$$

$$\begin{bmatrix} M_{i,k} & C_{1,i}^{(j)}Q_{i,k} \\ * & Q_{i,k} \end{bmatrix} > 0,$$

$$Trace\{M_{i,k}\}) \leq \gamma^2, \quad k = 0, \ldots K$$

$$\begin{bmatrix} Q_{i,K}A_i^{(j)'} + A_i^{(j)}Q_{i,K} & B_i^{(j)} \\ * & -I \end{bmatrix} < 0$$

and

$$Q_{i,K} - Q_{l,0} \leq 0, \quad \forall l \in L_i. \tag{5.10}$$

The covariance of the state vector x is then bounded in the time interval $[\tau_s, \tau_{s+1})$ by Q_i where

$$Q_i = \begin{cases} Q_{i,k} + (Q_{i,k+1} - Q_{i,k})\frac{t-\tau_{s,k}}{T/K} & t \in [\tau_{s,k}, \tau_{s,k+1}) \\ Q_{i,K} & t \in [\tau_{s,K}, \tau_{s+1,0}) \quad .s = 1, 2, \ldots \\ Q_{i_0,K} & t \in [0, \tau_1) \end{cases} \tag{5.11}$$

Proof We assume that at the switching instant τ_s the system switched to the ith subsystem and that it switches again, at τ_{s+1}, to lth subsystem.

In order to find a bound that satisfies (5.7), we take Q to be dependent on σ so that it is piecewise linear in time during the dwell time, and constant after the dwell and before the next switching occurs. This leads to Q_i of the structure described in (5.11).

The positive definiteness of $Q_{i,k}$ guarantees that Q_i is positive definite over the above interval. To show that Q_i satisfies (5.7) during the dwell time, we divide the time interval $[\tau_s, \tau_s + T]$ into K subintervals $[\tau_{s,k}, \tau_{s,k+1}], k = 0, 1, \ldots K - 1$. For each of these subintervals we have that $\dot{Q}_i = \frac{Q_{i,k+1}-Q_{i,k}}{T/K}$, and since Q_i changes linearly from $Q_{i,k}$ to $Q_{i,k+1}$, the matrix Q_i is a convex combination of $Q_{i,k}$ and $Q_{i,k+1}$. We demand that condition (5.7) holds for both $Q_{i,k}$ and $Q_{i,k+1}$, and therefore for any convex combination of these matrices. This leads to conditions (5.10a) which guarantee that (5.7) holds for any value of Q_i at the time interval $t \in [\tau_{s,k}, \tau_{s,k+1})$, for any $k < K$, and therefore during the entire time interval $[\tau_s, \tau_s + T)$.

After time $\tau_s + T$, and before the next switching occurs, we take Q_i to be constant. Since then $Q_i = Q_{i,K}$, condition (5.10d) guarantees that (5.7) holds. At the switching instant τ_{s+1}, the covariance matrix of the state vector $\bar{Q}$ is bounded by $Q_{i,K}$. The latter does not change at the switching instant, and therefore any matrix that is greater than or equal to $Q_{i,K}$, is a bound on $\bar{Q}$ at the switching instant τ_{s+1}. The bound on $\bar{Q}$ just after the switching occurs is $Q_{l,0}$, and therefore condition (5.10e) guarantees the Q_i is a bound on $\bar{Q}$ also at the switching instant τ_{s+1}.

A bound on the variance of the objective vector z is given for each vertex of the polytope that corresponds to the ith subsystem by $Tr(C_{1,i}^{(j)}Q_iC_{1,i}^{(j)'})$. Condition (5.10b) thus implies that $Trace\{M_{i,k}\}$ is a bound on the latter variance when $Q_i = Q_{i,k}$, $k = 0, 1, \ldots K$, and (5.10c) then guarantees the required bound.   ∎

5.4   Filtering

We apply in this section the covariance analysis of Sect. 5.3 to the problem of finding the best bound on the filtering error of the system (5.4). To simplify the derivation, we first consider the filtering problem of the system (5.3). Recall that A_c, B_c, and C_c denote the state-space representation of the filter, then for the system (5.3), the filtering problem is characterized by the augmented system which is given by $\bar{A}, \bar{B}, \bar{C}$, where

$$\bar{A} = \begin{bmatrix} A & 0 \\ B_cC_2 & A_c \end{bmatrix}, \quad \bar{B} = \begin{bmatrix} B_1 \\ B_cD_{21} \end{bmatrix}, \quad \text{and} \quad \bar{C} = \begin{bmatrix} C_1 & -C_c \end{bmatrix}.$$

We denote

$$P = \begin{bmatrix} X & \bar{M} \\ * & U \end{bmatrix}, \quad Q = \begin{bmatrix} Y & \bar{N} \\ * & V \end{bmatrix}, \quad R = Y^{-1}, \quad \text{and} \quad J = \begin{bmatrix} I & I \\ 0 & \bar{N}'R \end{bmatrix}. \quad (5.12)$$

We seek a condition equivalent to (5.8), that guarantees that $Q = P^{-1}$ is a bound on the covariance of the state, and a condition equivalent to (5.9) that guarantees that M is a bound on the covariance of the objective vector. We obtain these following conditions by multiplying conditions (5.8) and (5.9) by $diag(J, I, I)$, from the right, and its transpose, from the left, and taking $\bar{M} = I$, where $A_f = A_cN'R$, $B_f = B_c$, $C_f = C_cN'R$.

$$\begin{bmatrix} \dot{X} + A'X + XA + B_fC_2 + C_2'B_f' & * & * \\ \dot{X} + A'X + A_f' + C_2'B_f' + RA & \dot{R} + RA + A'R & * \\ B_1'X + D_{21}'B_f' & B_1'R & -I \end{bmatrix} < 0,$$

$$(5.13)$$

$$\begin{bmatrix} M & C_1 & C_1 - C_f \\ * & X & R \\ * & * & R \end{bmatrix} > 0.$$

The condition for γ^2 to be a bound on the variance of the objective vector remains $\gamma^2 > Trace\{(M\}$.

Next, we apply these conditions to the system (5.4) to obtain LMI conditions for filtering of the system (5.4) with a bound on the error covariance γ^2.

Theorem 5.2 *If there exists a collection of matrices $X_{i,k} > 0$, $R_{i,k} > 0$, $A_{f,i}$, $B_{f,i}$, $C_{f,i}$, $D_{f,i}$, $i = 1, \ldots \bar\sigma$, $k = 0, \ldots K$ of compatible dimensions, where K is a prescribed integer, such that, for all $i = 1, \ldots \bar\sigma$, and $j = 1, \ldots N$ the following holds:*

$$\begin{bmatrix} \frac{X_{i,k+1}-X_{i,k}}{T/K} + A_i^{(j)'}X_{i,h} + X_{i,h}A_i^{(j)} + B_{f,i}C_{2,i}^{(j)} + C_{2,i}^{(j)'}B_{f,i}' & * & * \\ \frac{X_{i,k+1}-X_{i,k}}{T/K} + A_i^{(j)'}X_{i,h} + A_{f,i}' + C_{2,i}^{(j)'}B_{f,i}' + R_{i,h}A_i^{(j)} & \tilde\Gamma_{2,2} & * \\ B_{1,i}^{(j)'}X_{i,h} + D_{21,i}^{(j)'}B_{f,i}' & B_{1,i}^{(j)'}R_{i,h}I \end{bmatrix} < 0$$

$$h = k, k+1, \quad k = 0, \ldots K-1,$$

$$\tilde\Gamma_{2,2} = \frac{R_{i,k+1} - R_{i,k}}{T/K} + R_{i,h}A_i^{(j)} + A_i^{(j)'}R_{i,h},$$

$$\begin{bmatrix} A_i^{(j)'}X_{i,K} + X_{i,K}A_i^{(j)} + B_{f,i}C_{2,i}^{(j)} + C_{2,i}^{(j)'}B_{f,i}' & * & * \\ A_i^{(j)'}X_{i,K} + A_{f,i}' + C_{2,i}^{(j)'}B_{f,i}' + R_{i,K}A_i^{(j)} & R_{i,K}A_i^{(j)} + A_i^{(j)'}R_{i,K} & * \\ B_{1,i}^{(j)'}X_{i,K} + D_{21,i}^{(j)'}B_{f,i}' & B_{1,i}^{(j)'}R_{i,K} & -I \end{bmatrix} < 0$$

$$\begin{bmatrix} M_{i,k} & C_{1,i}^{(i)} & C_{1,i}^{(i)} - C_{f,i} \\ * & X_{i,k} & R_{i,k} \\ * & * & R_{i,k} \end{bmatrix} > 0,$$

$$Trace\{M_{i,k}\} \leq \gamma^2, \quad k = 0, \ldots K$$

and

$$\begin{bmatrix} R_{l,0} - R_{i,K} & X_{l,0} - X_{i,K} \\ X_{l,0} - X_{i,K} & X_{l,0} - X_{i,K} \end{bmatrix} \leq 0, \quad \forall l \in L_i, \tag{5.14}$$

then, for a dwell time of $\bar T \geq T$, γ^2 is a bound on the variance of the estimation error of the filter given by the state-space representation

$$A_{c,i} = A_{f,i}(R_i - X_i)^{-1}, \quad B_{c,i} = B_{f,i}, \quad C_{c,i} = C_{f,i}(R_i - X_i)^{-1}, \tag{5.15}$$

where

$$X_i = \begin{cases} X_{i,k} + (X_{i,k+1} - X_{i,k})\frac{t-\tau_{s,k}}{T/K} & t \in [\tau_{s,k}, \tau_{s,k+1}) \\ X_{i,K} & t \in [\tau_{s,K}, \tau_{s+1,0}) \quad , s = 1, 2, \ldots \\ X_{i_0,K} & t \in [0, \tau_1) \end{cases}$$

and $\tag{5.16}$

$$R_i = \begin{cases} R_{i,k} + (R_{i,k+1} - R_{i,k})\frac{t-\tau_{s,k}}{T/K} & t \in [\tau_{s,k}, \tau_{s,k+1}) \\ R_{i,K} & t \in [\tau_{s,K}, \tau_{s+1,0}) \quad . s = 1, 2, \ldots \\ R_{i_0,K} & t \in [0, \tau_1) \end{cases}$$

Proof Let X, R be as is (5.16), then conditions (5.13) are piecewise linear in time between the switching instances.

For (5.13) to hold during the dwell time, we consider the time intervals $t \in [\tau_{s,k}, \tau_{s,k+1})$. In each section, X, R, M are linear in time, similarly to Q in Theorem 5.1. During those intervals we have that $\dot{R} = \frac{R_{i,k+1}-R_{i,k}}{T/K}, \dot{X} = \frac{X_{i,k+1}-X_{i,k}}{T/K}$, and that X, R, M are convex combinations of $X_{i,k}, R_{i,k}, M_{i,k}$ and $X_{i,k+1}, R_{i,k+1}, M_{i,k+1}$, respectively. Therefore, if conditions (5.13) hold for both $X = X_{i,k}, R = R_{i,k}, M = M_{i,k}$ and $X = X_{i,k+1}, R = R_{i,k+1}, M = M_{i,k+1}$ then these conditions are satisfied for $t \in [\tau_{s,k}, \tau_{s,k+1})$. This leads to condition (5.14c) for $k = 0, \ldots K - 1$, and (5.14a).

After the dwell time and before the next switching occurs, the conditions (5.13) are time independent. Since at these times $X = X_{i,K}, R = R_{i,K}$ and $M = M_{i,K}$, conditions (5.14c) for $k = K$ and (5.14b) guarantee that condition (5.13) are satisfied for $t \in [\tau_s + T, \tau_{s,k+1})$.

To address the bound on the covariance at the switching instances, we first note that from the definitions of (5.12), and from $\bar{M} = I$ we have that $R\bar{N} = R - X$, and $U = (X - R)^{-1}$. These equalities can be verified from the equality $PQ = QP = I$. At the switching instant τ_{s+1}, the covariance matrix of the state vector $\bar{Q}$ is bounded by $P_{i,K}^{-1} = Q_{i,K}$. The covariance matrix does not change at the switching instant, and therefore a valid bound on P just after the switching occurs is any matrix that is less or equal to $P_{i,K}$. The bound on $\bar{Q}$ after the switching is $P_{l,0}^{-1}$, and therefore, in order to guarantee that P^{-1} is a bound on $\bar{Q}$ at the switching instant, we demand $P_{l,0} - P_{i,K} \leq 0$. Substitution of the definitions of (5.12), and $\bar{M} = I$ conditions $X_{l,0} - X_{i,K} \leq 0$, and $(X_{l,0} - R_{l,0})^{-1} \leq (X_{i,K} - R_{i,K})^{-1}$, where the latter is equivalent to the condition $R_{l,0} - X_{l,0} \leq R_{i,K} - X_{i,K}$. These conditions are rewritten as

$$\begin{bmatrix} R_{l,0} - R_{i,K} - (X_{l,0} - X_{i,K}) & 0 \\ 0 & X_{l,0} - X_{i,K} \end{bmatrix} \leq 0.$$

Applying Schur complements [9] conditions (5.14d) are obtained.

To find a bound on the variance of the objective vector z, we take M to be a bound on the covariance of z. Condition (5.9) guarantees that M is a bound on the covariance of z. This leads to conditions (5.10b). Since the variance is the trace of the covariance, we have that $\gamma^2 \geq Trace\{(M\}$ implies that γ^2 is a bound on the variance of z. This leads to conditions (5.10c).

A bound on the variance of the objective vector z is obtained by demanding that (5.9) holds for each vertex of the polytope that corresponds to the ith subsystem. Condition (5.14e) thus implies that $Trace\{M_{i,k}\}$ is a bound on the latter variance when $P_i = P_{i,k}, k = 0, 1, \ldots K$, and (5.14d) then guarantees the required bound. ∎

The latter result may be somewhat conservative, since it applies the same decision variables R_i and X_i over the whole uncertainty polytope of each subsystem. Since it is only required that in (5.15) the matrix $R_i - X_i$ will be constant over the whole polytope, we can define $\Psi_i = R_i - X_i$ and $\Phi_i = R_i + X_i$, and let Φ_i be vertex dependent $\Phi_i = \sum_{j=1}^{N} \eta_j(t)\Phi_i^{(j)}$, and therefore $X_i^{(j)} = 0.5\Phi_i^{(j)} - 0.5\Psi_i, R_i^{(j)} = 0.5\Phi_i^{(j)} + 0.5\Psi_i$

for $j = 1, \ldots N$. Note that $R_i^{(j)} - X_i^{(j)} = \Psi_i$ for all $j = 1, \ldots N$. Thus we obtain the following:

Corollary 5.1 *If there exists a collection of matrices $\Phi_{i,k}^{(j)} > 0$, $\Psi_{i,k} > 0$, $A_{f,i}$, $B_{f,i}$, $C_{f,i}$, $D_{f,i}$, $i = 1, \ldots \bar{\sigma}$, $k = 0, \ldots K, j = 1, \ldots N$ of compatible dimensions, where K is a prescribed integer, such that $X_i^{(j)} = 0.5\Phi_i^{(j)} - 0.5\Psi_i$, $R_i^{(j)} = 0.5\Phi_i^{(j)} + 0.5\Psi_i$, and for all $i = 1, \ldots \bar{\sigma}$, and $j = 1, \ldots N$ the following holds:*

$$\begin{bmatrix} \Gamma_{11} & * & * \\ \Gamma_{21} & \frac{R_{i,k+1}^{(j)} - R_{i,k}^{(j)}}{T/K} + R_{i,h}^{(j)} A_i^{(j)} + A_i^{(j)'} R_{i,h}^{(j)} & * \\ B_{1,i}^{(j)'} X_{i,h}^{(j)} + D_{21,i}^{(j)'} B_{f,i}' & B_{1,i}^{(j)'} R_{i,h}^{(j)} & -I \end{bmatrix} < 0$$

$$h = k, k+1, \quad k = 0, \ldots K-1,$$

where

$$\Gamma_{11} = \frac{X_{i,k+1}^{(j)} - X_{i,k}^{(j)}}{T/K} + A_i^{(j)'} X_{i,h}^{(j)} + X_{i,h}^{(j)} A_i^{(j)} + B_{f,i} C_{2,i}^{(j)} + C_{2,i}^{(j)'} B_{f,i}',$$

$$\Gamma_{21} = \frac{X_{i,k+1}^{(j)} - X_{i,k}^{(j)}}{T/K} + A_i^{(j)'} X_{i,h}^{(j)} + A_{f,i}' + C_{2,i}^{(j)'} B_{f,i}' + R_{i,h}^{(j)} A_i^{(j)},$$

$$\begin{bmatrix} A_i^{(j)'} X_{i,K}^{(j)} + X_{i,K}^{(j)} A_i^{(j)} + B_{f,i} C_{2,i}^{(j)} + C_{2,i}^{(j)'} B_{f,i}' & * & * \\ A_i^{(j)'} X_{i,K}^{(j)} + A_{f,i}' + C_{2,i}^{(j)'} B_{f,i}' + R_{i,K}^{(j)} A_i^{(j)} & R_{i,K}^{(j)} A_i^{(j)} + A_i^{(j)'} R_{i,K}^{(j)} & * \\ B_{1,i}^{(j)'} X_{i,K}^{(j)} + D_{21,i}^{(j)'} B_{f,i}' & B_{1,i}^{(j)'} R_{i,K}^{(j)} & -I \end{bmatrix} < 0$$

and

$$\begin{bmatrix} M_{i,k} & C_{1,i}^{(j)} & C_{1,i}^{(j)} - C_{f,i} \\ * & X_{i,k}^{(j)} & R_{i,k}^{(j)} \\ * & * & R_{i,k}^{(j)} \end{bmatrix} > 0,$$

$$Trace\{(M_{i,k}\} \le \gamma^2, \quad k = 0, \ldots K$$

and

$$\begin{bmatrix} R_{l,0}^{(j)} - R_{i,K}^{(j)} & X_{l,0}^{(j)} - X_{i,K}^{(j)} \\ X_{l,0}^{(j)} - X_{i,K}^{(j)} & X_{l,0}^{(j)} - X_{i,K}^{(j)} \end{bmatrix} \le 0, \quad \forall l \in L_i, \tag{5.17}$$

then, for a dwell time of $\bar{T} \ge T$, γ^2 is a bound on the variance of the estimation error of the filter, and a state-space realization of the filter is the same as (5.15).

We note that this method assumes that systems parameters change very slowly within the uncertainty polytopes. If they change fast, a bound on the derivative of the matrices R, X due to their dependence of η_j should be added to these LMIs, as in [10].

5.5　Application to Robust Filtering

Consider the system (5.1) with the measurement

$$y(t) = C_2 x(t) + D_{21} w(t). \tag{5.18}$$

The systems parameters are uncertain and they are assumed to reside in the following polytope:

$$\Omega = \begin{bmatrix} A & B_1 \\ C_1 & 0 \\ C_2 & D_{21} \end{bmatrix} = \sum_{j=1}^{N} \eta^j(t) \Omega^{(j)}, \quad \sum_{j=1}^{N} \eta^j(t) = 1, \ \eta^j(t) \geq 0, \ \text{and}$$

$$\Omega^{(j)} = \begin{bmatrix} A^{(j)} & B_1^{(j)} \\ C_1^{(j)} & 0 \\ C_2^{(j)} & D_{21}^{(j)} \end{bmatrix}, \ j = 1, \ldots N. \tag{5.19}$$

In the literature, this uncertainty polytope has been treated as a single region and various filtering methods have been proposed which require conditions to be satisfied over the entire uncertainty polytope [3, 9, 11]. In order to achieve improved performance, we use the theory of Sect. 5.4 and introduce an alternative approach where the polytope Ω is divided into several overlapping sub-polytopes:

$$\Omega = \bigcup_{i=1,\ldots\bar{\sigma}} \Omega_i$$
where

$$\Omega_i = \begin{bmatrix} A_i & B_{1,i} \\ C_{1,i} & 0 \\ C_{2,i} & D_{21,i} \end{bmatrix} = \sum_{j=1}^{N} \eta^j(t) \Omega_i^{(j)}, \quad \sum_{j=1}^{N} \eta_i^j(t) = 1, \ \eta_i^j(t) \geq 0, \ \text{and}$$

$$\Omega_i^{(j)} = \begin{bmatrix} A_i^{(j)} & B_{1,i}^{(j)} \\ C_{1,i}^{(j)} & 0 \\ C_{2,i}^{(j)} & D_{21,i}^{(j)} \end{bmatrix}. \tag{5.20}$$

The whole system is thus treated as a switched system, where each subsystem is defined by a different sub-polytope. The overlap between the sub-polytopes, along with a given bound on the rate of change of the system parameters, impose a dwell time constraint on the switching between the subsystems. We assume that the system parameters can be grossly measured, or estimated with sufficient accuracy in real-time, so that it is known in which the sub-polytope system parameters reside at any given time. Dissimilar to the Gain Scheduling method (GS) [12], the exact location of the parameters inside Ω is not required.

Figure 5.1 illustrates the split of Ω for a single uncertain parameter ($N = 2$). The overlap of the right and the left regions is apparent. Each of these regions defines and corresponds to a subsystem of the switched system. The size of the solid line and the bound on the rate of change of the single parameter determine the dwell time.

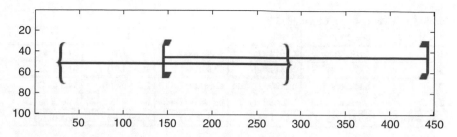

Fig. 5.1 The polytope Ω, divided into 2 overlapping regions

Once the polytope is divided into sub-polytopes, the system is treated as a switched system, where each sub-polytope defines a subsystem, and a switching-dependent filter is obtained for the system. Note that the filter depends on the switching, and that after a switch occurs, during the dwell time, the filter is time dependent. After the dwell time, the filter becomes time invariant until the next switching occurs.

5.6   Examples

Example 5.1   (Switched System)
We consider the filtering problem of [13],

$$A = \begin{bmatrix} 0 & -1 + 0.3\alpha \\ 1 & -0.5 \end{bmatrix}, \quad B_1 = \begin{bmatrix} -2 & 0 \\ 1 & 0 \end{bmatrix}, \quad C_1 = \begin{bmatrix} 1 & 0 \end{bmatrix},$$
$$C_2 = \begin{bmatrix} -100 + 10\beta & 100 \end{bmatrix}, \quad D_{11} = \begin{bmatrix} 0 & 0 \end{bmatrix}, \quad D_{21} = \begin{bmatrix} 0 & 1 \\ 0 & 0 \end{bmatrix}.$$

The parameters (α, β) switch, with a dwell time of 20 seconds, between the following 4 combinations: $(-2.5, -1), (-2.5, 1), (2.5, -1)$, and $(2.5, 1)$.

Using the result of Theorem 5.2, a bound of 2.95 on the estimation error variance ($Trace\{Q\}$) is guaranteed for $K = 1$. A gain scheduled filter [12] for this system does not exist.

Seeking an individual filter for each of the subsystems, estimation error variances of 0.0126, 0.0146, 5.257, and 7.164 are obtained for the parameters above. The bound of 7.165 that is obtained by Theorem 5.2 is very close to the minimal possible error variance of 7.164 that is achieved for the fourth subsystem.

Example 5.2   (Robust Filtering)
We consider again the filtering problem of [13], with $|\alpha| < 2.3$ and $|\beta| < 1$. We divide the parameter regions into four overlapping subregions that are described by

Table 5.1 Values of γ^2 for Example 5.2

Dwell time	K = 1	K = 2	K = 4	K = 10	K = 20
10	36 (9.77)	35.6 (9.33)	34.8 (8.82)	32.6 (8.24)	31.3 (7.8)
5	67.3 (10.4)	45.4 (9.99)	38.8 (9.34)	35.7 (8.37)	31.5 (7.72)
3	Infeasible (13.2)	Infeasible (11.9)	Infeasible (11)	138.5 (9.77)	54.3 (8.42)
2	Infeasible (22.6)	Infeasible (17)	Infeasible (14.7)	Infeasible (13.1)	Infeasible (10.6)
Min. dwell time	3.94 (1.74)	3.49 (1.51)	3.21 (1.37)	2.84 (1.09)	2.35 (0.98)

$$\alpha \in [-2.3, 0.115], \beta \in [-1, 0.05];$$
$$\alpha \in [-2.3, 0.115], \beta \in [-0.05, 1];$$
$$\alpha \in [-0.115, 2.3], \beta \in [-1, 0.05];$$
and
$$\alpha \in [-0.115, 2.3], \beta \in [-0.05, 1].$$

We assume that the dwell time, during which the values of the parameters will remain in the same subregion is 20 seconds.

Robust quadratic and GS filters could not be found that solve the estimation problem over the above range of uncertainty. The relevant LMIs turn out to be infeasible.

Using the result of Theorem 5.2 (Corollary 5.1), a switched filter that guarantees a variance bound of γ^2 is obtained for this system. In Table 5.1, we show the bound on the variance for different values of K for different dwell times, as well as the minimal dwell time for which we could obtain a solution. The results in brackets there are obtained using Corollary 5.1.

5.7 Conclusions

Two filtering problems have been solved. The first achieves a minimum upper-bound on the estimation error variance of an uncertain switched system. The second problem is robust filtering in uncertain linear systems, where a deliberate switching with dwell is introduced in order to improve the estimation. In both the problems, a set of switched filters has been derived with switching that depends on the information on which subsystem is currently active. The solution to these problems is based on a new covariance analysis of switched systems with dwell-time constraint. This analysis can also be applied to other filtering methods that will be developed for switched systems in the future.

References

1. Jazwinsky, A.H.: Stochastic Processes and Filtering Theory. Academic Process, New York (1970)
2. Kwakernaak, H., Sivan, R.: Linear Optimal Control Systems. Wiley, New York (1972)
3. Scherer, C.: Theory of Robust Control. Lecture Note, Mechanical Engineering Systems and Control Group, Delft University of Technology, The Netherlands (2001)
4. Gershon, E., Shaked, U., Yaesh, I.: H_∞ control and filtering of discrete-time stochastic systems with multiplicative noise. Automatica **37**, 409–417 (2001)
5. Anderson, B.D.O., Moore, J.B.: Optimal Filtering. Prentice-hall, Englewood (1979)
6. Allerhand, L.I., Shaked, U.: Robust stability and stabilization of linear switched systems with dwell time. IEEE Trans. Autom. Control **56**, 381–386 (2011)
7. Geromel, J., Colaneri, P.: Stability and stabilization of continuous-time switched linear systems. SIAM J. Control Optim. **45**(5), 1915–1930 (2006)
8. Geromel, J., Colaneri, P.: H_∞ and dwell time specifications of continuous-time switched linear systems. IEEE Trans. Autom. Control **55**(1), 207–212 (2010)
9. Boyd, S., El Ghaoui, L., Feron, E., Balakrishnan, V.: Linear Matrix Inequality in Systems and Control Theory. SIAM Frontier Series, Philadelphia (1994)
10. de Oliveira, M.C., Skelton, R.E.: Stability test for constrained linear systems. In: Reza Moheimani, S.O. (eds.) Perspectives in Robust Control. Lecture Notes in Control and Information Sciences vol. 268. Springer, London (2001)
11. Green, M., Limebeer, D.J.N.: Linear Robust Control. Prentice Hall, NJ (1995)
12. Apkarian, P., Gahinet, P.: A convex characterization of gain-scheduled H_∞ controllers. IEEE Trans. Autom. Control **40**(5), 853–864 (1995)
13. Barbosa, K., de Souza, C., Tronifno, A.: Robust H_2 filtering for uncertain linear systems: LMI based methods with parametric Lyapunov functions. Syst. Control Lett. **54**(3), 251–262 (2005)

Chapter 6
Stability and Controller Synthesis of Discrete Linear-Switched Systems

Abstract Sufficient conditions are presented for the robust stability of discrete-time, switched, linear systems with dwell time in the presence of polytopic-type parameter uncertainty. A Lyapunov function, in quadratic form, is assigned to each of the subsystems. This function is allowed to be time-varying and piecewise linear during the dwell time and it becomes time invariant afterward. Asymptotic stability conditions are obtained in terms of linear matrix inequalities for the nominal set of subsystems. These conditions are then extended to the case, where the subsystems encounter polytopic type parameter uncertainties. The developed method is applied to l_2-gain analysis where a bounded real lemma is derived, and to H_∞ control and estimation, both for the nominal and the uncertain cases.

6.1 Introduction

The stability analysis of continuous-time, linear, switched systems with dwell time has received considerable attention in the past two decades (see, for example, [1–5]). When dealing with switched systems, a variety of constraints may be imposed on the switching law; an overview of these can be found in [6]. In [7], the variational approach is utilized to analyze the stability of switched linear systems, and in [8] the approach is utilized to analyze the l_2-gain of switched systems. In [9], analysis and controller synthesis for discrete-time switched systems is addressed, and necessary and sufficient conditions are obtained by using path-dependent Lyapunov functions. In [10] sufficient conditions for the stability of discrete-time switched system, and stabilization via static output feedback, are obtained in terms of quadratic Lyapunov functions. The topic of switched systems controllability and observability is considered in [11] and the references therein. Robust filter synthesis for switched systems with polytopic type parameter uncertainties in discrete-time is investigated in [12]. The invertibility of switched linear systems has been studied in [13], and the references therein.

© Springer Nature Switzerland AG 2019

E. Gershon and U. Shaked, *Advances in H_∞ Control Theory*,
Lecture Notes in Control and Information Sciences 481,
https://doi.org/10.1007/978-3-030-16008-1_6

In the present chapter, we treat the discrete-time case. By applying a quadratic Lyapunov function, sufficient conditions are first obtained for the stability of linear-switched systems with dwell time and with polytopic-type parameter uncertainties. The least conservative results obtained so far for these systems, in terms of quadratic Lyapunov function, seem to be those of [14]. These results are obtained, however, for the case where the systems parameters are perfectly known (namely for nominal systems). The condition of [14] depends, however, on an exponent of the dynamics matrix of the switched system which prevents its extension to the uncertain case and to control synthesis. We are not aware of other works on discrete-time switched systems with dwell time.

In [15], the method of [14] has been extended to find an upper-bound of the l_2-gain of a switched linear system under a dwell time constraint. The method of [15] still depends on an exponent of the dynamics matrix of the switched system, a fact that prevents its extension to the uncertain case and to control synthesis.

In [3], a piecewise linear Lyapunov function is used to derive sufficient conditions for the stability of continuous-time linear switched systems. We apply the concept of piecewise linear Lyapunov functions in the present chapter to discrete-time linear switched systems with dwell time. The Lyapunov function is assigned separately to each subsystem and it is time-varying during the dwell time in a piecewise linear manner. After the dwell time, this function becomes time invariant. This Lyapunov function allows derivation of sufficient conditions via linear matrix inequalities (LMIs) that are affine in the systems dynamics matrix, thus allowing extension to the case with polytopic- type uncertainties.

In Sect. 6.2, the mathematical formulation of discrete-time switched systems with dwell time is presented, and some results from the literature are discussed. In Sect. 6.3, piecewise linear Lyapunov functions are presented which are applied to the stability analysis of switched systems. The same Lyapunov function is applied, in Sect. 6.4, to derive bounds on the l_2-gain of switched systems, and a Bounded Real Lemma (BRL) is derived. In Sect. 6.5, the BRL is used in the synthesis of state-feedback controllers. In Sect. 6.6 the Luenberger-type filters are considered, and the latter BRL is applied there to the synthesis of H_∞ filters. Finally, in Sect. 6.7 illustrative examples are given that demonstrate the applicability of the developed theory.

Notation: Throughout this chapter the superscript $'$ stands for matrix transposition, $\mathcal{N}$ denotes the set of natural numbers, $\mathcal{R}^n$ denotes the n-dimensional Euclidean space and $\mathcal{R}^{n \times m}$ is the set of all $n \times m$ real matrices. l_2 is the set of all sequences with finite energy. For a symmetric $P \in \mathcal{R}^{n \times n}$, $P > 0$ means that it is positive definite. Matrices with eigenvalues in the open unit circle are refereed to as stability matrices. Switching law of a switched system is denoted by $\sigma(t)$. A symmetric matrix $\begin{bmatrix} Q & R \\ R^T & P \end{bmatrix}$ is denoted by $\begin{bmatrix} Q & R \\ * & P \end{bmatrix}$.

6.2 Problem Formulation

In the context of stability, we consider the following linear-switched system:

$$x(t + 1) = A_{\sigma(t)}(t)x(t), \quad x(0) = x_0 \tag{6.1}$$

defined for all $t \in \mathcal{N}$, where $x \in \mathcal{R}^n$ is the state vector. The switching rule $\sigma(t) \in \{1 \ldots M\}$, for each $t \in \mathcal{N}$, is such that $A_{\sigma(t)} \in \{A_1, \ldots, A_M\}$, where $A_i \in \mathcal{R}^{n \times n}$, $i = 1, \ldots M$ are stability matrices. In most practical cases the parameters of the subsystems are not precisely known. One of the most useful descriptions of the uncertainty is one where the systems parameters are assumed to reside in a prespecified polytope [16, 17]. Each of the matrices A_i is assumed to reside within a polytope with N vertices $A_i^{(j)}$, $j = 1 \ldots N$:

$$\Omega_i = \{A_i(t) | A_i(t) = \sum_{j=1}^{N} \eta^j(t) A_i^{(j)}, \quad \sum_{j=1}^{N} \eta^j(t) = 1, \quad \eta^j(t) \geq 0\}, \quad i = 1, \ldots, M. \tag{6.2}$$

Note that in this configuration $A_{\sigma(t)}(t)$ can vary in time between the switching instances, but only within the polytop whose vertices are $A_i^{(1)}, \ldots A_i^{(N)}$, where $\sigma(t){=}i$.

A dwell time constraint is imposed on the switching rule, which means that if the dwell time is T, and the switching instants are $\tau_1, \tau_2, \ldots$, then $\tau_{h+1} - \tau_h > T$, $\forall h \geq 1$.

In [14], a simple sufficient condition for the stability of the nominal system is introduced.

Lemma 6.1 ([14]) *Assume that for some $T \geq 0$ there exists a collection of positive definite matrices $P_1, P_2, \ldots P_M$ of compatible dimensions such that*

$$A_m' P_m A_m - P_m < 0, \quad A_m'^{T+1} P_q A_m^{T+1} - P_m < 0, \quad \forall m = 1, \ldots M,$$

$$q \neq m = 1, \ldots M. \tag{6.3}$$

Then, the system is globally asymptotically stable for dwell time greater than or equal to T.

The following is a conservative generalization of the result of Lemma 6.1 to the polytopic-type uncertain case. It is the only generalization we know of for the later result. It is brought here for comparison sake in the examples section.

Lemma 6.2 *Assume that for some scalars $T \geq 0$, and $0 < \lambda_i < 1$, $i = 1, \ldots, M$ there exists a collection of symmetric matrices $P_1, P_2, \ldots P_M$ of compatible dimensions such that*

$$P_m > 0, \quad \begin{bmatrix} -\lambda_m P_m & A_m^{'(j)} P_m \\ P_m A_m^{(j)} & -P_m \end{bmatrix} < 0,$$

$$\lambda_q^{T+1} P_q - P_m < 0, \quad \forall\, m = 1, \dots M, \quad q \neq m = 1, \dots M \;\; \forall j = 1, \dots N. \tag{6.4}$$

Then, the system is globally asymptotically stable for dwell time greater than or equal to T.

Proof Consider the Lyapunov function $V(t) = x'(t) P_{\sigma(t)} x(t)$. Condition (6.4a) guarantees that the value of the Lyapunov function is positive for any value of $x(t) \neq 0$. When the mth subsystem is active, $T - 1$ time steps after switching, we have from (6.4b) for all $j = 1 \dots N$ that

$$\begin{bmatrix} -\lambda_m P_m & A_m' P_m \\ P_m A_m & -P_m \end{bmatrix} < 0$$

for every A_m within the polytope, and hence that $A_m' P_m A_m - \lambda_m P_m < 0$. Applying this inequality to the Lyapunov function we find that: $V(t+1) = x'(t + 1) P_{\sigma(t+1)} x(t+1) = x(t) A_m' P_{\sigma(t)} A_m x(t) < -x'(t) \lambda_m P_m x(t) = \lambda_m V(t)$. This guarantees that the subsystem is stable with a decay rate of at least λ_m even if no further switching occurs. Now, assume that at a certain time instance the system switches from the mth to the qth subsystem. At the switching instant, the value of the Lyapunov function is $V(t) = x'(t) P_q x(t)$. T steps later, when the dwell time ends, we have $V(t + T + 1) = x'(t + T + 1) P_q x(t + T + 1) = x(t) A_q^{'T+1} P_q A_q^{T+1} x(t)$, and from (6.4b) we have that $V(t + T + 1) = x(t) A_q^{'T+1} P_q A_q^{T+1} x(t) < \lambda_q^{T+1} x(t) P_q x(t)$. Invoking (6.4b), we find that $\lambda_q^{T+1} x(t) P_q x(t) < x(t) P_m x(t) = V(t)$, and hence $V(t + T + 1) < V(t)$. This assures that the value of V, dwell time steps after switching, is reduced, and consequently the asymptotic stability of the switched system is guaranteed. ∎

In the context of l_2-gain analysis, we consider the following system:

$$\begin{aligned} x(t+1) &= A_{\sigma(t)}(t) x(t) + B_{\sigma(t)}(t) w(t), \quad x(0) = 0, \\ z(t) &= C_{\sigma(t)}(t) x(t) + D_{\sigma(t)}(t) w(t), \end{aligned} \tag{6.5}$$

which is defined for all $t \geq 0$, where $\sigma(t)$ is as defined above, $x \in \mathcal{R}^n$ is the state, $w \in \mathcal{R}^q$ is the exogenous disturbance in l_2, and $z \in \mathcal{R}^r$ is the objective vector. The parameters of the subsystems are assumed to reside in the following polytope:

$$\tilde{\Omega}_i(t) = \sum_{j=1}^{N_i} \eta^j(t) \Omega_i^{(j)}, \quad \sum_{j=1}^{N_i} \eta^j(t) = 1, \quad \eta^j(t) \geq 0, \tag{6.6}$$

where $\tilde{\Omega}_i^{(j)} = \begin{bmatrix} A_i^{(j)} & B_{1,i}^{(j)} \\ C_{1,i}^{(j)} & D_{11,\,i}^{(j)} \end{bmatrix}$.

In the feedback control and filtering settings, the following linear-switched system is considered.

$$
\begin{aligned}
x(t+1) &= A_{\sigma(t)}(t)x(t) + B_{1,\sigma(t)}w(t) + B_{2,\sigma(t)}(t)u(t), \quad x(0) = 0, \\
z(t) &= C_{1,\sigma(t)}(t)x(t) + D_{11,\sigma(t)}(t)w(t) + D_{12,\sigma(t)}(t)u(t), \\
y(t) &= C_{2,\sigma(t)}x(t) + D_{21,\sigma(t)}w(t),
\end{aligned}
\tag{6.7}
$$

where x, w, z, and σ are defined above, $u \in \mathcal{R}^p$ is the control signal, $y \in \mathcal{R}^s$ is the measurement. The subsystems are assumed to reside in the following polytope:

$$
\bar{\Omega}_i(t) = \sum_{j=1}^{N_i} \eta^j(t)\bar{\Omega}_i^{(j)}, \quad \sum_{j=1}^{N_i} \eta^j(t) = 1, \quad \eta^j(t) \ge 0,
\tag{6.8}
$$

where

$$
\bar{\Omega}_i(t) = \begin{bmatrix} A_i(t) & B_{1,i}(t) & B_{2,i}(t) \\ C_{1,i}(t) & D_{11,i}(t) & D_{12,i}(t) \end{bmatrix}, \quad \text{and} \quad \bar{\Omega}_i^{(j)} = \begin{bmatrix} A_i^{(j)} & B_{1,i}^{(j)} & B_{2,i}^{(j)} \\ C_{1,i}^{(j)} & D_{11,i}^{(j)} & D_{12,i}^{(j)}, \end{bmatrix}
\tag{6.9}
$$

$$
, i = 1, \ldots M, \ j = 1, \ldots N_i.
$$

Note that A_i is no longer required to be a stability matrix, and also that $C_{2,i}$ and $D_{21,i}$ are assumed to be known and constant.

6.3 Time-Switching Stability

We introduce a Lyapunov function which, although is more conservative than the one of [14], in the nominal case, it provides an efficient way to deal with uncertainties and with controller synthesis. In order to derive the Lyapunov function, the following result, inspired by [18], is first applied.

Lemma 6.3 *Assume that for some time interval $t \in [t_0, t_f]$, where $\delta = t_f - t_0$ there exist two symmetric matrices P_1 and P_2 of compatible dimensions that satisfy the following:*

$$
P_1, P_2 > 0, \quad \begin{bmatrix} -P_1 & * \\ P_1 A + \frac{P_2-P_1}{\delta} A - P_1 & -\frac{P_2-P_1}{\delta} \end{bmatrix} < 0, \quad \begin{bmatrix} -P_2 + \frac{P_2-P_1}{\delta} & * \\ P_2 A & -P_2 \end{bmatrix} < 0.
\tag{6.10}
$$

Then, for the system $x(t+1) = Ax(t)$ the Lyapunov function

$$
V(t) = x'(t)P(t)x(t), \quad \text{with } P(t) = \alpha(t)P_1 + (1 - \alpha(t))P_2, \quad \alpha(t) = 1 - \frac{t - t_0}{\delta}
\tag{6.11}
$$

is strictly decreasing over the time interval $t \in [t_0, t_f]$.

Proof The condition for $V(t)$ to decrease for every value of $x(t)$ is $A'P(t+1)A - P(t) < 0$. Using Schurs complements [19], this condition is equivalent to

$$\begin{bmatrix} -P(t) & * \\ P(t+1)A & -P(t+1) \end{bmatrix} < 0. \tag{6.12}$$

Let $P(t)$ and $\alpha(t)$ be as in (6.11), then the condition (6.12) is affine in $\alpha(t)$. Hence, (6.10b–c) which correspond to $\alpha(t) = 1$ and $\alpha(t) = \frac{1}{\delta}$, respectively, imply that (6.12) is verified for any $\alpha(t)$ in the interval $[\frac{1}{\delta}, 1]$, i.e., for any time instant t in the considered interval.   ∎

We note that the choice of the Lyapunov function of (6.11) introduces an extra degree of freedom in comparison to the quadratic structure Lyapunov function with a constant P. The time affinity of $P(t)$ allows us to solve the stability problem here over the time interval $[t_0, t_f - 1]$.

Choosing below K to be an integer in $[1, T]$ apriori, according to the allowed computational complexity, the following is obtained.

Theorem 6.1 *The nominal system (6.1) is globally asymptotically stable for any switching law with dwell time greater than or equal to $T > 0$ if there exist a collection of symmetric matrices $P_{i,k}$, $i = 1, \ldots M$, $k = 0, \ldots K$ of compatible dimensions, and a sequence of integers $\{\delta_k \geq 1, \quad k = 1, \ldots K, \quad \sum_{k=1}^{K} \delta_k = T\}$ such that, for all $i = 1, \ldots M$ the following holds:*

$$P_{i,k} > 0, \quad \begin{bmatrix} -P_{i,k} & * \\ P_{i,k}A_i + \frac{P_{i,k+1}-P_{i,k}}{\delta_{k+1}}A_i & -P_{i,k} - \frac{P_{i,k+1}-P_{i,k}}{\delta_{k+1}} \end{bmatrix} < 0,$$

$$\begin{bmatrix} -P_{i,k+1} + \frac{P_{i,k+1}-P_{i,k}}{\delta_{k+1}} & * \\ P_{i,k+1}A_i & -P_{i,k+1} \end{bmatrix} < 0, \quad \begin{bmatrix} -P_{i,K} & * \\ P_{i,K}A_i & -P_{i,K} \end{bmatrix} < 0, \tag{6.13}$$

$$\begin{bmatrix} -P_{q,K} & * \\ P_{i,0}A_q & -P_{i,0} \end{bmatrix} < 0, \quad k = 0, \ldots K-1, \quad i = 1 \ldots M,$$

$$q = 1, \ldots i-1, i+1, \ldots M.$$

Proof Let $i_0 = \sigma(0)$, and let $\tau_1, \tau_2, \ldots$ be the switching instants, where $\tau_{h+1} - \tau_h > T$, $\forall h = 1, 2, \ldots$. Define $\tau_{h,k} = \tau_h + \sum_{j=1}^{k} \delta_j$ for $k \geq 1$, and $\tau_{h,0} = \tau_h$. Note that the dwell time constraint implies $\tau_{h,K} + 1 \leq \tau_{h+1,0} = \tau_{h+1}$. Choose the Lyapunov function $V(t) = x'(t)P(t)x(t)$, where

$$P(t) = \begin{cases} P_{i,k} + (P_{i,k+1} - P_{i,k})\frac{t-\tau_{h,k}}{\delta_{k+1}} & t \in [\tau_{h,k}, \tau_{h,k+1}) \\ P_{i,K} & t \in [\tau_{h,K}, \tau_{h+1,0}), \\ P_{i_0,K} & t \in [0, \tau_1) \end{cases} \quad h = 1, 2, \ldots \tag{6.14}$$

and where i is the index of the subsystem that is active at time t.

Assume that at some switching instant τ_h, the systems switches from the qth subsystem to the ith subsystem, meaning that τ_h is the last time step in which the system follows the qth dynamics. At the switching instant τ_h, we have

$$V(\tau_h) = x(\tau_h)' P_{q,K} x(\tau_h), \quad V(\tau_h + 1) = x(\tau_h)' A_q' P_{i,0} A_q x(\tau_h).$$

Therefore, for the Lyapunov function to decrease at the switching instants for any $x(\tau_h)$, we require that $A_q' P_{i,0} A_q - P_{q,K} < 0$ which is condition (6.13e). After the dwell time, and before the next switching occurs, we have that $V(t) = x(t)' P_{i,K} x(t)$, where $x(t + 1) = A_i x(t)$. Therefore, $V(t + 1) - V(t) = x(t)'[A_i' P_{i,K} A_i - P_{i,K}]$ $x(t)$ and (6.13d) guarantees that this expression is negative for any $x(t) \neq 0$. During the dwell time, we consider the time intervals $t \in [\tau_{h,k}, \tau_{h,k+1} - 1]$ where $\delta_{k+1} = \tau_{h,k+1} - \tau_{h,k}$. The matrix $P(t)$ changes then linearly from $P_{i,k}$ to $P_{i,k+1}$, and applying the conditions of Lemma 6.3, conditions (6.13b–c) guarantee that in these intervals $V(t)$ is strictly decreasing. This implies the asymptotic stability of the system. ∎

Remark 6.1 We note that the computation burden is increasing with K. In most cases, one may apply $K \ll T$. In the case where $K = T$, we have $\delta_k = 1$, $\forall k = 1 \ldots K$. In this case, $P_{i,k}$ are decision variables at each step during the dwell time. Unless the computational cost is too high, this choice of K should provide the least conservative result.

The LMIs in Theorem 6.1 are affine in the system matrices. Therefore, if the subsystems entail polytopic-type uncertainty, this theorem can provide a solution if its conditions hold at the vertices of all the subsystems. Robust stability in the case of uncertain subsystems with polytopic uncertainties is thus given by the following.

Corollary 6.1 *The system (6.1) with the uncertainty as in (6.2) is globally asymptotically stable for any switching law with dwell time greater than or equal to $T > 0$ if there exist a collection of symmetric matrices $P_{i,k}, i = 1, \ldots M, k = 0, \ldots K$ of compatible dimensions, where K is a prechosen integer in $[1, T]$, and a sequence of integers $\{\delta_k \geq 1, \quad k = 1, \ldots K, \quad \sum_{k=1}^{K} \delta_k = T\}$ such that, for all $i = 1, \ldots M$, the following holds.*

$$P_{i,k} > 0, \quad \begin{bmatrix} -P_{i,k} & * \\ P_{i,k} A_i^{(j)} + \frac{P_{i,k+1}-P_{i,k}}{\delta_{k+1}} A_i^{(j)} & -P_{i,k} - \frac{P_{i,k+1}-P_{i,k}}{\delta_{k+1}} \end{bmatrix} < 0,$$

$$\begin{bmatrix} -P_{i,k+1} + \frac{P_{i,k+1}-P_{i,k}}{\delta_{k+1}} & * \\ P_{i,k+1} A_i^{(j)} & -P_{i,k+1} \end{bmatrix} < 0, \quad \begin{bmatrix} -P_{i,K} & * \\ P_{i,K} A_i^{(j)} & -P_{i,K} \end{bmatrix} < 0,$$

and

$$\begin{bmatrix} -P_{q,K} & * \\ P_{i,0} A_q^{(j)} & -P_{i,0} \end{bmatrix} < 0,$$

$$k = 0, \ldots K - 1, \quad i = 1 \ldots M, \quad q = 1, \ldots i - 1, i + 1, \ldots M, \quad j = 1, \ldots N.$$

$$(6.15)$$

6.4 l_2-gain Analysis

In order to analyze the l_2-gain of a switched system, we first consider the l_2-gain of a general LTV system:

$$
\begin{aligned}
x(t+1) &= A(t)x(t) + B(t)w(t), \quad x(0) = 0 \\
z(t) &= C(t)x(t) + D(t)w(t),
\end{aligned}
\tag{6.16}
$$

with the cost function:

$$
J(\bar{t}) = \sum_{k=0}^{\bar{t}} [z'(t)z(t) - \gamma^2 w'(t)w(t)],
\tag{6.17}
$$

where it is required that $J(\bar{t}) < 0$ for $\bar{t} \to \infty$. It is well known that the condition for this requirement to hold is the existence of $P(t) > 0$, $t \in \mathcal{N}$ that satisfies the following [20]:

$$
\begin{bmatrix}
-P(t) & * & * & * \\
0 & -\gamma^2 I & * & * \\
P(t+1)A(t) & P(t+1)B(t) & -P(t+1) & * \\
C(t) & D(t) & 0 & -I
\end{bmatrix} < 0
$$

or equivalently that there exists $P(t) > 0$, $t \in \mathcal{N}$ that satisfies the following [20]:

$$
\begin{bmatrix}
-Q(t) & * & * & * \\
0 & -\gamma^2 I & * & * \\
A(t)Q(t) & B(t) & -Q(t+1) & * \\
C(t)Q(t) & D(t) & 0 & -I
\end{bmatrix} < 0,
\tag{6.18}
$$

where $Q(t) = P^{-1}(t)$.

Taking $P(t)$ to be as in (6.14), the following conditions are obtained.

Theorem 6.2 *The l_2-gain of the system (6.5) with the uncertainty (6.6) is less than a prescribed scalar γ, for any switching law with dwell time greater than or equal to $T > 0$, if there exist a collection of symmetric matrices $P_{i,k}, i = 1, \ldots M, k = 0, \ldots K$ of compatible dimensions, where K is a prechosen integer in $[1, T]$, and a sequence of integers $\{\delta_k \geq 1, \quad k = 1, \ldots K, \quad \sum_{k=1}^{K} \delta_k = T\}$ such that, for all $i = 1, \ldots M$ the following holds:*

$$
P_{i,k} > 0,
$$

$$
\begin{bmatrix}
-P_{i,k} & * & & * & * \\
0 & -\gamma^2 I & & * & * \\
\Gamma_{3,1}(i,k,j) & P_{i,k}B_i^{(j)} + \frac{P_{i,k+1}-P_{i,k}}{\delta_{k+1}}B_i^{(j)} & -P_{i,k} - \frac{P_{i,k+1}-P_{i,k}}{\delta_{k+1}} & * \\
C_i^{(j)} & D_i^{(j)} & 0 & -I
\end{bmatrix} < 0,
$$

$$
\Gamma_{3,1}(i,k,j) = P_{i,k}A_i^{(j)} + \frac{P_{i,k+1}-P_{i,k}}{\delta_{k+1}}A_i^{(j)},
$$

$$
\begin{bmatrix}
-P_{i,k+1} + \frac{P_{i,k+1}-P_{i,k}}{\delta_{k+1}} & * & * & * \\
0 & -\gamma^2 I & * & * \\
P_{i,k+1}A_i^{(j)} & P_{i,k+1}B_i^{(j)} & -P_{i,k+1} & * \\
C_i^{(j)} & D_i^{(j)} & 0 & -I
\end{bmatrix} < 0
$$

$$
\begin{bmatrix}
-P_{i,K} & * & * & * \\
0 & -\gamma^2 I & * & * \\
P_{i,K}A_i^{(j)} & P_{i,K}B_i^{(j)} & -P_{i,K} & * \\
C_i^{(j)} & D_i^{(j)} & 0 & -I
\end{bmatrix} < 0, \; and \;
\begin{bmatrix}
-P_{q,K} & * & * & * \\
0 & -\gamma^2 I & * & * \\
P_{i,0}A_q^{(j)} & P_{i,0}B_q^{(j)} & -P_{i,0} & * \\
C_q^{(j)} & D_q^{(j)} & 0 & -I
\end{bmatrix} < 0
$$

$$
k = 0,\ldots K-1, \quad i = 1\ldots M, \quad q = 1,\ldots i-1, i+1, \ldots M, \quad j = 1 \ldots N.
$$
(6.19)

Proof Let $P(t)$ be as in (6.14), and assume that at some switching instant τ_h the systems switches from the qth subsystem to the ith subsystem. At the switching instant τ_h we have

$$
P(t) = P_{q,K}, \quad P(t+1) = P_{i,0},
$$

and the dynamics matrices are those at time t, namely those of the qth subsystem. Substituting into (6.18a) yields (6.19e). After the dwell time, and before the next switching instant, we have that $P(t) = P_{i,K}$, and the dynamics matrices are those of the ith subsystem. Substitution in (6.18a) leads then to (6.19d). During the dwell time, we consider the time intervals $t \in [\tau_{h,k}, \tau_{h,k+1} - 1]$, where $\delta_{k+1} = \tau_{h,k+1} - \tau_{h,k}$. The matrix $P(t)$ changes then linearly from $P_{i,k}$ to $P_{i,k+1}$, and the dynamics matrices are those of the ith subsystem. Substitution in (6.18a) yields (6.19b–c). ∎

Equivalently, it is possible to write the conditions of Theorem 6.2 in terms of $Q = P^{-1}$:

Corollary 6.2 *The l_2-gain of the system (6.5) with the uncertainty (6.6) is less than a prescribed positive γ, for any switching law with dwell time greater than or equal to $T > 0$, if there exist a collection of symmetric matrices $Q_{i,k}, i = 1, \ldots M, k = 0, \ldots K$ of compatible dimensions, where $K \in [1, T]$, and a sequence of integers $\{\delta_k \geq 1, \quad k = 1, \ldots K, \quad \sum_{k=1}^{K} \delta_k = T\}$ such that, for all $i = 1, \ldots M$ the following holds.*

$$Q_{i,k} > 0, \quad \begin{bmatrix} -Q_{i,k} & * & * & * \\ 0 & -\gamma^2 I & * & * \\ A_i^{(j)} Q_{i,k} & B_i^{(j)} & -Q_{i,k} - \frac{Q_{i,k+1}-Q_{i,k}}{\delta_{k+1}} & * \\ C_i^{(j)} Q_{i,k} & D_i^{(j)} & 0 & -I \end{bmatrix} < 0,$$

$$\begin{bmatrix} -Q_{i,k+1} + \frac{Q_{i,k+1}-Q_{i,k}}{\delta_{k+1}} & * & * & * \\ 0 & -\gamma^2 I & * & * \\ A_i^{(j)} Q_{i,k+1} - A_i^{(j)}\frac{Q_{i,k+1}-Q_{i,k}}{\delta_{k+1}} & B_i^{(j)} & -Q_{i,k+1} & * \\ C_i^{(j)} Q_{i,k+1} - C_i^{(j)}\frac{Q_{i,k+1}-Q_{i,k}}{\delta_{k+1}} & D_i^{(j)} & 0 & -I \end{bmatrix} < 0,$$

and

$$\begin{bmatrix} -Q_{i,K} & * & * & * \\ 0 & -\gamma^2 I & * & * \\ A_i^{(j)} Q_{i,K} & B_i^{(j)} & -Q_{i,K} & * \\ C_i^{(j)} Q_{i,K} & D_i^{(j)} & 0 & -I \end{bmatrix} < 0, \quad \begin{bmatrix} -Q_{q,K} & * & * & * \\ 0 & -\gamma^2 I & * & * \\ A_q^{(j)} Q_{q,K} & B_q^{(j)} & -Q_{i,0} & * \\ C_q^{(j)} Q_{q,K} & D_q^{(j)} & 0 & -I \end{bmatrix} < 0,$$

$$k = 0,\ldots K-1, \quad i = 1\ldots M, \quad q = 1,\ldots i-1, i+1, \ldots M, \quad j = 1 \ldots N. \tag{6.20}$$

6.5   State Feedback

We restrict our discussion to controllers of the form

$$u(t) = G(t)x(t). \tag{6.21}$$

The system (6.7) with the controller (6.21) is equivalent to the one of (6.5), where $A_{\sigma(t)}$, and $C_{1,\sigma(t)}$ are replaced by $A_{\sigma(t)} + B_{2,\sigma(t)}G(t)$ and $C_{1,\sigma(t)} + D_{12,\sigma(t)}G(t)$, respectively. Therefore, if $G(t)$ is taken to be constant, denoting $A_{cl,i}^{(j)} = A_i^{(j)} + B_{2,i}^{(j)}G$, $C_{cl,1,i}^{(j)} = C_{1,i}^{(j)} + D_{12,i}^{(j)}G$, the system (6.7) with the controller (6.21) is equivalent to the one of (6.5), where $A_{\sigma(t)}$, and $C_{1,\sigma(t)}$ are replaced by $A_{cl,i}^{(j)}$ and $C_{cl,1,i}^{(j)}$ respectively. Applying this substitution to the conditions of Corollary 6.2, a set of Bilinear Matrix Inequality (BMI) conditions is obtained, a solution to which, if exists, guarantees an l_2-gain bound of γ for the closed-loop system (6.7). These conditions are obtained via BMIs because the same state-feedback gain should be applied to all the subsystems.

Remark 6.2 BMIs are known to be non-convex. They can be solved either by using a solver that is able to solve non-convex problems, or by applying local iterations, where at each step LMIs are solved, which allows the use of standard solvers.

The case where the switching signal is measured online can be readily solved by replacing the constant state-feedback gain matrix G by switching dependent G_i, $i = 1, \ldots M$. In this case, a computationally efficient result can be obtained that applies time-varying gains via solutions of LMIs. The latter is achieved by allowing the gains to be time-varying during the dwell time and by seeking a matrix $Q(t)$ as in Corollary 6.2. In the theorem below, we use a time-varying gain $G_{\sigma(t)}(t)$, apply Corollary 6.2, and in the resulting inequalities denote $Y_{i,k} = G_{i,k}Q_{i,k}$.

Theorem 6.3 *The l_2-gain of the system (6.7) with the uncertainty (6.8), and (6.9), and the controller (6.21) is less than a prescribed positive scalar γ, for any switching law with dwell time greater than or equal to $T > 0$, if there exist: a collection of matrices $Q_{i,k} = Q'_{i,k}, Y_{i,k}i = 1,\ldots M, k = 0,\ldots K$ of compatible dimensions, where $K \in [1,\ T]$, and a sequence of integers $\{\delta_k \geq 1, \quad k = 1,\ldots K, \quad \sum_{k=1}^{K}\delta_k = T\}$ such that, for all $i = 1,\ldots M$ the following holds.*

$$Q_{i,k} > 0, \quad \begin{bmatrix} -Q_{i,k} & * & * & * \\ 0 & -\gamma^2 I & * & * \\ A_i^{(j)}Q_{i,k} + B_{2,i}^{(j)}Y_{i,k} & B_{1,i}^{(j)} & -Q_{i,k} - \frac{Q_{i,k+1}-Q_{i,k}}{\delta_{k+1}} & * \\ C_{1,i}^{(j)}Q_{i,k} + D_{12,i}^{(j)}Y_{i,k} & D_{11,i}^{(j)} & 0 & -I \end{bmatrix} < 0,$$

$$\begin{bmatrix} -Q_{i,k+1} + \frac{Q_{i,k+1}-Q_{i,k}}{\delta_{k+1}} & * & * & * \\ 0 & -\gamma^2 I & * & * \\ \bar{\Gamma}_{3,1}(i,k,j) & B_{1,i}^{(j)} & -Q_{i,k+1} & * \\ \bar{\Gamma}_{4,1}(i,k,j) & D_{11,i}^{(j)} & 0 & -I \end{bmatrix} < 0,$$

$$\begin{bmatrix} -Q_{i,K} & * & * & * \\ 0 & -\gamma^2 I & * & * \\ A_i^{(j)}Q_{i,K} + B_{2,i}^{(j)}Y_{i,K} & B_{1,i}^{(j)} & -Q_{i,K} & * \\ C_{1,i}^{(j)}Q_{i,K} + D_{12,i}^{(j)}Y_{i,K} & D_{11,i}^{(j)} & 0 & -I \end{bmatrix} < 0,$$

and

$$\begin{bmatrix} -Q_{q,K} & * & * & * \\ 0 & -\gamma^2 I & * & * \\ A_q^{(j)}Q_{q,K} + B_{2,q}^{(j)}Y_{q,k} & B_{1,q}^{(j)} & -Q_{i,0} & * \\ C_{1,q}^{(j)}Q_{q,K} + D_{12,q}^{(j)}Y_{q,k} & D_{11,q}^{(j)} & 0 & -I \end{bmatrix} < 0,$$

$$k = 0,\ldots K-1, \quad i = 1\ldots M, \quad q = 1,\ldots i-1, i+1,\ldots M, \quad j = 1\ldots N,$$
(6.22)

where

$$\bar{\Gamma}_{3,1}(i,k,j) = A_i^{(j)}Q_{i,k+1} - A_i^{(j)}\frac{Q_{i,k+1} - Q_{i,k}}{\delta_{k+1}} + B_{2,i}^{(j)}Y_{i,k+1} - B_{2,i}^{(j)}\frac{Y_{i,k+1} - Y_{i,k}}{\delta_{k+1}},$$

and

$$\bar{\Gamma}_{4,1} = (i, k, j)C_{1,i}^{(j)}Q_{i,k+1} - C_i^{(j)}\frac{Q_{i,k+1} - Q_{i,k}}{\delta_{k+1}} + D_{12,i}^{(j)}Y_{i,k+1} - D_{12,i}^{(j)}\frac{Y_{i,k+1} - Y_{i,k}}{\delta_{k+1}}.$$

We note that in order to preserve the convexity for time values in the interval $[\tau_{h,k}, \tau_{h,k+1})$, $Y_\sigma(t)$ must change linearly over this time interval, which implies $Y_\sigma(t) = Y_{i,k} + \frac{t-\tau_{h,k}}{\delta_{k+1}}(Y_{i,k+1} - Y_{i,k})$. Therefore, the controller gain $G_\sigma(t)$ will be

$$G_{\sigma(t)} =$$

$$\begin{cases} [Y_{i,k} + \frac{t-\tau_{h,k}}{\delta_{k+1}}(Y_{i,k+1} - Y_{i,k})][Q_{i,k} + (Q_{i,k+1} - Q_{i,k})\frac{t-\tau_{h,k}}{\delta_{k+1}}]^{-1} & t \in [\tau_{h,k}, \tau_{h,k+1}) \\ Y_{i,K}Q_{i,K}^{-1} & t \in [\tau_{h,K}, \tau_{h+1,0}). \\ Y_{i_0,K}Q_{i_0,K}^{-1} & t \in [0, \tau_1) \end{cases}$$

$$(6.23)$$

A simpler controller that is constant between the switching instances can be obtained by applying the method of [21]. This controller leads, however, to more conservative results.

6.6 Filtering

We consider the estimation problem in switched systems with dwell time. We treat here the case where the switching rule is known online. We consider the system (6.7) with $B_{2,i} = 0$, $D_{11,i} = 0$, $D_{12,i} = 0 \ \forall i = 1 \dots M$, where A_i, $C_{1,i}$, $C_{2,i} \ \forall i = 1 \dots M$ are fully known, and the switching signal is measured online. The cases where either A_i, $C_{2,i} \ \forall i = 1 \dots M$ entail uncertainty, or the switching signal is unknown, may be solved using the results obtained for the output feedback in the proceeding section.

We consider the following filter:

$$\hat{x}(t+1) = A_{\sigma(t)}\hat{x}(t) + L_{t,\sigma(t)}(y(t) - C_{2,\sigma(t)}\hat{x}(t)). \qquad (6.24)$$

Defining $e \equiv x - \hat{x}$, we have

$$e(t+1) = x(t+1) - \hat{x}(t+1) = \underline{A}_{\sigma(t)}e + \underline{B}_{1,\sigma(t)}w, \qquad (6.25)$$

where

$$\underline{A}_{\sigma(t)} = A_{\sigma(t)} - L_{t,\sigma(t)}C_{2,\sigma(t)} \text{ and } \underline{B}_{1,\sigma(t)} = B_{1,\sigma(t)} - L_{t,\sigma(t)}D_{21,\sigma(t)} \qquad (6.26)$$

and the objective vector is $z_e = C_{1,\sigma(t)}e$.

Replacing A_i and $B_{1,i}$ of Theorem 6.1 by $\underline{A}_i$ and $\underline{B}_{1,i}$, of (6.26a–b), respectively, and denoting $Y_{i,k} = -P_{i,k}L_{i,k}$, a sufficient condition is obtained for the stabilization and the bound on the closed-loop L_2-gain of the estimation error system.

Theorem 6.4 *The l_2-gain from w to the estimation error of z, which is denoted by z_e, is less than a prescribed positive γ, for any switching law with dwell time greater than or equal to $T > 0$, if there exist a collection of matrices $P_{i,k} = P'_{i,k}, Yi, k, i = 1, \ldots M, k = 0, \ldots K$ of compatible dimensions, where $K \in [1, \ T]$, and a sequence of integers $\{\delta_k \geq 1, \ \ k = 1, \ldots K, \ \ \sum_{k=1}^{K} \delta_k = T\}$ such that, for all $i = 1, \ldots M$ the following holds.*

$P_{i,k} > 0,$

$$\begin{bmatrix} -P_{i,k} & * & * & * \\ 0 & -\gamma^2 I & * & * \\ \Psi_{3,1}^{(j)}(i,k) & \Psi_{3,2}^{(j)}(i,k) & -P_{i,k} - \frac{P_{i,k+1}-P_{i,k}}{\delta_{k+1}} & * \\ C_{1,i} & 0 & 0 & -I \end{bmatrix} < 0,$$

$$\begin{bmatrix} -P_{i,k+1} + \frac{P_{i,k+1}-P_{i,k}}{\delta_{k+1}} & * & * & * \\ 0 & -\gamma^2 I & * & * \\ P_{i,k+1}A_i + Y_{i,k+1}C_{2,i} & P_{i,k+1}B_{1,i}^{(j)} + Y_{i,k+1}D_{21,i}^{(j)} & -P_{i,k+1} & * \\ C_{1,i} & 0 & 0 & -I \end{bmatrix} < 0,$$

$$\begin{bmatrix} -P_{i,K} & * & * & * \\ 0 & -\gamma^2 I & * & * \\ P_{i,K}A_i + Y_{i,K}C_{2,i} & P_{i,K}B_{1,i}^{(j)} + Y_{i,K}D_{21,i}^{(j)} & -P_{i,K} & * \\ C_{1,i} & 0 & 0 & -I \end{bmatrix} < 0,$$

and

$$\begin{bmatrix} -P_{q,K} & * & * & * \\ 0 & -\gamma^2 I & * & * \\ P_{i,0}A_q + Y_{i,0}C_{2,q} & P_{i,0}B_{1,q}^{(j)} + Y_{i,0}D_{21,q}^{(j)} & -P_{i,0} & * \\ C_q & 0 & 0 & -I \end{bmatrix} < 0,$$

$$k = 0, \ldots K-1, \quad i = 1 \ldots M, \quad q = 1, \ldots i-1, i+1, \ldots M, \quad j = 1 \ldots N,$$

where:

$$\Psi_{3,1}^{(j)}(i,k) = P_{i,k}A_i + \frac{P_{i,k+1}-P_{i,k}}{\delta_{k+1}}A_i + Y_{i,k}C_{2,i} + \frac{Y_{i,k+1}-Y_{i,k}}{\delta_{k+1}}C_{2,i},$$

and

$$\Psi_{3,2}^{(j)}(i,k) = P_{i,k}B_{1,i}^{(j)} + \frac{P_{i,k+1}-P_{i,k}}{\delta_{k+1}}B_{1,i}^{(j)} + Y_{i,k}D_{21,i}^{(j)} + \frac{Y_{i,k+1}-Y_{i,k}}{\delta_{k+1}}D_{21,i}^{(i)}.$$

$$(6.27)$$

Similar to the state-feedback case, the convexity of $Y_\sigma(t)$ must be preserved for time values in the interval $[\tau_{h,k}, \tau_{h,k+1})$, and therefore it must change linearly in time, which implies that $Y_\sigma(t) = Y_{i,k} + \frac{t-\tau_{h,k}}{\delta_{k+1}}(Y_{i,k+1} - Y_{i,k})$. The filter gain $L_\sigma(t)$ will, therefore, be

$$L_\sigma(t) =$$

$$\begin{cases} -[P_{i,k} + (P_{i,k+1}-P_{i,k})\frac{t-\tau_{h,k}}{\delta_{k+1}}]^{-1}[Y_{i,k} + \frac{t-\tau_{h,k}}{\delta_{k+1}}(Y_{i,k+1}-Y_{i,k})] & t \in [\tau_{h,k}, \tau_{h,k+1}) \\ -P_{i,K}^{-1}Y_{i,K} & t \in [\tau_{h,K}, \tau_{h+1,0}). \\ -P_{i_0,K}^{-1}Y_{i_0,K} & t \in [0, \tau_1) \end{cases}$$

$$(6.28)$$

Table 6.1 Minimum dwell time for Example 6.1

Method	Lemma 6.1 [14]	Lemma 6.2	$K = 1$	$K = 2$	$K = 3$	$K = 4$	$K = 5$
Min T	5	7	10	7	6	5	5

6.7 Examples

In the following, we solve five examples to demonstrate the application of the theory developed in the previous sections. We chose simple second-order examples and compared our results with those that were reported in the literature.

Example 6.1 (Stability) We consider the system taken from [14] with:

$$A_1 = exp\left(0.5\begin{bmatrix} 0 & 1 \\ -10 & -1 \end{bmatrix}\right), \quad A_2 = exp\left(0.5\begin{bmatrix} 0 & 1 \\ -0.5 & -0.1 \end{bmatrix}\right).$$

Note that this system is not quadratically stable. Table 6.1 shows the minimal dwell time for which the system is proved to be stable, calculated by Lemmas 6.1, 6.2 and Theorem 6.1.

It is readily seen that the result of [14] is recovered already for $K = 4$. The table also demonstrates the conservatism of Lemma 6.2.

We note that the finding minimal dwell time is accomplished by performing a line-search, where, at each step, the feasibility problem is solved. In the line-search, one may apply a binary search, thus obtaining the minimal dwell time by solving a number of LMIs which is proportional to $log(T_{min})$. This method is considered to be, therefore, computationally efficient.

Example 6.2 (l_2-gain analysis) We consider the discrete-time equivalent of the system treated in [22]. Denoting the sampling time as t_s, and discretizing via Zero-Order Hold (ZOH) ([23]), the following matrices of the state-space model are obtained.

$$A_1 = exp\left(\begin{bmatrix} 0 & 1 \\ -10 & -1 \end{bmatrix} t_s\right), \quad A_2 = exp\left(\begin{bmatrix} 0 & 1 \\ -0.5 & -0.1 \end{bmatrix} t_s\right),$$

$$B_1 = \int_0^{t_s} exp\left(\begin{bmatrix} 0 & 1 \\ -10 & -1 \end{bmatrix} \tau\right)\begin{bmatrix} 0 \\ 1 \end{bmatrix} d\tau, \quad B_2 = \int_0^{t_s} exp\left(\begin{bmatrix} 0 & 1 \\ -0.5 & -0.1 \end{bmatrix} \tau\right)\begin{bmatrix} 0 \\ 1 \end{bmatrix} d\tau,$$

$$C_1 = \begin{bmatrix} 0.8715 & 0 \end{bmatrix}, \quad C_2 = \begin{bmatrix} 0 & 0.335 \end{bmatrix}, \quad D_1 = \begin{bmatrix} -0.8715 \end{bmatrix} \text{ and } D_2 = \begin{bmatrix} 0.335 \end{bmatrix}.$$

In [22], bounds on the l_2-gain of the continuous-time system were calculated for various dwell times. The minimum dwell time that preserves stability is found there to be in the interval [2.71, 2.76]. For $t_s = 0.01$, the corresponding dwell time should be around $T \geq 270$, which means that a very large number of inequalities should be solved if one applies $K = T$. Using Theorem 6.2, this system is found to be stable for, say, $T = 300$ with $K = 30$. This result indicates the efficiency of our method

Table 6.2 Minimal γ for Example 6.2

T	275	300	500
$K = 10$	Infeasible	Infeasible	3.69
$K = 30$	Infeasible	441	3.06
$K = T$	163	8.1	2.73

in reducing considerably the computational burden that is encountered when taking $K = T$, while slightly sacrificing the value of the minimal dwell time.

Applying Theorem 6.2, bounds on the l_2-gain of the system are given in Table 6.2 for various values of T and K. We note that the minimum value of T for which stability is guaranteed by Theorem 6.2 is found to be 275.

The table shows that for large dwell times, it is possible to apply smaller values of K and still obtain good results. For $T = 700$, a 5 percent difference in the values of γ for $K = 700$ and $K = 30$ can be found. The results become highly sensitive to changes in K the closer T is to the critical value of $T = 275$.

Example 6.3 (l_2-gain analysis) We consider the discrete-time system treated in [15].

$$A_1 = \begin{bmatrix} 0 & 0.25 & 0 \\ 1 & 0 & 0 \\ 0 & 1 & 0 \end{bmatrix}, \quad A_2 = \begin{bmatrix} -2 & -1.5625 & -0.4063 \\ 1 & 0 & 0 \\ 0 & 1 & 0 \end{bmatrix}, \quad A_3 = \begin{bmatrix} 1 & -0.5625 & 0.1563 \\ 1 & 0 & 0 \\ 0 & 1 & 0 \end{bmatrix},$$

$$B_1 = B_2 = B_3 = \begin{bmatrix} 1 \\ 0 \\ 0 \end{bmatrix} \quad C_1 = [0\ 0\ 0.7491], \quad C_2 = [0.0964\ 0.0964\ 0.0964],$$

$$C_3 = [0.2031\ 0.0444\ 0.1174], \quad D_1 = [0] \quad D_2 = [0] \text{ and } D_3 = [0.1015].$$

In [15], bounds on the l_2-gain of the system were calculated for various values of the dwell time. The minimum dwell time for which a bound on the l_2-gain is found there is 5. The bound for $T = 5$ is found there to be around $\gamma = 10.5$. Using our Theorem 6.2, with $K = T$, we are able to show that the system is stable with dwell time $T = 4$ and for this dwell time the bound on the l_2-gain for the system is found to be $\gamma = 9.44$. Using our Theorem 6.2 with $K = T = 5$ we obtain an l_2-gain bound of $\gamma = 8.32$.

Example 6.4 (State feedback) We consider the system:

$$A_1 = \begin{bmatrix} 0 & 1 \\ 0.9 & 0 \end{bmatrix}, \quad A_2 = \begin{bmatrix} 0 & 0.9 \\ 1 & 0 \end{bmatrix}, \quad B_{1,1} = \begin{bmatrix} 10 \\ 0 \end{bmatrix}, \quad B_{1,2} = \begin{bmatrix} 0 \\ 10 \end{bmatrix},$$

$$B_{2,1} = \begin{bmatrix} 0 \\ 1 \end{bmatrix}, \quad B_{2,2} = \begin{bmatrix} 1 \\ 0 \end{bmatrix}, \quad C_{1,1} = C_{1,2} = [1\ 1], \text{ and } D_{11,1} = D_{11,2} = 0,$$

$$D_{12,1} = D_{12,2} = 0.1.$$

Table 6.3 Minimal γ for Example 6.4

T	1	3	5
$K = 1$	15.15	12.73	12.09
$K = 2$	–	12.08	11.45
$K = 5$	–	–	11.23

Table 6.4 Minimal γ for Example 6.5

T	Robust	GS	$K = 1$	$K = 2$
$\alpha = 3$	94.9	10.79	9.14	8.36
$\alpha = 5$	Infeasible	11.68	9.65	7.42
$\alpha = 10$	Infeasible	15.28	12.2	9.35

It is assumed that the switching signal is known online, and Theorem 6.3 is invoked in order to design a state-feedback control that minimizes the l_2-gain of the system. The minimal γ obtained for this system, with various values of the dwell time and of K, is described in Table 6.3.

It is evident from the table, once again, that for shorter dwell times, changes in K have a bigger effect on the guaranteed l_2-gain.

Example 6.5 (filtering) We consider the system taken from [24] with:

$$A = \begin{bmatrix} 0.9 & 0.1 + 0.06\alpha \\ 0.01 + 0.05\beta & 0.9 \end{bmatrix}, \ B_1 = \begin{bmatrix} 1 & 0 & 0 \\ 0 & 1 & 0 \end{bmatrix}, \ C_1 = \begin{bmatrix} 1 & 1 \end{bmatrix}, \ C_2 = \begin{bmatrix} 1 & 0 \end{bmatrix},$$

and $D_{21} = \begin{bmatrix} 0 & 0 & \sqrt{2} \end{bmatrix}$, where the parameters (α, β) switch, with dwell time of 10 seconds, between the following combinations: $(\alpha, \alpha), (-\alpha, \alpha), (\alpha, -\alpha), (-\alpha, -\alpha)$. Three methods of filtering are compared in Table 6.4. The first is a standard robust filter design [25] that treats the above four combinations as vertices of an uncertain polytope. The second is gain-scheduled filtering [26], where different filters are obtained for the four vertices of the latter polytope, applying the same Lyapunov matrix P all over the polytope. The third method is the one of Theorem 6.4.

The advantage of the design that is based on Theorem 6.4 is evident.

6.8 Conclusions

Sufficient conditions are obtained for robust stability of discrete-time switched systems with dwell time. These conditions, which apply Lyapunov functions that are piecewise linear in time, are given in terms of LMIs that are affine in the system parameters. They are thus applicable to polytopic-type uncertainties and to controller and filtering syntheses. The results obtained depend on the number of intervals K used during the dwell time. This number should be determined by the allowed computational complexity. It is possible to apply the above theory to systems with different number of vertices of the uncertainty polytope for each subsystem.

References

1. Liberzon, D.: Switching in Systems and Control. Birkhauser, Boston, PA (2003)
2. Colaneri, P.: Dwell time analysis of deterministic and stochastic switched systems. Eur. J. Autom. Control **15**, 228–249 (2009)
3. Allerhand, L.I., Shaked, U.: Robust stability and stabilization of linear switched systems with dwell time. IEEE Trans. Autom. Control **56**, 381–386 (2011)
4. Sun, Z., Ge, S.S.: Analysis and synthesis of switched linear control systems. Automatica **41**, 181–195 (2005)
5. Savkin, A.V., Evans, R.J.: Hybrid Dynamical Systems - Controller and Sensor Switching Problems. Birkhauser, Boston, PA (2002)
6. Hespanha, J.P.: Uniform stability of switched linear systems: extensions of LaSalle's invariance principle. IEEE Trans. Autom. Control **49**, 470–482 (2004)
7. Margaliot, M., Langholz, G.: Necessary and sufficient conditions for absolute stability: the case of second-order systems. IEEE Trans. Circuits Syst. Fundam. Theory Appl. **50**, 227–234 (2003)
8. Margaliot, M., Hespanha, J.: Root-mean-square gains of switched linear systems: a variational approach. Automatica **44**, 2398–2402 (2008)
9. Lee, J.W., Dullerud, G.E.: Optimal disturbance attenuation for discrete-time switched and markovian jump linear systems. SIAM J. Control Optim. **45**, 1915–1930 (2006)
10. Daafouz, J., Riedinger, P., Iung, C.: Stability analysis and control synthesis for switched systems: a switched Lyapunov function approach. IEEE Trans. Autom. Control **47**, 1883–1887 (2002)
11. Zhao, S., Sun, J.: Controllability and observability for time-varying switched impulsive controlled systems. Int. J. Robust Nonlinear Control **20**, 1313–1325 (2010)
12. Zhang, L., Shi, P., Wang, C., Gao, H.: Robust H_∞ filtering for switched linear discrete-time systems with polytopic uncertainties. Int. J. Adapt. Control Signal Process. **20**, 291–304 (2006)
13. Vu, L., Liberzon, D.: Invertibility of switched linear systems. Automatica **44**, 949–958 (2008)
14. Geromel, J., Colaneri, P.: Stability and stabilization of discrete time switched systems. Int. J. Control **79**, 719–728 (2006)
15. Colaneri, P., Bolzern, P., Geromel, J.C.: Root mean square gain of discrete-time switched linear systems under dwell time constraints. Automatica **47**, 1677–1684 (2001)
16. Petersen, I.R.: Quadratic Stabilizability of uncertain linear systems: existence of a nonlinear stabilizing control does not imply existence of a stabilizing control. IEEE Trans. Autom. Control **30**, 291–293 (1985)
17. Ackermann, J.: Longitudinal control of fighter aircraft F4E with additional canards. In: Sondergeld K.P. (compiler), Germany: A Collection of Plant Models and Design Specifications for Robust Control. DFVLR, Oberpfaffenhofen
18. Boyarski, S., Shaked, U.: Time-convexity and time-gain-scheduling in finite-horizon robust H_∞-control. In: Proceedings of the 48th CDC09. Shanghai, China (2009)
19. Boyd, S., El Ghaoui, L., Feron, E., Balakrishnan, V.: Linear Matrix Inequality in Systems and Control Theory. SIAM Frontier Series, Philadelphia (1994)
20. Green, M., Limebeer, D.J.N.: Linear Robust Control. Prentice Hall, Englewood Cliffs (1995)
21. de Oliveira, M.C., Skelton, R.E.: Stability test for constrained linear systems. In: Reza Moheimani S.O., (eds.) Perspectives in robust control. Lecture Notes in Control and Information Sciences 268. Springer, London (2001)

22. Geromel, J., Colaneri, P.: H_∞ and dwell time specifications of continuous-time switched linear systems. IEEE Trans. Autom. Control **55**, 207–212 (2010)
23. Ogata, K.: Discrete Time Control Systems. Prentice Hall, New Jersey, PA (1995)
24. Xie, L., Lu, L., Zhang, D., Zhang, H.: Improved robust H_2 and H_∞ filtering for uncertain discrete-time systems. Automatica **40**, 873–880 (2004)
25. Scherer, C., Weiland, S.: Linear Matrix Inequalities in Control (2004). Ebook: http://www.dcsc.tudelft.nl/~cscherer/lmi/notes05.pdf
26. Apkarian, P., Gahinet, P.: A convex characterization of gain-scheduled H_∞ conterollers. IEEE Trans. Autom. Control **40**, 853–864 (1995)

Chapter 7
Output-Dependent Switching Control of Linear Systems with Dwell Time

Abstract A unified approach to output-dependent switching law synthesis with dwell time is presented. This approach guarantees the asymptotic stability and achieves a prescribed upper-bound on the $\mathcal{L}_2$-gain of switched linear systems. When the dwell time is zero, the approach recovers, as special cases, the results of the known methods of minimizing the Lyapunov function and its derivative. Two switching laws emerge from the developed theory. The first switching law depends directly on the measurements, without applying any filter. At the cost of an added conservatism, this switching law can be designed without applying tuning parameters. The second switching law applies a filter; the switching law depends then on the filter's states. Two examples are given that demonstrate the application of the proposed design methods.

7.1 Introduction

In the present chapter, we consider switching in the case where the state vector of the system is not fully accessible. We introduce output-dependent switching laws that obey a dwell-time constraint with a prescribed dwell time. Two different families of output-dependent switching laws are investigated. The first family of laws is based directly on the measurement vector without applying any filter. We refer to this case is as a static output-dependent switching. This kind of switching law is easy to apply. It has a special merit in controller switching where an estimate of the systems state is already available through the controller and the designer usually does not wish to add more states. When the dwell time is zero (no dwell-time constraint), our approach recovers, as special cases, the known methods of (a) minimizing the Lyapunov function [1] and (b) minimizing the derivative of the Lyapunov function [2].

The second family of switching laws applies filters. Filter-based output-dependent switching laws that are based on minimizing the LF have been considered in [3] and have been later extended in [4] to guarantee a bound on the $\mathcal{L}_2$-gain and to simultaneously design the switching rule and a linear feedback controller for every subsystem. These switching laws are designed by adding states to the system (a filter) and using these states to determine the switching law. The latter method does not deal with uncertainty in the system's parameters and it does not cope with dwell time.

© Springer Nature Switzerland AG 2019

E. Gershon and U. Shaked, *Advances in H∞ Control Theory*,
Lecture Notes in Control and Information Sciences 481,
https://doi.org/10.1007/978-3-030-16008-1_7

In practice, the systems parameters entail uncertainty. In the present chapter, we obtain laws for output-dependent switching with dwell that are affine in the system matrices. These laws do not require the explicit knowledge of the system matrices for implementation. Our result can, therefore, be trivially extended to switched systems with polytopic type uncertainty. To the best of our knowledge, this is the first time that robust output-dependent switching laws are introduced. A piecewise-linear approach is used to solve the resulting Differential Linear Matrix Inequalities (DLMIs).

We formulate the problem in Sect. 7.2 and introduce the synthesis method for static output-dependent switching in Sect. 7.3. In Sect. 7.4, the result of the previous section is used to design filter-based switching. Two illustrative examples are given in Sect. 7.5, where various types of switching laws are considered.

7.2 Problem Formulation

We consider the following linear-switched system:

$$
\begin{aligned}
\dot{x} &= A_\sigma x + B_{1,\sigma} w, \quad x(0) = 0, \\
z &= C_{1,\sigma} x + D_{11,\sigma} w, \\
y &= C_{2,\sigma} x + D_{21,\sigma} v,
\end{aligned}
\tag{7.1}
$$

where $x \in \mathcal{R}^n$ is the state, $z \in \mathcal{R}^r$ is the objective vector, $y \in \mathcal{R}^s$ is the measurement, and $w \in \mathcal{R}^q$ is an exogenous disturbance in $\mathcal{L}_2$. This system is defined for all $t \geq 0$. The system switches between M subsystems and the right continuous switching law $\sigma(t) \in \{1 \ldots M\}$, for each $t \geq 0$, is such that $\Omega_{\sigma(t)} \in \{\Omega_1, \ldots, \Omega_M\}$, where

$$
\Omega_{\sigma(t)} = \begin{bmatrix} A_{\sigma(t)} & B_{1,\sigma(t)} \\ C_{1,\sigma(t)} & D_{11,\sigma(t)} \\ C_{2,\sigma(t)} & D_{21,\sigma(t)} \end{bmatrix}.
$$

Conditions will be derived below for the stability and the performance of the above-switched system. These conditions are all affine in the system matrices (except for $C_{2,i}$ and $D_{21,i}$) and they can, therefore, be trivially extended to incorporate polytopic-type parameter uncertainties. For each subsystem, a polytopic description can be adapted for Ω_i, and the conditions below must hold then for all the vertices of the uncertainty polytope. For switching without a filter, it is also assumed that all $D_{21,i} = 0$, meaning that the exogenous disturbances do not affect the measurement directly.

If $C_{2,i}$ and $D_{21,i}$ encounter uncertainty, or if a static switching law is sought when $D_{21,i} \neq 0$, a standard routine, which is used in gain-scheduling design [5], can be applied to obtain $D_{21,i} = 0$ and to transfer the uncertainty of $C_{2,i}$ and $D_{21,i}$ into A_i or $B_{1,i}$, respectively.

The present chapter is focused on stabilizing the system via output-dependent switching and on achieving a bound on the $\mathcal{L}_2$-gain of the system. It is assumed that the switching law satisfies a well-time constraint. The latter means that if the dwell time is T, and the switching instants are $\tau_1, \tau_2, \ldots$, then $\tau_{h+1} - \tau_h \geq T$, $\forall h \geq 1$. In practice, a dwell-time constraint is always imposed due to physical limitations of the system components, such as switches and valves. Therefore, some (small) dwell time should always be included.

7.3 Switching Without a Filter

Let τ_h and τ_{h+1} be the last, and the next switching instances, respectively, where τ_{h+1} will be determined by the switching law. Denoting, for a finite set of integers Ξ, $argmin_{q \in \Xi}(\psi_q)$ to be the $min(\{q \in \Xi | \forall h \in \Xi, \psi_q \leq \psi_h\})$, we consider the switching law that determines the next switching by:

$$
\begin{aligned}
\sigma(t) &= i \ \forall t \in [\tau_h, \tau_h + T), \\
\sigma(t) &= i \ \forall t > \tau_h + T, \ \text{if} \ y'S_{i,q}y \geq 0, \ \forall q \in L_i, \\
\sigma(\tau_{h+1}^+) &= argmin_{q \in L_i} \ \{y(\tau_{h+1})'S_{i,q}y(\tau_{h+1})\}, \ otherwise,
\end{aligned}
\tag{7.2}
$$

for some set of symmetric matrices $S_{i,q}, i = 1 \ldots M, q \in L_i$.

Theorem 7.1 *The system (7.1) with $D_{21,i} = 0, i = 1, \ldots M$ is stabilized by the switching law (7.2), with a dwell time T, and attains a prescribed $\mathcal{L}_2$-gain bound γ, if there exist: a collection of absolutely continuous matrix-valued functions $P_{i,q}(\theta) : [0, T] \to \mathcal{R}^{n \times n}$, a collection of matrices $P_i^{(f)} \in \mathcal{R}^{n \times n}, S_{i,q} \in \mathcal{R}^{s \times s}$ and non-negative scalars $\lambda_{i,q}, \beta_{i,q}, i = 1, \ldots M, q \in L_i$ that satisfy:*

$$
\begin{bmatrix}
\dot{P}_{i,q}(\theta) + P_{i,q}(\theta)A_q + A_q'P_{i,q}(\theta) & P_{i,q}(\theta)B_{1,q} & C_{1,q}' \\
* & -\gamma^2 I & D_{11,q}' \\
* & * & -I
\end{bmatrix} < 0,
$$

$$
\begin{bmatrix}
P_i^{(f)}A_i + A_i'P_i^{(f)} + \Psi_i & P_i^{(f)}B_{1,i} & C_{1,i}' \\
* & -\gamma^2 I_i & D_{11,i}' \\
* & * & -I
\end{bmatrix} < 0,
\tag{7.3}
$$

$$
P_{i,q}(0) - P_i^{(f)} - \beta_{i,q}C_{2,i}'S_{i,q}C_{2,i} \leq 0,
$$

and

$$
P_{i,q}(\theta) > 0, \ \text{and} \ P_{i,q}(T) = P_q^{(f)} \geq 0, \forall i = 1, \ldots M, \ q \in L_i, \theta \in [0, T],
$$

where $\dot{P}_{i,q}(\theta) = \frac{dP_{i,q}(\theta)}{d\theta}$ *and* $\Psi_i = \Sigma_{q \in L_i} \lambda_{i,q}C_{2,i}'S_{i,q}C_{2,i}$.

Proof We consider the LF:

$$
V(t) = x'(t)\bar{P}(t)x(t),
$$

where

$$\bar{P}(t) = \begin{cases} P_{\sigma(\tau_{h-1}),\sigma(t)}(t-\tau_h) & t-\tau_h < T \\ P_\sigma^{(f)} & t-\tau_h \geq T \end{cases} \tag{7.4}$$

and where τ_h is the last switching instant before time t. To further clarify this choice, we choose $\bar{P} = P_{i,q}(t - \tau_h)$ during the dwell time after the system has switched from the ith dynamics to the qth dynamics, and $\bar{P} = P_i^{(f)}$ after the dwell when $\sigma = i$.

Conditions (7.3d) imply that $V > 0 \ \forall x \neq 0$ and they guarantee that the LF is continuous between the switching instances.

To prove the stability and the bound on the $\mathcal{L}_2$-gain we define:

$$J(t_f) = \int_0^{t_f} [z'z - \gamma^2 w'(\bar{\tau})w(\bar{\tau})]d\tau.$$

We first show that the LF is nonincreasing if, $w=0$, and that $J(t_f) \leq 0$ for every feasible trajectory and for every $t_f \geq 0$, including $\lim t_f \to \infty$, Let $\tilde{J}(t_f) = J(t_f) + V(t_f)$. We have already shown that $V(t_f) \geq 0$ and, therefore, $\tilde{J}(t_f) \geq J(t_f)$. For $w = 0$, $J(t_f) \geq 0$ which implies that $\tilde{J}(t_f) \geq V(t_f)$. It is, therefore, sufficient to show that $\tilde{J}(t_f) \leq 0$ in order to guarantee the stability and the $\mathcal{L}_2$-gain of the system.

The LF is absolutely continuous anywhere but at the switching instances. Conditions (7.3c) guarantee that $V(t)$ is nonincreasing at the switching instances, which implies that $V(t) \leq \int_0^t \dot{V}(\bar{\tau})d\bar{\tau}$. We write:

$$\tilde{J}(t_f) = V(t_f) + \int_0^{t_f} \bar{L}(\bar{\tau})d\bar{\tau} \leq \sum_{s=0}^{\bar{s}(t_f)-1} \int_{\tau_s}^{\tau_{s+1}} \bar{L}(\bar{\tau})\,d\bar{\tau} + \int_{\tau_{\bar{s}(t_f)}}^{t_f} \bar{L}(\bar{\tau})\,d\bar{\tau}, \tag{7.5}$$

where $\bar{s}(t_f)$ is the number of switchings that preceded the one at t_f, and

$$\bar{L}(t) = \dot{V}(t) + z'(t)z(t) - \gamma^2 w'(t)w(t).$$

To show that $J(t_f) \leq 0, \forall t_f \geq 0$, it is sufficient to show that: $\bar{L} \leq 0$ between the switching instances, V is nonincreasing at these instances and $J(0) = 0$. From (7.4) and from the system dynamics (7.1) the following expression is obtained for the integrand between the switching instances:

$$\bar{L} = \begin{bmatrix} x' & w' & z' \end{bmatrix} \begin{bmatrix} \dot{\bar{P}} + \bar{P}A_\sigma + A'_\sigma\bar{P} & \bar{P}B_{1,\sigma} & C'_{1,\sigma} \\ * & -\gamma^2 I & D'_{1,\sigma} \\ * & * & -I \end{bmatrix} \begin{bmatrix} x \\ w \\ z \end{bmatrix}.$$

Conditions (7.3a) then readily guarantee that $\bar{L} \leq 0$ between the switching instances.

We show next that when the switching law (7.2) is applied, the LF is nonincreasing at the switching instances. Applying (7.2) means that once the dwell time has elapsed

$$x'C'_{2,\sigma}S_{\sigma,q}C_{2,\sigma}x=y'S_{\sigma,q}y \geq 0 \ \forall q \in L_\sigma.$$

Conditions (7.3b), in conjunction with the nonnegativity of the above term, imply that $\bar{L} \leq 0$ after the dwell and before the next switching occurs. The initial condition $x(0) = 0$ implies $J(0) = 0$. Since J is bounded by zero, from above, and is nonincreasing, $\lim \sup_{t_f \to \infty} J(t_f)$ exists and it satisfies $\lim \sup_{t_f \to \infty} J(t_f) \leq 0$. ■

The above conditions are affine in the matrices $A_i, B_{1,i}, C_{1,i}, D_{11,i}$, $i = 1\ldots M$, and the knowledge of these matrices is not required in order to determine the switching law. The above results can, therefore, be trivially generalized to systems with polytopic type parameter uncertainties by solving the Linear Matrix Inequalities (LMIs) for all the vertices of the uncertainty polytope applying the same $P_{i,q}(\theta), i = 1\ldots M, q \in L_i$.

7.3.1 The Solution Method

The conditions of Theorem 7.1 entail two difficulties. The first is that (7.3a) are DLMIs and the second difficulty is the need to choose the scalars $\beta_{i,q}$ and $\lambda_{i,q}$. The first difficulty is easy to overcome since efficient methods for solving DLMIs, by converting them to LMIs, exist in the literature, see for example, [6–8]. When using the method of [6] P is defined as an integral of a piecewise constant right continuous function and its derivative is, therefore, well defined.

The choice of both sets $\{\beta_{i,q}\}$ and $\{\lambda_{i,q}\}$ is clearly redundant. In order to avoid overcomplex solutions and to obtain a solution with a reasonable computational cost, we restrict all $\beta_{i,q}$, $i = 1\ldots M$, $q \in L_i$ to have the same value β. One then faces two options: choosing $\beta = 0$ or $\beta = 1$.

Standard methods of state-dependent switching [1, 9] and filter-dependent switching [3] are special cases of Theorem 7.1 with $\beta = 1$, where the $S_{i,q}$ matrices are taken to be $S_{i,q} = P_q(0) - P_i(T)$. To obtain a solution with a reasonable computational effort for $\beta = 1$, one may restrict all $\lambda_{i,q}$ to be of the same value λ, and we are thus left with a single-tuning parameter to search for. By using a line-search for this tuning parameter, the resulting inequalities can be solved via standard LMI solvers with a single tuning parameter λ. We include, in Sect. 7.7, a detailed algorithm for finding a solution.

For $\beta=0$, one can choose, without loss of generality, $\lambda=1$. In this case, a solution is obtained without tuning parameters, and the LF is nonincreasing at the switching instances regardless of the measurements. Conditions (7.3c) will then be replaced by $P_{i,q}(0) \leq P_i^{(f)}$. The resulting conditions are explicitly given in Procedure 7.1 in Sect. 7.7 below.

Remark 7.1 Switching law synthesis that is based on minimizing the derivative of the LF, as presented in [2], is a special case of Theorem 7.1 with $\beta = 0$. To see this, consider the case of [2] where full state measurements are available and the dwell time is zero. Choosing $P_{i,q}(\theta) = P_q^{(f)} = P$, $\forall \theta \in [0, T]$, $i = 1, \ldots M$, $q \in L_i$, a possible solution for $S_{i,q}$ is then $S_{i,q} = \eta_q (PA_q + A_q'P - PA_i - A_i'P)$ with some positive scalars η_q and $\sum_q \eta_q \leq 1$. This solution recovers the derivative minimizing conditions of [2]. While the solution of [2] is restricted to nominal systems, our solution can easily treat the case where the system matrices encounter polytopic type uncertainty.

7.4   Filter-Based Switching

In order achieve better performance, one may add additional states, "a filter", to the system. The new states may be a part of the controller and they are available for measurement. We note that the added states should not necessarily provide an estimate of the system's states.

Adding the filter

$$\dot{\hat{x}} = A_{c,\sigma}\hat{x} + B_{c,\sigma}y, \qquad (7.6)$$

an augmented system is obtained with the state vector $\xi = col\{x, \hat{x}\}$ and the measurement $y_f = col\{y, \hat{x}\}$. In order to avoid the computational issues that were explained in Sect. 7.3, we apply $\hat{x}$ to determine the switching law. We therefore treat $y_a = \hat{x} = [0, \ I]\xi$ as the measurement signal for the design of the switching signal.

Consider the following switching law:

Let τ_h be the last switching instant, and τ_{h+1}, $\sigma(\tau_{h+1})$ to be determined by

$$
\begin{aligned}
&\sigma(t) = i \ \forall t \in [\tau_h, \tau_h + T), \\
&\sigma(t) = i \ \forall t \geq \tau_h + T, \ \text{if} \ \hat{x}(t)'[\Delta_i^{-1} - \Delta_q^{-1}]\hat{x}(t) \leq 0, \ \forall q \in L_i, \qquad (7.7) \\
&\sigma(\tau_{h+1}) = argmin_q\{\hat{x}(\tau_{h+1})'(\Delta_q^{-1} - \Delta_i^{-1})\hat{x}(\tau_{h+1})\}, \ otherwise,
\end{aligned}
$$

for some set of symmetric matrices Δ_i, $i = 1 \ldots M$, to be chosen by the designer. Applying the conditions of Theorem 7.1 with $\beta = 1$ to the augmented system, with the above choice for the switching conditions, the following result is obtained.

Theorem 7.2 *The system (7.1) with the filter (7.6) is stabilized by the switching law (7.7) with an $\mathcal{L}_2$-gain bound γ and dwell time T if there exist absolutely continuous matrix-valued functions $X_{i,q} : [0, T] \to \mathcal{R}^{n \times n}$, a collection of matrices $\Delta_i, R_i^{(f)}, A_{f,i} \in \mathcal{R}^{n \times n}, B_{f,i} \in \mathcal{R}^{n \times s}, i = 1, \ldots M$, a set of scalars $\lambda_{i,q} \geq 0$, $i = 1 \ldots M$, $q \in L_i$ and positive scalars ϵ_p, ρ_p, such that*

$$
\left[
\begin{array}{cccccccc}
\Phi_i^{(f)} & * & * & * & * & * & & * \\
\bar{\Phi}_i^{(f)} & \hat{\Phi}_i^{(f)} - \Sigma_{q \in L_i} \lambda_{i,q} \Delta_i & * & * & * & * & & * \\
\tilde{\Phi}_i^{(f)} & B_{1,i}' R_i^{(f)} & -\gamma^2 I & * & * & * & & * \\
C_{1,i} & C_{1,i} & D_{11,i} & -I & * & * & & * \\
0 & \lambda_{i,q_1^{(i)}} \Delta_{q_1^{(i)}} & 0 & 0 & -\lambda_{i,q_1^{(i)}} \Delta_{q_1^{(i)}} & * & & * \\
0 & \vdots & & \vdots & \vdots & \ddots & & \vdots \\
0 & \lambda_{i,q_{M-1}^{(i)}} \Delta_{q_{M-1}^{(i)}} & 0 & 0 & 0 & \ldots & & -\lambda_{i,q_{M-1}^{(i)}} \Delta_{q_{M-1}^{(i)}}
\end{array}
\right] < 0,
$$

$$
\left[
\begin{array}{cccc}
\Phi_{i,q}(\theta) + \dot{X}_{i,q}(\theta) & * & * & * \\
\bar{\Phi}_{i,q}(\theta) + \dot{X}_{i,q}(\theta) & \hat{\Phi}_{i,q}(\theta) + \dot{R}_{i,q}(\theta) & * & * \\
\tilde{\Phi}_{i,q}(\theta) & B_{1,q}' R_{i,q}(\theta) & -\gamma^2 I & * \\
C_{1,q} & C_{1,q} & D_{11,q} & -I
\end{array}
\right] < 0,
$$

$$
\begin{aligned}
& R_{i,q}(\theta) \geq 0, \; \Delta_i \geq 0, \; R_{i,q}(0) = R_q^{(f)} \text{ and} \\
& R_{i,q}(T) = R_i^{(f)}, \; \forall i = 1, \ldots, M, \, q \in L_i, \, \theta \in [0, T],
\end{aligned}
\tag{7.8}
$$

where

$$
X_{i,q}(\theta) \triangleq \Delta_i + R_{i,q}(\theta), \, X_i^{(f)} \triangleq \Delta_i + R_i^{(f)}, \, \dot{X}_{i,q}(\theta) = \frac{dX_{i,q}(\theta)}{d\theta}, \, \dot{R}_{i,q}(\theta) = \frac{dR_{i,q}(\theta)}{d\theta},
$$

and

$$
\begin{aligned}
\Phi_{i,q}(\theta) &= A_q' X_{i,q}(\theta) + X_{i,q}(\theta) A_q + B_{f,q} C_{2,q} + C_{2,q}' B_{f,q}', \\
\bar{\Phi}_{i,q}(\theta) &= A_q' X_{i,q}(\theta) + A_{f,q}' + C_{2,q}' B_{f,q}' + R_{i,q}(\theta) A_q, \\
\tilde{\Phi}_{i,q}(\theta) &= B_{1,q}' X_{i,q}(\theta) + D_{21,q}' B_{f,q}', \\
\hat{\Phi}_{i,q}(\theta) &= R_{i,q}(\theta) A_q + A_q' R_{i,q}(\theta), \\
\Phi_i^{(f)} &= A_i' X_i^{(f)} + X_i^{(f)} A_i + B_{f,i} C_{2,i} + C_{2,i}' B_{f,i}', \\
\bar{\Phi}_i^{(f)} &= A_i' X_i^{(f)} + A_{f,i}' + C_{2,i}' B_{f,i}' + R_i^{(f)} A_i, \\
\tilde{\Phi}_i^{(f)} &= B_{1,i}' X_i^{(f)} + D_{21,i}' B_{f,i}', \\
\hat{\Phi}_i^{(f)} &= R_i^{(f)} A_i + A_i' R_i^{(f)}.
\end{aligned}
\tag{7.9}
$$

Proof The augmented system is described by

$$
\begin{aligned}
\dot{\xi} &= \begin{bmatrix} A_\sigma & 0 \\ B_{c,\sigma} C_{2\sigma} & A_{c,\sigma} \end{bmatrix} \xi + \begin{bmatrix} B_{1,\sigma} \\ B_{c,\sigma} D_{21,\sigma} \end{bmatrix} w, \\
z &= \begin{bmatrix} C_{1,\sigma} & 0 \end{bmatrix} \xi + D_{11,\sigma} w, \\
y_a &= \begin{bmatrix} 0 & I \end{bmatrix} \xi,
\end{aligned}
\tag{7.10}
$$

where $A_{c,i} = -A_{f,i}(\Delta_i(0))^{-1}$ and $B_{c,\sigma} = B_{f,\sigma}$.

We choose $P_{i,q}(\theta) \triangleq \begin{bmatrix} X_{i,q}(\theta) & I \\ I & U_i \end{bmatrix}$, denote $Q_{i,q}(\theta) = P_{i,q}^{-1}(\theta)$, partition $Q_{i,q}(\theta) =$

$\begin{bmatrix} R_{i,q}^{-1}(\theta) & N_{i,q}(\theta) \\ N_{i,q}(\theta)' & V_{i,q}(\theta) \end{bmatrix}$, and denote $R_{i,q}(T) = R_q^{(f)}$, $N_{i,q}(T) = N_q^{(f)}$, $V_{i,q}(T) = V_q^{(f)}$, $X_{i,q}(\theta) =$

$\Delta_q + R_{i,q}(\theta)$, $X_{i,q}(T) = X_q^{(f)}$, $J_{i,q}(\theta) = \begin{bmatrix} I & I \\ 0 & N_{i,q}(\theta)'R_{i,q}(\theta) \end{bmatrix}$, $J_{i,q}(T) = J_i^{(f)}$.

Applying the conditions of (7.3) to the augmented system, and choosing $y_a' S_{i,q} y_a$ to be $\hat{x}'(\Delta_i^{-1} - \Delta_q^{-1})\hat{x}$ with $\beta_{i,q} = 1$, we multiply (7.3a) and (7.3b, c), from the right by $diag(J_i(\theta), I, I)$ and $diag(J_i(T), I, I)$, respectively, and by their transposes from the left, and we readily obtain inequalities (7.8), invoking the property $Q_i(\theta)P_i(\theta) = I$. ∎

We note that the constraints $X_{i,q}(\theta) \triangleq \Delta_i + R_{i,q}(\theta)$, where Δ_i is independent of q and θ, is required for the filters to be time-invariant between the switching instances.

As mentioned above, the conditions of Theorem 7.2 can be trivially generalized to the case where the system matrices encounter polytopic-type uncertainties.

7.5 Examples

Two examples are given where we compare our methods with those of [4]. The work in [4] treats simultaneous output-feedback control and switching law synthesis, where the switching depends on the controller states. In order to apply it only for switching law synthesis, we take all the matrices that multiply the control signal there to be zero. Since the method of [4] does not treat dwell-time constraints, we also apply a zero dwell. In both examples, we choose Π of [4] to be $\begin{bmatrix} -\lambda & \lambda \\ \lambda & -\lambda \end{bmatrix}$ and the optimal values of λ is then found by a line-search. Without dwell time the result of [4] and our Theorem 7.2 are analytically the same, for any choice the multiplier matrix Π. We have also verified this claim numerically and it holds in all of the examples that we checked.

When DLMIs are solved below, we use the method of [6] where, in order to minimize the computational burden, the matrix-valued functions are taken to be linear during the dwell time. The latter means that the derivative is constant and therefore, the matrix derivatives can be replaced by constant matrices. Due to the affinity of the conditions in the Lyapunov matrices, we can then solve the DLMIs at the switching instances and at the end of the dwell time. Between these points in time, the Lyapunov matrices are convex combinations of the extreme values and the DLMIs conditions are, therefore, satisfied during the dwell time by solving for the vertices. It is possible to further improve the performance by allowing more degrees of freedom in the time-dependent Lyapunov matrices. The results below use the simplified version in order to show the merit of our method even for this computationally simple approach.

We apply, in Theorem 7.2, $X_{i,q}(\theta)$ that are independent of i, q. This choice does not affect the results in these two examples, and it significantly simplifies the numerical computations.

Example 7.1 We consider a modified version of the nominal system of [1] with added measurement noise, and partial information:

$$A_1 = \begin{bmatrix} 0 & 1 \\ 2 & -9 \end{bmatrix}, \; A_2 = \begin{bmatrix} 0 & 1 \\ -2 & 2 \end{bmatrix}, \; B_{1,i} = \begin{bmatrix} 0 & 0 \\ 1 & 0 \end{bmatrix}, \; C_{1,i} = [1 \; 0],$$
$$D_{11,i} = 0, \; C_{2,i} = [1 \; 0], \; D_{21,i} = [0 \; \rho], \; i = 1, 2.$$

The two subsystems are both unstable. We compare the state-dependent switching method of [9], the method of [4] and the one of Theorem 7.2, all of which stabilize the system and guarantee performance. The results are summarized in Table 7.1 for different values of ρ and T. The values of λ are given in brackets.

It is clear from Table 7.1 that without measurement noise the filter-based switching laws almost recover the result of the state-dependent switching of [9]. We note that, in spite of the fact that [4] does not apply dwell and is limited to nominal systems, the results of Theorem 7.2 recover the result of [4] without dwell time, with a relatively small difference when a dwell time is imposed.

The results of our Theorem 7.2 are identical to those of [4], when both dwell and uncertainty are absent. This example could not be solved for static output-dependent switching and it seems that a filter is necessary to stabilize this system.

For completeness, we include simulation results. We show, in Figs. 7.1 and 7.2 below, the state trajectories of the two unstable subsystems without switching. We then show, in Fig. 7.3, the state trajectories of the system and the control signal with the proposed switching law. The simulations were performed with zero initial state with time in seconds, and with a unit pulse signal as the exogenous disturbance.

Example 7.2 (A physical system with uncertainty and dwell): We consider the shock absorption by the semi-active suspension system of a road vehicle given in [3]. The system is described by

$$A_i = \begin{bmatrix} 0 & 1 & 0 & 0 \\ -\frac{k}{M} & -\frac{c_i}{M} & \frac{k}{M} & \frac{c_i}{M} \\ 0 & 0 & 0 & 1 \\ -\frac{k}{m} & -\frac{c_i}{m} & \frac{k+k_t}{m} & \frac{c_i}{M} \end{bmatrix}, \; C_{2,i} = \begin{bmatrix} 1 & 0 \\ 0 & 1 \\ -1 & 0 \\ 0 & -1 \end{bmatrix}',$$

$$C_{1,i} = \left[-\frac{k}{M} \; -\frac{c_i}{M} \; \frac{k}{M} \; \frac{c_i}{M} \right], \; D_{11,i} = 0,$$

Table 7.1 $\mathcal{L}_2$-gain bounds for Example 7.1

Method	$\rho = 0, \; T = 0$	$\rho = 0.5, \; T = 0$	$\rho = 0.5, \; T = 0.1$
[9]	3.5 (10)	3.5 (10)	7.36 (8)
[4]	3.64 (12)	17.08 (16)	–
Theorem 7.2	3.64 (12)	17.08 (16)	19.73 (16)

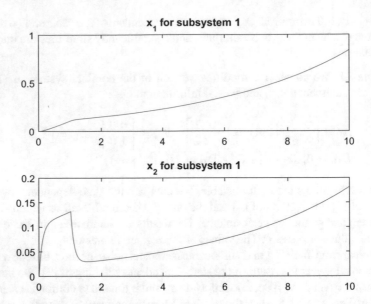

Fig. 7.1 State trajectories for subsystems 1 without switching

with $M = 400$, $m = 50$, $k = 2 \cdot 10^4$, $k_t = 2.5 \cdot 10^5$, $c_1 \in 300 \cdot [1 - \eta, 1 + \eta]$, and $c_2 \in 3900 \cdot [1 - \eta, 1 + \eta]$,
where η is an uncertain parameter.

For static switching: $B_{1,i} = \begin{bmatrix} 0 & -1 & 0 & -1 \end{bmatrix}'$, and $D_{21,i} = \begin{bmatrix} 0 & 0 \end{bmatrix}'$. For filter-based switching measurement disturbances are also included and then:

$$B_{1,i} = \begin{bmatrix} 0 & 0 & 0 \\ -1 & 0 & 0 \\ 0 & 0 & 0 \\ -1 & 0 & 0 \end{bmatrix} \text{ and } D_{21,i} = \begin{bmatrix} 0 & 0.1 & 0 \\ 0 & 0 & 0.5 \end{bmatrix}.$$

Switching is made between the two values c_i in A_i, $C_{1,i}$. In [3], it is assumed that the damping coefficient of the passive shock absorber switches between c_1 and c_2 instantaneously. Here, we assume that the disturbance induced by the transient behavior is modeled as uncertainty in the value of c, where η determines the size of the uncertainty.

Once again, we compare the state-dependent switching method of [9], the method of [4] (without dwell and uncertainty), Theorems 7.1 and 7.2. The results are summarized in Table 7.2, for different values of η and T. The value of λ is given in brackets.

Clearly, the results of Theorem 7.2 cannot compete with those obtained by state-dependent switching [9]. The results of Theorem 7.1 are similar to those found for the filter-based methods with zero dwell. When the dwell time increases, and uncertainty is introduced, the results of Theorem 7.2 change only slightly, whereas the results of

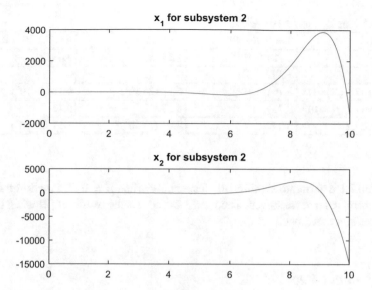

Fig. 7.2 State trajectories for subsystems 2 without switching

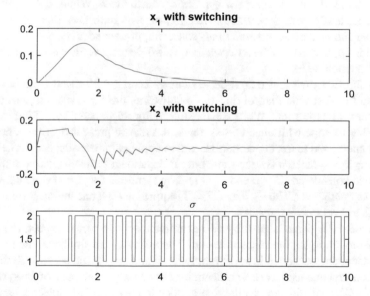

Fig. 7.3 State trajectories and control with switching

Table 7.2   The $\mathcal{L}_2$-gain for Example 7.2

Method	$\eta = 0, T = 0,$	$\eta = 0, T = 0.1,$	$\eta = 0.1, T = 0.1,$
[9]	2.12 $(2 \cdot 10^5)$	3.25 (5)	3.4 (5)
[4]	3.3 $(2 \cdot 10^5)$	–	–
Theorem 7.1 $(\beta = 1)$	3.76 $(2 \cdot 10^5)$	5.72 $(2 \cdot 10^5)$	6.15 $(2 \cdot 10^5)$
Theorem 7.1 $(\beta = 0)$	3.76	8.23	9.12
Theorem 7.2	3.3 $(2 \cdot 10^5)$	3.47 $(2 \cdot 10^5)$	3.57 $(2 \cdot 10^5)$

Theorem 7.1 deteriorate significantly, especially when $\beta = 0$. This was the case in other examples that we solved. Theorem 7.2 recovers the results of [4] when there is no uncertainty and dwell.

7.6   Conclusions

A novel method for switching law synthesis is introduced. While existing methods in the past apply state-dependent switching, or switching laws that are based on full-order filters, the new method treats static output-dependent switching. The new method is applied also to the case where a full-order robust filter is designed along with the switching law.

It is the first time that output-dependent switching with dwell time is treated, and also the first time that, during the dwell time, the LF takes into account the subsystem it has just left. While the existing output-dependent switching methods are unable to cope with uncertain systems, the theory presented in this chapter is easily generalized to treat uncertainties and to apply parameter-dependent LF in the switching law synthesis. For the static output-dependent switching, two possibilities have been considered ($\beta = 0$ and $\beta = 1$). The synthesis for $\beta = 0$ is made without tuning any parameter. The choice of $\beta = 1$ requires at least one tuning parameter but it seems to lead to better results.

The theory is demonstrated by two examples. As one would expect, the results obtained by applying the robust filters of Theorem 7.2, are similar to the results of state-dependent switching. From Example 7.1, it appears that both the dwell time and the measurement noise have a significant impact on the performance of the system. In Example 7.2, switching law synthesis is applied to the control of shock absorption by semi-active suspension in cars. In this example, static and filter-based switching are compared. The advantage of applying a filter there is clear; it significantly reduces the impact of the uncertainty and the dwell time on the performance.

7.7 Appendix: The Solution Methods

We present here the suggested solution method for the conditions of Theorem 7.1. We first bring Algorithm 7.1 that treats the case of $\beta = 1$. This method is based on a line-search for λ. Procedure 7.1 that describes how to solve the conditions of Theorem 7.1 for $\beta = 0$ then follows.

- -

Algorithm 7.1
$\quad \beta \leftarrow 1 \; \forall i = 1, \dots M, \; q \in L_i$
$\quad \gamma_{min} \leftarrow \infty$
for λ in $logspace(\lambda_{min}, \lambda_{max}, \lambda_{numpoints})$ **do**
$\quad |\qquad \gamma, \; P_i^{(f)}, \; S_{i,q}, \; P_{i,q}(\theta), \; \forall i = 1, \dots M, \; q \in L_i, \; \theta \in [0, T] \leftarrow argumin$
$(eq.3a - e)$
$\quad |$ **if** $\gamma < \gamma_{min}$ **then**
$\quad |\qquad \gamma_{min} \leftarrow \gamma$
$\quad |\quad$ **end**
end

- -

Algorithm 7.1: Solving Theorem 7.1 with $\beta = 1$

The next procedure for $\beta = 0$ applies conditions (7.3b, d, e) without changes (in condition (7.3c) terms with β are zeroed out) and all the $\lambda_{i,j}$ are taken to be 1.

Procedure 7.1 *The system (7.1) with* $D_{21,i} = 0, i = 1, \dots M$ *is stabilized by the switching law (7.2), with a dwell time* T, *and attains a prescribed* $\mathcal{L}_2$-*gain bound* γ, *if there exist: a collection of absolutely continuous matrix-valued functions* $P_{i,q}(\theta)$: $[0, T] \rightarrow \mathcal{R}^{n \times n}$, *a collection of matrices* $P_i^{(f)} \in \mathcal{R}^{n \times n}$, $S_{i,q} \in \mathcal{R}^{s \times s}$, *and nonnegative scalars* $\lambda_{i,q}, \beta_{i,q}, i = 1, \dots M, q \in L_i$ *that satisfy:*

$$\begin{bmatrix} \dot{P}_{i,q}(\theta) + P_{i,q}(\theta)A_q + A_q'P_{i,q}(\theta) & P_{i,q}(\theta)B_{1,q} & C_{1,q}' \\ * & -\gamma^2 I & D_{11,q}' \\ * & * & -I \end{bmatrix} < 0,$$

$$\begin{bmatrix} P_i^{(f)}A_i + A_i'P_i^{(f)} + \bar{\Psi}_i & P_i^{(f)}B_{1,i} & C_{1,i}' \\ * & -\gamma^2 I_i & D_{11,i}' \\ * & * & -I \end{bmatrix} < 0,$$

$$P_{i,q}(0) \le P_i^{(f)},$$

$$P_{i,q}(\theta) > 0, \text{ and } P_{i,q}(T) = P_q^{(f)} \ge 0, \forall i = 1, \dots M, \; q \in L_i, \theta \in [0, T],$$

(7.11)

where: $\dot{P}_{i,q}(\theta) = \frac{dP_{i,q}(\theta)}{d\theta}$ *and* $\bar{\Psi}_i = \Sigma_{q \in L_i} C_{2,i}' S_{i,q} C_{2,i}$.

References

1. Geromel, J., Colaneri, P.: Stability and stabilization of continuous-time switched linear systems. SIAM J. Control Optim. **45**(5), 1915–1930 (2006)
2. Zhai, G., Lin, H., Antsaklis, P.J.: Quadratic stabilizability of switched linear systems. Int. J. Control **77**(6), 598–605 (2004)
3. Geromel, J., Colaneri, P., Bolzern, P.: Dynamic output feedback control of switched linear systems. IEEE Trans. Autom. Control **53**(3), 720–733 (2008)
4. Deaecto, G.S., Geromel, J., Daafouz, J.: Dynamic output feedback H_∞ control of switched linear systems. Automatica **47**, 1713–1720 (2011)
5. Gahinet, P., Nemirovski, A., Laub, A.J., Chilali, M.: LMI Control Toolbox. The MathWorks Inc., Natick, MA, USA (1995)
6. Amato, F., Ariola, M., Carbone, M., Cosentino, C.: Finite-time output feedback control of linear systems via differential linear matrix conditions. In: Proceedings of the 45th CDC06, Sain-Diego, USA (2006)
7. Briat, C.: Convex conditions for robust stabilization of uncertain switched systems with guaranteed minimum and mode-dependent dwell-time. Syst. Control Lett. **78**, 63–72 (2015)
8. Valmorbida, G., Ahmadi, M., Papachristodoulou, A.: Stability analysis for a class of partial differential equations via semidefinite programming. IEEE Trans. Autom. Control **61**(6), 1649–1654 (2016)
9. Allerhand, L.I., Shaked, U.: Robust state dependent switching of linear systems with dwell time. IEEE Trans. Autom. Control **58**(4), 994–1001 (2013)

Chapter 8
Robust Switching-Based Fault-Tolerant Control

Abstract A new method for fault-tolerant control is presented. The fault is considered as switching in the system dynamics. A switching-based approach is then adopted for the fault-tolerant control of the system. This approach can be used to model multiplicative faults, such as actuator and sensor faults, as well as component faults. It is applied to passive fault-tolerant control, where the fault is never detected, and to active fault-tolerant control, where the fault is detected and isolated within a given time limit. The fault may not be known accurately and some uncertainty in the fault can be considered as an uncertainty in the systems dynamics after the fault has occurred. A controller that stabilizes the system, and achieves guaranteed performance is obtained by solving linear matrix inequalities. An illustrative example of aircraft control is given.

8.1 Introduction

Fault-tolerant Control (FTC) deals with systems whose standard operation is interrupted by a malfunction in actuators, sensors, or other components of the system [1, 2]. The standard approach for modeling such a malfunction is to consider it to be an instantaneous shift in the systems dynamics. Faults can thus be treated as switching in the system dynamics where it is assumed that once a fault has occurred, the system does not return to its original mode of operation.

FTC is achieved either by Passive Fault-tolerant Control (PFTC) or by Active Fault-tolerant Control (AFTC) [1–3]. In the former, a single controller is designed that is able to stabilize the system and achieve the required performance, regardless of the fault. This controller does not require knowledge of the fault. AFTC, on the other hand, reconfigures the controller by detecting and isolating the fault that occurred in the system. Fault detection and isolation have been treated by many ([2] and the references therein, [4, 5]).

© Springer Nature Switzerland AG 2019
E. Gershon and U. Shaked, *Advances in H∞ Control Theory*,
Lecture Notes in Control and Information Sciences 481,
https://doi.org/10.1007/978-3-030-16008-1_8

107

A variety of control methods have been applied to FTC, for example, [6–8]. For AFTC, it is assumed that faults can be detected with sufficient accuracy, usually as they happen [9]. In [10], AFTC with delayed detection and isolation has been considered, where an approximation of the performance is obtained. The case where the fault parameters are only partially known has been treated in [11]. In practice, faults are detected sometime after they occurred. Once the fault is detected, the controller is reconfigured to compensate for the fault, which allows the controller to be specifically designed to the system with the fault dynamics. The disadvantage of AFTC is that the resulting controller is more complex. Another drawback of AFTC is that the detection of the fault is a hard problem and the fault detection algorithms cannot guarantee a correct detection in the presence of external disturbances. Despite these drawbacks, AFTC provides sophisticated controllers that potentially achieve better performance compared to the PFTC.

Obtaining a bound on the performance of a passive linear controller can be formulated as a robust control problem. This problem can be solved by applying existing methods, see for example and [12]. Many FTC problems are solved in practice by applying Gain-scheduled Control (GSC) [13] where the GSC adapts the controller to cope with the fault by applying a single Lyapunov Function (LF) [14].

Switched systems models have already been applied to AFTC [2], for example by [15], using the supervisory control methodology [16]. The main drawback of the latter is that it is unable, so far, to guarantee the input–output properties of the closed-loop, such as a bound on the L_2-gain.

In this chapter, we present new methods for the synthesis of AFTC and PFTC that guarantee a bound on the L_2-gain of the closed-loop and both stability and performance requirements [12]. We consider FTC to be a special case of a switched system. An important concept in the area of switched system is the dwell time [17]. The dwell time is a lower bound on the time length that the system stays in one of its possible dynamics. This is related to, but is not the same as, the time it takes a control system to detect and isolate a fault which we refer to as the Detection and Isolation Interval (DII). The DII is a lower bound on the time length where the controller does not change.

A method that treats switched systems with parameter uncertainty has recently been developed [18], where both systems with and without dwell time are treated. This method has been extended to guarantee a bound on the performance of the system in the L_2-gain sense [19]. A similar Lyapunov function has also been applied to weighted performance for systems with average dwell time [20–22]. In [22], output-feedback control for switched systems is considered and a bound on a weighted norm is achieved. We adopt the method of [18, 19] to the special kind of switching that occurs in FTC. For AFTC, we treat both the case of Instantaneous Detection and Isolation (IDI) and the case where only an upper-bound on the DII is known. We focus on the design of Static Output Feedback (SOF) controllers. The method of [23] is used to obtain a fault-tolerant controller that guarantees a prescribed bound on the system's performance.

8.2 Problem Formulation

We consider the following linear system:

$$
\begin{aligned}
\dot{x}(t) &= A_0 x(t) + B_{1,0} w(t) + B_{2,0} u(t), \quad x(0) = 0, \\
z(t) &= C_{1,0} x(t) + D_{11,0} w(t) + D_{12,0} u(t), \\
y(t) &= C_{2,0} x(t) + D_{21,0} w(t),
\end{aligned}
\tag{8.1}
$$

where $x \in \mathcal{R}^n$ is the state, $w \in \mathcal{R}^q$ is an exogenous disturbance in $\mathcal{L}_2$, $z \in \mathcal{R}^r$ is the objective vector, $u \in \mathcal{R}^p$ is the control signal, and $y \in \mathcal{R}^s$ is the measurement. The systems parameters are uncertain and they are assumed to reside within the following polytope:

$$
\Omega_0 = \begin{bmatrix} A_0 & B_{1,0} & B_{2,0} \\ C_{1,0} & D_{11,0} & D_{12,0} \\ C_{2,0} & D_{21,0} & 0 \end{bmatrix} = \sum_{j=1}^{N} \eta_j(t) \Omega_0^{(j)}, \; \sum_{j=1}^{N} \eta_j(t) = 1, \; \eta_j(t) \geq 0,
$$

where

$$
\Omega_0^{(j)} = \begin{bmatrix} A_0^{(j)} & B_{1,0}^{(j)} & B_{2,0}^{(j)} \\ C_{1,0}^{(j)} & D_{11,0}^{(j)} & D_{12,0}^{(j)} \\ C_{2,0} & D_{21,0}^{(j)} & 0 \end{bmatrix}, \; j = 1, \ldots N.
\tag{8.2}
$$

We assume that $C_{2,0}$ is known. In the case where the latter matrix encounters parameter uncertainty, a standard routine, which is used in GSC [24], can be applied and the parameter dependance of $C_{2,0}$ is then inserted into an augmented A_0.

When the fault occurs, the system dynamics is changed according to the fault. For example, an actuator fault may cause one of the columns of $B_{2,0}$ to become zero. We assume that once the fault occurs, this fault remains in effect until the system is shut down for maintenance. We consider a system where one of N_f different faults may occur. Therefore, we assume that after the fault occurs, the system parameters reside within one of the following polytopes:

$$
\Omega_i = \begin{bmatrix} A_i & B_{1,i} & B_{2,i} \\ C_{1,i} & D_{11,i} & D_{12,i} \\ C_{2,i} & D_{21,i} & 0 \end{bmatrix} = \sum_{j=1}^{N} \eta_j(t) \Omega_i^{(j)}, \; \sum_{j=1}^{N} \eta_j(t) = 1, \; \eta_j(t) \geq 0,
$$

where

$$
\Omega_i^{(j)} = \begin{bmatrix} A_i^{(j)} & B_{1,i}^{(j)} & B_{2,i}^{(j)} \\ C_{1,i}^{(j)} & D_{11,i}^{(j)} & D_{12,i}^{(j)} \\ C_{2,i} & D_{21,i}^{(j)} & 0 \end{bmatrix}, \; \begin{array}{l} j = 1, \ldots N \\ i = 1, \ldots N_f \end{array}.
\tag{8.3}
$$

We assume, as explained above, that $C_{2,i}$, $i = 0, \ldots N_f$ are known.

The above description can be used to model systems with multiplicative faults such as actuator, sensor, and component faults. Multiplicative faults are modeled by multiplying the relevant system matrices by the appropriate matrix. For example, a sensor fault in the fth sensor can be modeled as $C_{2,i} = M_i C_{2,0}$ with $M_i = I - g_{fault} e_f e'_f$ where g_{fault} is the reduction in the sensors effectiveness and e_f is a unit vector whose elements are all zero except for the fth element which is 1. Component faults that affect the system's parameters, including both gain factor and fixed value parameter change, can be described using the above method. This modeling approach is common in the literature, see for example [16, 25–27].

The system is treated as a switched system. In the PFTC framework, the system has one switching instant, t_{fault}, when the fault occurs. At this time instant, the system dynamics switches from the normal mode of operation, $i = 0$, to the dynamics of one of the faults $i = 1, \ldots N_f$. In PFTC, the control law does not change. In the AFTC framework, the system possesses two switching instances. The first is the time when the fault occurs t_{fault}. At this time instant, the system dynamics switches from the normal mode of operation to the dynamics of the fault but the controller remains the one designed for the normal mode of operation. The second switch occurs at $t_{detection}$ when the fault is detected and isolated and the controller switches then to the one designed for the appropriate fault dynamics. It is assumed that the detection and isolation of the fault occur at the same instant. This is a reasonable assumption when the fault detection and isolation algorithm is based on fault reconstruction. If this is not the case, it is still possible to keep using the nominal controller until the fault is isolated.

The above model does not consider state and input constraints. Since the system may be open-loop unstable, input constraints may lead to an unstabilizable system. State constraints can, however, be added to our model by using the S-procedure [28]. An alternative method for dealing with these constraints can be found in [29].

The one-directional switching of the LF and the method for AFTC H_∞-design with a strict bound on the L_2-gain of the closed-loop and with an unknown delay in the detection of the fault is the main novelty of the present chapter.

The switched version of (8.1a–d) is the following:

$$
\begin{aligned}
\dot{x}(t) &= A_{\sigma(t)} x(t) + B_{1,\sigma(t)} w(t) + B_{2,\sigma(t)} u(t), \quad x(0) = 0, \\
z(t) &= C_{1,\sigma(t)} x(t) + D_{11,\sigma(t)} w(t) + D_{12,\sigma(t)} u(t), \\
y(t) &= C_{2,\sigma(t)} x(t) + D_{21,\sigma(t)} w(t).
\end{aligned}
\tag{8.4}
$$

To guarantee a given bound on the L_2-gain of the closed-loop system, the conditions of the Bounded Real Lemma (BRL) [12] should be satisfied. It has been shown in [19] and [30] that sufficient conditions for the L_2-gain of a switched system to be less than a prescribed value require that the conditions of the BRL are satisfied between the switching instances and that the LF is nonincreasing at the switching instances.

Practical FTC applies output-feedback controllers. It is well known, however, that the problem of finding robust dynamical output-feedback controllers is nonconvex and very difficult to obtain. In order to robustly control the system with fault, we concentrate on SOF control design. The SOF design is based on assigning a special structure to the LF [23].

Lemma 8.1 *The L_2-gain of the system (8.1) with the output feedback $u(t) = Gy(t)$ is less than $\gamma > 0$ if, for a tuning parameter α, there exist matrices $Q > 0 \in \mathcal{R}^{n \times n}$, $\hat{Q} \in \mathcal{R}^{s \times s}$, and $Y \in \mathcal{R}^{p \times s}$ that, for $j = 1, \ldots N$, satisfy*

$$\begin{bmatrix} \Xi(j) & \bar{\Xi}^{(j)} & B_1^{(j)} & QC_1^{(j)'} + C_2'Y'D_{12}^{(j)'} \\ * & -\alpha(\hat{Q} + \hat{Q}') & D_{21}^{(j)} & \alpha Y'D_{12}^{(j)'} \\ * & * & -\gamma^2 I & D_{11}' \\ * & * & * & -I \end{bmatrix} < 0, \qquad (8.5)$$

where

$$\Xi^{(j)} = A^{(j)}Q + Q'A^{(j)'} + B_2^{(j)}YC_2 + C_2'Y'B_2^{(j)'}$$
and
$$\bar{\Xi}^{(j)} = \alpha B_2^{(j)}Y + QC_2' - C_2'\hat{Q}'G = Y\hat{Q}^{-1}.$$

We note that this method is by no means optimal, and other methods are available in the literature for SOF. We adopt this method because it is given in terms of LMIs, it is based on a simple LF, and it provides good performance.

This method can be readily applied to State-feedback (SF) control by taking $C_{2,i} = I$ and $D_{21,i} = 0$. Using this method to design an SF controller, the conservatism that stems from the constraints on the structure of the LF is removed. This applies to all the theorems below as well.

Since the objective of the optimization depends on the application, we consider a multi-objective optimization problem. For the case where no fault occurs, the L_2-gain of the closed-loop is bounded by γ_n. If a fault does occur, we guarantee a different bound on the L_2-gain $\gamma_f \geq \gamma_n$. In the results given below, γ_n^2 and γ_f^2 can be taken as decision variables, in which case any linear combination of γ_n^2 and γ_f^2 can be minimized. The designer can also add strict constraints on γ_n^2 and γ_f^2. Alternatively, the designer can choose either γ_n^2 or γ_f^2 to be a prescribed scalar and then minimize the another.

8.3 Passive Fault-Tolerant Control

In order to design PFTC that guarantees a bound on the L_2-gain of the closed-loop system, a single controller must satisfy the conditions of the BRL for both the normal mode of operation and the fault dynamics. The simplest way to obtain a controller that complies with this demand is to apply the same LF for both modes [3]. This is sometimes referred to as the quadratic approach. While such a controller achieves an upper-bound on the L_2-gain of the closed-loop [3], the demand for the same LF for the normal mode of operation and the fault dynamics is conservative. A common LF has been used to obtain full-order output-feedback PFTC for systems without uncertainty in [31, 32]. It was applied to design gain-scheduled controllers [14] for AFTC assuming IDI [13]. It has been applied in conjunction with adaptive control

[33] and it was also used to guarantee some input-to-state properties of supervisory control for FTC [34].

We bring next a result that applies the method of [3] to the robust SOF synthesis method of Lemma 8.1:

Lemma 8.2 [3] *The L_2-gain of the system (8.4) with the output feedback $u(t) = Gy(t)$ is less than $\gamma_n > 0$ without fault, and less than $\gamma_f \geq \gamma_n$ if a fault occurs, if, $\forall i = 1, \ldots N_f$, $C_{2,i} = C_{2,0} = C_2$, and if for a tuning parameter α, there exist matrices $Q > 0 \in \mathcal{R}^{n \times n}$, $\hat{Q} \in \mathcal{R}^{s \times s}$, and $Y \in \mathcal{R}^{p \times s}$, that, for $i = 0, \ldots N_f$, $j = 1, \ldots N$, satisfy*

$$\begin{bmatrix} \Xi_i^{(j)} & \bar{\Xi}_i^{(j)} & B_{1,i}^{(j)} & QC_{1,i}^{(j)'} + C_2'Y'D_{12,i}^{(j)'} \\ * & -\alpha(\hat{Q} + \hat{Q}') & D_{21,i}^{(j)} & \alpha Y'D_{12,i}^{(j)'} \\ * & * & -\gamma_i^2 I & D_{11,i}' \\ * & * & * & -I \end{bmatrix} < 0, \tag{8.6}$$

where

$$\gamma_0 = \gamma_n; \quad \gamma_i = \gamma_f, \ i = 1 \ldots N_f;$$
$$\Xi_i^{(j)} = A_i^{(j)}Q + Q'A_i^{(j)'} + B_{2,i}^{(j)}YC_2 + C_2'Y'B_{2,i}^{(j)'};$$
$$\bar{\Xi}_i^{(j)} = \alpha B_{2,i}^{(j)}Y + QC_2' - C_2'\hat{Q}'.$$

We note that because of the terms that include multiplication of C_2 by $B_{2,i}^{(j)}$ and $D_{12,i}^{(j)}$, C_2 must be independent of the fault. This assumption is relaxed in the theorems below. In this configuration, the SOF gain G is given by $G = Y\hat{Q}^{-1}$.

Theorem 8.1 *The L_2-gain of the system (8.4) with the output feedback $u(t) = Gy(t)$ is less than $\gamma_n > 0$ without fault, and less than $\gamma_f \geq \gamma_n$ if a fault occurs, if, for a prescribed scalar α, there exist matrices $Q_i > 0 \in \mathcal{R}^{n \times n}$, $\hat{Q} \in \mathcal{R}^{s \times s}$, and $Y \in \mathcal{R}^{p \times s}$ that, for $i = 0, \ldots N_f$, and $j = 1, \ldots N$, satisfy*

$$\begin{bmatrix} \Xi_i(j) & \bar{\Xi}_i^{(j)} & B_{1,i}^{(j)} & Q_iC_{1,i}^{(j)'} + C_{2,i}'Y'D_{12,i}^{(j)'} \\ * & -\alpha(\hat{Q} + \hat{Q}') & D_{21,i}^{(j)} & \alpha Y'D_{12,i}^{(j)'} \\ * & * & -\gamma_i^2 I & D_{11,i}' \\ * & * & * & -I \end{bmatrix} < 0, \tag{8.7}$$
$$and \quad Q_0 \leq Q_i,$$

where

$$\gamma_0 = \gamma_n; \quad \gamma_i = \gamma_f, \ i = 1 \ldots N_f;$$
$$\Xi_i^{(j)} = A_i^{(j)}Q_i + Q_i'A_i^{(j)'} + B_{2,i}^{(j)}YC_{2,i} + C_{2,i}'Y'B_{2,i}^{(j)'};$$
$$\bar{\Xi}_i^{(j)} = \alpha B_{2,i}^{(j)}Y + Q_iC_{2,i}' - C_{2,i}'\hat{Q}',$$

and where $G = Y\hat{Q}^{-1}$.

Proof From [19], we know that for the L_2-gain of the system to be less than γ_f, the conditions of the BRL must be satisfied before and after the switching, and that the LF must be nonincreasing at the switching instances. Condition (8.7a) for $i = 0$ guarantees that the BRL holds before the fault occurs, and conditions (8.7a) for $i = 1, \ldots N_f$ guarantees that the conditions of the BRL hold after the fault occurs, for each of the possible faults. If a fault does not occur, conditions (8.7a) for $i = 0$ is the standard condition of the BRL for an uncertain system without faults, and γ_n is therefore a bound on the L_2-gain in this case.   ∎

The conservatism of the Lemma 8.2 stems from the fact that a single LF is assigned to the normal mode of operation and to the fault dynamics. This result is improved in Theorem 8.1 by assigning different LF to the different modes. If the system encounters parameter uncertainty, a further reduction of the conservatism can be achieved by selecting the LF to depend on the vertices polytope that describes the parameter uncertainty [35]. The vertex-dependent LF can be used, however, only if there is a known bound on the derivative of the parameters. If this is the case, the vertex dependence can be introduced to all the results below.

8.4 Active FTC with Delayed Detection and Isolation

In this section, we assume that the Detection and Isolation Interval (the DII) is bounded from above by a known constant T. In practice, the DII is unknown, and even an upper-bound on the DII cannot be a priori determined. The model that we apply is more practical than the commonly used IDI assumption since, in practice, the DII cannot be neglected. The practicality of this approach also stems from the fact that it is possible to guarantee that the fault will be detected, within a given time frame, with some desired probability. For example, if we know that the fault will be detected in less than 10 s. For 95% of the cases, then we can guarantee that the given performance will hold in 95% of the cases. The time it takes to detect the fault with some given probability can be easily found by using Monte Carlo simulations. The results obtained then will be different from those obtained by Markov jumps since the Markov jump models guarantee a bound on the mean of the L_2-gain while our approach guarantees a bound on the L_2-gain with some probability.

The LF $x'Q_\sigma^{-1}x$ with the piecewise linear Q_σ that is introduced below is found by using a prescribed integer K that is chosen according to the allowed computational burden. A larger K means that the LF is more complex but it may lead to less conservative results. It appears, however, that in many cases the results are insensitive to K.

During the DII, the system operates with the controller that was designed for the normal mode of operation, although the system has already encountered a fault. During this time interval, it may be beneficial to allow the LF to be time dependent. This is done in the theorem below.

We divide the time interval into three subintervals. The first interval, denoted by $s = 1$, is the time before any fault has occurred. The second time interval, denoted by $s = 2$, is the detection and isolation interval. The third time interval, denoted by $s = 3$, corresponds to the time when the fault has been isolated. We apply this division, using s as an index, in the theorem below. Note that j denotes the vertex of the uncertainty polytope, as in (8.3).

Theorem 8.2 *The L_2-gain of the controlled system (8.4) is less than $\gamma_n > 0$ without fault, and less than $\gamma_f \geq \gamma_n$ if a fault occurs, for a fault with a DII of T sec. if, for prescribed scalars $\beta > 0$ and α and an integer K, there exist symmetric matrices $Q_1 = Q_{1,0}$, $Q_{2,i,k}$, $Q_{3,i} = Q_{3,i,0}$, and matrices $\hat{Q}_1 = \hat{Q}_{1,0}$, $\hat{Q}_{3,i}$, $Y_1 = Y_{1,0}$, $Y_{3,i}$, $i = 1, \ldots N_f$, $k = 0 \ldots K$, of the appropriate dimensions, that satisfy*

$$\begin{bmatrix} \varXi_{1,0}^{(j)} & \varXi_{1,0}^{(j)} & B_{1,0}^{(j)} & Q_{1,0}C_{1,0}^{(j)'}+C_{2,0}'Y_1'D_{12,0}^{(j)'} \\ * & -\alpha(\hat{Q}_{1,0}+\hat{Q}_{1,0}') & D_{21,0}^{(j)} & \alpha Y_{1,0}'D_{12,0}^{(j)'} \\ * & * & -\gamma_n^2 I & D_{11,0}' \\ * & * & * & -I \end{bmatrix} < 0,$$

$$Q_1 \leq Q_{2,i,0},$$

$$\begin{bmatrix} \Delta_{k,h,i}^{(j)} & \bar{\varXi}_{2,h,i}^{(j)} & B_{1,i}^{(j)} & Q_{2,i,h}C_{1,i}^{(j)'}+C_{2,i}'Y_2'D_{12,i}^{(j)'} \\ * & -\alpha(\hat{Q}_2+\hat{Q}_2') & D_{21,i}^{(j)} & \alpha Y_2'D_{12,i}^{(j)'} \\ * & * & -\gamma_f^2 I & D_{11,i}' \\ * & * & * & -I \end{bmatrix} < 0, \qquad (8.8)$$

$$Q_{2,i,k} \leq Q_{3,i}, \quad i = 1, \ldots N_f, \quad k = 0 \ldots K - 1, h = k, k + 1,$$

$$\begin{bmatrix} \varXi_{3,i}^{(j)} & \bar{\varXi}_{3,i}^{(j)} & B_{1,i}^{(j)} & Q_{3,i}C_{1,i}^{(j)'}+C_{3,i}'Y_{3,i}'D_{12,i}^{(j)'} \\ * & -\alpha(\hat{Q}_{3,i}+\hat{Q}_{3,i}') & D_{21,i}^{(j)} & \alpha Y_{3,i}'D_{12,1}^{(j)'} \\ * & * & -\gamma_f^2 I & D_{11,i}' \\ * & * & * & -I \end{bmatrix} < 0,$$

$$i = 1, \ldots N_f$$

where $Y_2 = \beta Y_1$, $\hat{Q}_2 = \beta \hat{Q}_1$,

$$\Delta_{k,h,i}^{(j)} = -\frac{Q_{2,i,k+1}-Q_{2,i,k}}{T/K} + \varXi_{2,h,i}^{(j)},$$
$$\varXi_{s,h,i}^{(j)} = A_i^{(j)}Q_{s,i,h} + Q_{s,i,h}'A_i^{(j)'}+B_{2,i}^{(j)}Y_{s,i}C_{2,i} + C_{2,i}'Y_{s,i}'B_{2,i}^{(j)'},$$
$$\bar{\varXi}_{s,h,i}^{(j)} = \alpha B_{2,i}^{(j)}Y_{s,i} + Q_{s,h}C_{2,i}' - C_{2,i}'\hat{Q}_{s,h}'.$$

In this configuration, the SOF gain G_σ is given by: $G_1 = G_2 = Y_1\hat{Q}_1^{-1} = Y_2\hat{Q}_2^{-1}$ for $t \leq f_{detection}$, and $G_3 = Y_{3,i}\hat{Q}_{3,i}^{-1}$ for $t > t_{detection}$, where i is the index of the subsystem which represents the fault.

Proof Over the time interval $t \in [0, t_{fault})$, the system dynamics corresponds to the matrix Ω_0. During this time interval, (8.8a) guarantees that the conditions of the BRL are satisfied. At the time instant t_{fault}, the system switches from the dynamics without faults to the fault dynamics (Ω_i, $i \in 1, \ldots N_f$) but the controller remains the one that was designed for the dynamics without fault. Condition (8.8b) guarantees that the LF is nonincreasing at the fault instant t_{fault}. During the time interval $t \in [t_{fault}, t_{detection})$, for $t \in [t_{fault} + \frac{k}{K/T}, t_{fault} + \frac{k+1}{K/T})$ let

$$Q_{\sigma(t)} = Q_{2,k} + (Q_{2,k+1} - Q_{2,k})(\frac{t - t_{fault}}{T/K} - k). \qquad (8.9)$$

Then, using the method of [36], conditions (8.8c) guarantee that the BRL is satisfied over this time interval. Condition (8.8d) guarantees that the LF is nonincreasing at the time instant of detection $t_{detection}$. Once the fault has been detected, condition (8.8e) guarantees that the LF is nonincreasing. ∎

In order to apply Theorem 8.2, two scalars, α and β, should be found. For simplicity, we choose $\beta = 1$ when Theorem 8.2 is applied. We then perform a line-search in order to find the best α. It is expected that a search on both parameters will lead to an even lower bound on the L_2-gain of the closed-loop system.

If the detection and isolation of the fault are instantaneous, simpler conditions can be derived.

Corollary 8.1 *The L_2-gain of the system (8.4) with IDI and SOF control is less than $\gamma_n > 0$ without fault, and less than $\gamma_f \geq \gamma_n$ if a fault occurs if conditions (8.8a) and (8.8e) of Theorem 8.2 are satisfied along with the condition $Q_{1,0} < Q_{3,i}$, $i = 1, \ldots N_f$.*

Proof Consider Theorem 8.2 where $T \to 0$, $K = 1$ and $Q_{2,i,0} = Q1, 0$, $Q_{2,i,1} = Q_{3,i}$. Conditions (8.8b,d) are satisfied by the definition of $Q_{2,i,k}$, $k = 0, 1$. The demand $Q_{1,0} < Q_{3,i}$ leads to $-\frac{Q_{2,i,k+1} - Q_{2,i,k}}{T/K} \to -\infty$.

8.5 Example

We consider the AC1 aircraft control problem of [37]. In the latter reference, the parameters of the aircraft are assumed to be perfectly known. An actuator fault is added where the gain of any of the control inputs may drop.

After the fault occurs, the matrix $B_{2,0}$ becomes one of the matrices: $B_{2,i}$, $i = 1, 2, 3$, where $B_{2,0}$ is the the B_2 matrix of [37], and where $B_{2,1} = B_{2,0} \cdot diag$ $(0.5, 1, 1)$, $B_{2,1} = B_{2,0} \cdot diag(1, 0, 1)$, and $B_{2,3} = B_{2,0} \cdot diag(1, 1, 0)$. This is done in order to consider both full and partial faults.

We solve the control problem by applying: Lemma 8.2 and Theorem 8.1 for PFTC and GSC, Corollary 8.1 for AFTC with IDI, and Theorem 8.2 for AFTC with delayed detection. We require $\gamma_f = 0.2$, and minimize for γ_n, for $\alpha = 0.6$. The resulting attenuation levels γ are given in Table 8.1 below. Results of using Theorem 8.2 for different DII, T, and different values of K are given in Table 8.2.

The method of Theorem 8.1 demonstrates an improvement of 26% for PFTC compared to the standard quadratic results (Lemma 8.2). For SOF, Theorem 8.2 provides an improvement of 38% to the result obtained by GSC for IDI. For $T = 10^4$ s, the minimum γ that is obtained by Theorem 8.2 is the same as the one found by Theorem 8.1 (PFTC). This result means that for a very large DII, the result of the passive controller is recovered.

Table 8.1 Minimum attenuation levels for PFTC and AFTC

PFTC	PFTC	AFTC	AFTC
Lemma 8.2	Theorem 8.1	GSC	Theorem 8.2
0.1858	0.1382	0.1602	0.0994

Table 8.2 Minimum attenuation levels for different T and K

K	$T = 0.1$	$T = 1$	$T = 10$	$T = 10^2$
K=1	0.0998	0.108	0.1147	0.1148
K=3	0.0997	0.105	0.1147	0.1148
K=30	0.0998	0.104	0.1146	0.1148

We note that, in this example, the result is insensitive to K. This behavior, which is encountered also in other examples, can be used to reduce the computational burden of the solution.

8.6 Conclusions

By considering the fault to be a switch, and by treating the system with a fault as a switched system, conditions are obtained for PFTC and AFTC to achieve a guaranteed bound on the L_2-gain of the closed-loop. These conditions are derived in terms of LMIs for the SOF control scheme (and thus also for the corresponding SF control), and the results are, therefore, readily applicable to systems with polytopic type parameter uncertainties.

The standard result of the PFTC is improved in this chapter by solving for a decision variable that depends on the fault. The new AFTC results treat the case where the fault detection and isolation occur within a bounded interval. They improve previous results even in the case where the fault is detected instantaneously, e.g., improvements of 26% for PFTC and 38% for AFTC are achieved in the above example. As one may expect, in the case where the fault takes very long time to detect, the results are the same as those obtained for a passive controller.

References

1. Patton, R.J.: Fault-tolerant control: the 1997 situation. In: Proceedings of the 3rd IFAC Symposium on Fault Detection, Supervision and Safety for Technical Processes (1997)
2. Zhang, Y., Jiang, J.: Bibliographical review on reconfigurable fault-tolerant control systems. Annu. Rev. Control **32**(2), 229–252 (2008)

References   117

3. Kanev, S.: Robust fault tolerant control. Ph.D. thesis, University of Twente, The Netherlands (2004)
4. Chen, R.H., Speyer, J.L.: Robust multiple-fault detection filter. Int. J. Robust Nonlinear Control **12**(8), 675–696 (2002)
5. Chen, R.H., Mingori, D.L., Speyer, J.L.: Optimal stochastic fault detection filter. Automatica **39**(3), 377–390 (2003)
6. Blanke, M., Kinnaert, M., Lunze, J., Staroswiecki, M.: Diagnosis and Fault-Tolerant Control. Springer (2010)
7. Sami, M., Patton, R.J.: Fault tolerant adaptive sliding mode controller for wind turbine power maximisation. In: Proceedings of the 7th IFAC Symposium on Robust Control Design (ROCOND12), Aalborg, Denmark (2012)
8. Rosa, P.A.N., Casau, P., Silvestre, C., Tabatabaeipour, S., Stoustrup, J.: A set-valued approach to FDI and FTC: theory and implementation issues. In: Proceedings of the 8th IFAC Symposium on Fault Detection. Supervision and Safety of Technical Processes, Mexico City, Mexico (2012)
9. Zolghadri, A.: The challange of advanced model-based FDIR techniques for aerospace systems: the 2011 situations. Prog. Flight Dyn. Guid. Navig. Control Fault Detect. Avion. **6**(12), 231–248 (2013)
10. Choukroun, D., Speyer, J.: Mode-estimator free quadratic control of jump-linear systems with mode-detection random delay. In: Proceedings of the AIAA Guidance, Navigation and Control Conference 2005 (GNC05), San-Francisco, California (2005)
11. Yang, Z., Stoustrup, J.: Robust reconfigurable control for parametric and additive faults with FDI uncertainties. In: Proceedings of the 39th Conference on Decision and Control, Australia, Sydney (2000)
12. Scherer, C.: Theory of robust control. Lecture Note, Mechanical Engineering Systems and Control Group, Delft University of Technology, The Netherlands (2001)
13. de Oca, S.M., Puig, V., Theilliol, D., Tornil-Sin, S.: Fault-tolerant control design using LPV admissible model matching with H2/H8 performance: application to a two-degree of freedom helicopter. In: Proceedings of the 1st Control and Fault-Tolerant Systems (SYSTOL10), Nice, France (2010)
14. Apkarian, P., Gahinet, P.: A convex characterization of gain-scheduled H_∞ controllers. IEEE Trans. Aotom. Control **40**(5), 853–864 (1995)
15. Yang, II., Jiang, B., Staroswiecki, M.: Supervisory fault tolerant control for a class of uncertain nonlinear systems. Automatica **45**(10), 2319–2324 (2009)
16. Yang, H., Jiang, B., Cocquempot, V.: Supervisory fault tolerant control design via switched system approach. In: Proceedings of the 1st Control and Fault-Tolerant Systems (SYSTOL10), Nice, France (2010)
17. Colaneri, P.: Dwell time analysis of deterministic and stochastic switched systems. Eur. J. Autom. Control **15**, 228–249 (2009)
18. Allerhand, L.I., Shaked, U.: Robust stability and stabilization of linear switched systems with dwell time. IEEE Trans. Autom. Control **56**, 381–386 (2011)
19. Allerhand, L.I., Shaked, U.: Robust control of linear systems via switching. IEEE Trans. Autom. Control **58**(2), 506–512 (2013)
20. Wu, L., Zheng, W., Gao, H.: Dissipativity-based sliding mode control of switched stochastic systems. IEEE Trans. Autom. Control **58**(3), 785–793 (2013)
21. Wu, L., Ho, D.W.C., Li, C.W.: Stabilisation and performance synthesis for switched stochastic systems. IET Control Theory Appl. **4**(10), 1877–1888 (2010)
22. Wu, L., Qi, T., Feng, Z.: Average dwell time approach to L_2–L_∞ control of switched delay systems via dynamic output feedback. IET Control Theory Appl. **3**(10), 1425–1436 (2009)
23. Shaked, U.: An LPD approch to robust H_2 and H_∞ static output-feedback design. IEEE Trans. Autom. Control **48**, 866–872 (2003)
24. Gahinet, P., Nemirovski, A., Laub, A.J., Chilali, M.: LMI Control Toolbox. The MathWorks Inc., Natick, Mass. USA (1995)
25. Odgaard, P.F., Stoustrup, J., Kinnaert, M.: Fault tolerant control of wind turbines—a benchmark model. In: Proceedings of the 7th IFAC Symposium on Fault Detection. Supervision and Safety of Technical Processes, Barcelona, Spain (2009)

26. Sloth, C., Esbensen, T., Stoustrup, J.: Robust and fault-tolerant linear parameter-varying control of wind turbines. Mechatronics **21**, 645–659 (2011)
27. Qi, R., Zhu, L., Jiang, B.: Fault-tolerant reconfigurable control for MIMO systems using online fuzzy identification. Int. J. Innov. Comput. Inf. Control **9**(10), 3915–3928 (2013)
28. Yakubovich, V.A.: S-procedure in nonlinear control theory. Vestn. Leningr. Univ. **1**, 62–77 (1971)
29. Raimondo, D.M., Marseglia, G.R., Braatz, R.D., Scott, J.K.: Fault-tolerant model predictive control with active fault isolation. In: Proceedings of the 2nd Control and Fault-Tolerant Systems (SYSTOL13), Nice, France (2013)
30. Allerhand, L.I., Shaked, U.: Robust state dependent switching of linear systems with dwell time. IEEE Trans. Autom. Control **58**(4), 994–1001 (2013)
31. Veillette, R.J., Medanid, J.V., Perkins, W.R.: Design of reliable control systems. IEEE Trans. Autom. Control **37**(3), 290–304 (1992)
32. Yang, G-H., Wang, J.L., Soh, Y.C.: Reliable H_∞ controller design for linear systems. Automatica **37**, 717–725 (2001)
33. Vargas-Martnez, A., Puig, V., Garza-Castan, L.E., Morales-Menendez, R.: MRAC + H_∞ fault tolerant control for linear parameter varying systems. In: Proceedings of the 1st Control and Fault-Tolerant Systems (SYSTOL10), Nice, France (2010)
34. Efimov, D., Cieslak, J., Henry, D.: Supervisory fault tolerant control via common Lyapunov function approach. In: Proceedings of the 1st Control and Fault-Tolerant Systems (SYSTOL10), Nice, France (2010)
35. de Oliveira, M.C., Skelton, R.E.: Stability test for constrained linear systems. In: Reza Moheimani, S.O. (ed.) Perspectives in Robust Control. Lecture Notes in Control and Information Sciences 268. Springer, London (2001)
36. Amato, F., Ariola, M., Carbone, M., Cosentino, C.: Finite-time output feedback control of linear systems via differential linear matrix conditions. In: Proceedings of the 45th CDC06, Sain-Diego, USA (2006)
37. Leibfritz, F.: COMP$l_e$ib: Constrained Matrix-optimization Problem Library—A Collection of Test Examples for Nonlinear Semidefinite Programs, Control System Design and Related Problems. Available from: http://www.compleib.de (2003)

Part II
Stochastic State-Multiplicative Systems

Chapter 9
Introduction and Literature Survey

Abstract This chapter is the introduction to Part II of the book. A short survey of the literature on delay-free and retarded systems with state- multiplicative noise is brought there. The robust control and estimation of delay-free systems with state-multiplicative noise is treated at the beginning of Part II. By applying Finsler's lemma, a considerable improvement in the control design and estimation of these systems is achieved in the presence of large parameter uncertainties. Stochastic systems with time delay are then treated. Two approaches are offered to their control design. The first transforms the system to non-retarded system with norm-bounded uncertainty, that is, readily treated by previously known methods. The second approach is one that applies a predictor-based state-feedback control to time-delayed stochastic systems. At the end of Part II, a robust static output-feedback control is applied to retarded uncertain stochastic systems and also the theory of switching systems with dwell that has been developed in Part I of the book is applied to systems with state-multiplicative noise.

Abbreviation

BLS	Bilinear System
BLSS	Bilinear Stochastic System
BRL	Bounded Real Lemma
BMI	Bilinear Matrix Inequality
CSTR	Continuous Stirred tank Reactor
DLMI	Difference Linear Matrix Inequality
DRE	Difference Riccati Equation
GBM	Geometrical Brownian Motion
GS	Gain Scheduling
LF	Lyapunov Function
LKF	Lyapunov Krasovskii
LMI	Linear Matrix Inequality
LPD	Lyapunov Parameter Dependent
LTI	Linear Time Invariant

© Springer Nature Switzerland AG 2019
E. Gershon and U. Shaked, *Advances in H∞ Control Theory*,
Lecture Notes in Control and Information Sciences 481,
https://doi.org/10.1007/978-3-030-16008-1_9

LTV   Linear Time Variant
OPs   Operating Points
PID   Proportion Integral Derivative (controller)
PI   Proportion Integral (controller)
OF   Output Feedback
P2P   Peak to Peak
SDN   State-Dependent Noise
SF   State Feedback
SNR   Signal-to-Noise Ratio
SOF   Static Output Feedback

9.1 Stochastic State-Multiplicative Noisy Systems

The theory of state-multiplicative noisy systems has been treated in our first two books. While in [1] the theory has been restricted to delay-free systems, the theory in [2] concerns both continuous- and discrete-time-delayed systems. In the current chapter, we bring some introductory material of stochastic control that may serve as a supplementary basis for Chaps. 10, 11, and 14 for delay-free stochastic systems, and in Chaps. 12 and 13 for systems with time delay. In the delayed time systems case, we apply the input–output approach to linear delayed systems. This approach transforms a given delayed system to a norm-bounded uncertain system that can be treated, in the stochastic context, by the various solutions methods that can be found in [1]. The major advantage of the input–output approach lies in its simplicity of use as the resulting inequalities that emerge are relatively tractable and simple for both: delay- dependent and delay-independent solutions. This technique entails, however, some degree of conservatism that can be compensated for by a clever choice of the relevant Lyapunov function.

In the following section, we bring a short survey of the literature on both delay-free and retarded stochastic state-multiplicative systems. More details on the research done in the field of stochastic state-multiplicative control can be found in the introduction chapters of [1] and [2]. Some introductory material of the basic issues contained in this part of the book is complemented by Appendices A and B.

9.2 Stochastic State-Multiplicative Time Delay Systems

The field of stochastic state-multiplicative control and filtering has greatly matured since its emergence in the 60s of the last century. The linear quadratic optimization problems (stochastic H_2) that were treated in the first two decades of its emergence ([3–15], see also [6, 10] and [12] for extensive review) cleared the way, in the mid-80s, to the H_∞ worst-case control strategy, resulting in a great expansion of research effort, aimed at the solution of related problems such as state-feedback control, estimation, dynamic output-feedback control, preview tracking control, and zero-order control among other problems for various types of nominal and uncertain systems (see [16–22] for linear systems [see also [1] for extensive review] and [23–29] for nonlinear stochastic systems).

In the last three decades, a great deal of the research in this field has been centered around time-delay systems for both linear systems and nonlinear ones, where solutions have been obtained for various problems in the H_∞ context. These problems concern mainly continuous-time- delayed systems and they include various stability issues, state-feedback control, estimation, and dynamic output-feedback control (for linear systems see [30–44] and [2] for a recent extensive review. For nonlinear systems see, for example, [45, 46]).

The latter linear problems are similar to the ones encountered in the delay-free stochastic systems and in the deterministic delayed systems counterpart [47–54]. The discrete-time stochastic setting for discrete-time retarded systems has been also tackled in the past, but to a lesser extent (see for example, [34] and the references therein).

The stability and control of deterministic delayed systems of various types (i.e., constant time delay, slow and fast varying delay, etc.) have been a central field in the system theory sciences. In the last four decades, systems with uncertain time-delay has been a subject of recurring interest, especially due to the emergence of the H_∞ control theory in the early 80s. Most of the research, within the control and estimation community, is based on the application of different types of Lyapunov Krasovskii Functionals (LKFs) (see for example, [47, 50]). Also continuous-time systems with *fast varying delays* (i.e., without any constraints on the delay derivative) were treated via descriptor type LKF [50], where the derivative of the LKF along the trajectories of the system depend on the state and the state derivative. The results that have been obtained for the stability and control of deterministic retarded systems have been naturally applied also to stochastic systems by large and in more than the last two decades to state-multiplicative noisy systems.

In our second monograph [2], an *input–output* approach [55] for mostly delay-dependent solutions of various control and the filtering stochastic problems was applied. By this approach, the system is replaced by one with norm-bounded operators without delays, which is treated by the standard theory of non-retarded systems with state-multiplicative noise [1].

In the systems treated in [2], we allowed for a slowly varying delay, (i.e., delays with the derivative less than 1) where the uncertain stochastic parameters multiply both the delayed and the non-delayed states in the state space model of the system. Thus, the filtering and dynamic output-feedback problems were solved by addressing first the stability issue of the stochastic state-multiplicative-delayed systems and subsequently addressing the corresponding Bounded Real Lemma (BRL). Based on the BRL derivation, various control and filtering problems were solved alongside many practical control examples taken from various fields of control engineering. The use of the input–output approach for the study of retarded stochastic system is also applied in this monograph and the interested reader may find the principles of this approach in Appendix B.

As mentioned above, the current part of the book (Part II) starts with extended topics in delay-free stochastic state-multiplicative systems. In Chap. 10, a recapitulation of the basic control and estimation problem is developed. Based on a simple version of the Finsler lemma, a stability condition is derived for uncertain polytopic systems which is, less conservative than that of the quadratic solution and which leads to an improved stochastic BRL condition. This, in turn, allows for better solutions of all the synthesis control problems that are based on the BRL, starting with the basic problem of state-feedback control and culminating in an improved robust solution of the general-type filtering problem. Chapter 10 starts with discrete-time systems and proceeds to similar solutions for continuous-time systems. In Chap. 11, a solution is developed for the problem of discrete-time dynamic output-feedback control where no preassigned Lyapunov function is used. Focusing on an observable canonical representation of the system under study, a simple LMI condition is derived for the latter problem. The theory developed is demonstrated by two examples; the first compares various solution methods and the second is engaged in a real-life robot manipulator system.

In Chap. 12, we turn to input delayed stochastic systems whose deterministic counterparts are abundant in process control, and we bring a predictor-based state-feedback control solution for systems that are subject to constant delay or to a delay that is fast varying in time. We obtain delay-dependent condition for the solution of the latter state-feedback control problem. The theory developed in Chap. 12 is applied in two examples where the second example is taken from the field of process control. A predictor-based state-feedback controller is derived there for a given input time-delay that is caused by a transport process.

In Chap. 13, the problem of zero-order output-feedback control is formulated and solved for retarded uncertain polytopic-type stochastic system based on transforming the retarded system to a norm-bounded uncertain system via the input–output approach. A simple LMI condition is derived there for the solution of the nominal system [i.e., with no polytopic uncertainties]. A solution for the robust case is also given there based on a version of the improved stability and the BRL conditions for discrete-time systems that have been given in Chap. 10, which are based on the application of the Finsler lemma.

Chapter 14 may be viewed as a direct continuation of the state-feedback control solution for delay-free stochastic switched systems with dwell time that have been brought in our second book [2]. While in [2], a solution is given for continuous-time systems (see there, Chap. 11), the solution in Chap. 14 focuses on discrete-time systems where the problems of the mean square stability, stochastic l_2-gain and state-feedback control are formulated and solved. In Chap. 14, again, a real practical control oriented problem is formulated and solved: the problem of stabilizing and attenuating disturbances acting on the longitudinal short period mode of the F4E fighter aircraft with additional canards.

References

1. Gershon, E., Shaked, U., Yaesh, I.: H_∞ Control and Estimation of State-Multiplicative Linear Systems. Lecture Notes in Control and Information Sciences, LNCIS, vol. 318. Springer (2005)
2. Gershon, E., Shaked, U.: Advanced Topics in Control and Estimation of State-Multiplicative Noisy Systems. LNCIS - Lecture Notes in Control and Information Sciences, vol. 439. Springer (2013)
3. Wonham, W.M.: Optimal stationary control of linear systems with state-dependent noise. SIAM J. Control **5**, 486–500 (1967)
4. Wonham, W.M.: On a matrix Riccati equation of stochastic control. SIAM J. Control **6**(4), 681–697 (1968)
5. Sagirow, P.: Stochastic Methods in the Dynamics of Satellites. Lecture Notes, Udine, CISM (1970)
6. Kozin, F.: A survey of stability of stochastic systems. Automatica **5**, 95–112 (1970)
7. Hussmann, U.G.: Optimal stationary control with state and control dependent noise. SIAM J. Control **9**, 184–198 (1971)
8. Mclane, P.J.: Optimal stochastic control of linear systems with state—and control-dependent disturbances. IEEE Trans. Autom. Control **16**, 793–798 (1971)
9. Willems, J.C.: Dissipative dynamical systems-part I: general theory. Arch. Rat. Mech. Anal. **45**, 321–351 (1972)
10. Bruni, C., Dipillo, G., Koch, G.: Bilinear systems: an appealing class of "nearly linear" systems in theory and applications. IEEE Trans. Autom. Control **19**, 334–348 (1974)
11. Pakshin, P.V.: State estimation and control synthesis for discrete linear systems with additive and multiplicative noise. Autom. Remote Control **4**, 526–535 (1978)
12. Mohler, M.M., Kolodziej, W.J.: An overview of stochastic bilinear control processes. IEEE Trans. Syst. Man Cybern. **10**, 913–919 (1980)
13. Milstein, G.N.: Design of stabilizing controller with incomplete state-data for linear stochastic system with multiplicative noise. Autom. Remote Control **5**, 653–659 (1982)
14. Willlems, J.L., Willems, J.C.: Robust stabilization of uncertain systems. SIAM J. Control and Optim. **21**, 352–373 (1983)
15. Phillis, Y.: Controller design of systems with multiplicative noise. IEEE Trans. Autom. Control **30**, 1017–1019 (1985)
16. Dragan, V., Halanay, A., Stoica, A.: A small gain theorem for linear stochastic system. Syst. Control Lett. **30**, 243–251 (1997)
17. Dragan, V., Morozan, T.: Mixed input-output optimization for time-varying Ito systems with state dependent noise. Dyn. Contin., Discret. Impuls. Syst. **3**, 317–333 (1997)
18. Dragan, V., Stoica, A.: A γ Attenuation Problem for Discrete-Time Time-Varying Stochastic Systems with Multiplicative Noise. Reprint series of the Institute of Mathematics of the Romanian Academy, No 10 (1997)
19. Bouhtouri, A.E., Hinricchsen, D., Pritchard, A.J.: H_∞-type control for discrete-time stochasic systems. Int. J. Robust Nonlinear Control **9**, 923–948 (1999)
20. Dragan, V., Morozan, T., Halanay, A.: Optimal stability compensator for linear systems with state dependent noise. Stoch. Anal. Appl. **10**, 557–572 (1997)
21. Dragan, V., Morozan, T.: The linear quadratic optimiziation problems for a class of linear stochastic systems with multiplicative white noise amd Markovian jumping. IEEE Trans. Autom. Control **49**(5), 665–675 (2004)
22. Gershon, E., Shaked, U.: H_∞ output-feedback of discrete-time systems with state-multiplicative noise. Automatica **44**, 574–579 (2008)
23. Florchinger, P.: A passive system approach to feedback stabilization of nonlinear control stochastic systems. SIAM J. Control Optim. **37**, 1848–1864 (1999)
24. Charalambous, C.D.: Stochastic nonlinear minmax dynamic games with noisy measurements. IEEE Trans. Autom. Control **48**, 261–266 (2003)
25. Berman, N., Shaked, U.: Stochastic H_∞ control for nonlinear stochastic systems. In: Proceeding of the Conference Decision Control, Maui, Hawaii (2003)

26. Zhang, W., Chen, B.S., Tseng, C.S.: Robust H_∞ filtering for nonlinear stochastic systems. IEEE Trans. Signal Process. **53**(2), 589–598 (2005)
27. Berman, N., Shaked, U.: H_∞-like control for nonlinear stochastic systems. Syst. Control Lett. **55**, 247–257 (2006)
28. Zhang, W., Chen, B.S.: State feedback H_∞ control for a class of nonlinear stochastic systems. SIAM J. Control Optim. **44**, 1973–1991 (2006)
29. Berman, N., Shaked, U.: H_∞ control for nonlinear stochastic systems, the output-feedback case. Int. J. Control **81**, 1733–1746 (2008)
30. Mao, X.: Robustness of exponential stability of stochastic differential delay equations. IEEE Trans. Autom. Control **41**, 442–447 (1996)
31. Mao, X., Shan, A.: Exponential stability of stochastic delay differential equations. Stoch. Stoch. Rep. **60**(1), 135–153 (1996)
32. Boukas, E.K., Liu, Z.K.: Deterministic and Stochastic Time Delay Systems. Birkhauser, Basel (2002)
33. Chen, W.H., Guan, Z., Lu, X.: Delay-dependent exponential stability of uncertain stochastic systems with multiple delays: an LMI approach. Syst. Control Lett. **54**, 547–555 (2005)
34. Gao, H., Chen, T.: New results on stability of discrete-time systems with time-varying state delay. IEEE Trans. Autom. Control **52**(2), 328–334 (2007)
35. Gershon, E., Shaked, U., Berman, N.: H_∞ control and estimation of state-multiplicative stochastic systems with delay. IEEE Trans. Autom. Control **52**, 329–334 (2007)
36. Gershon, E., Shaked, U., Berman, N.: H_∞ control and estimation of state-multiplicative stochastic systems with delay. In: Proceedings of the American Control Conference (ACC07), New York, USA (2007)
37. Gershon, E., Berman, N., Shaked, U.: Robust H_∞ control and estimation of state-multiplicative stochastic systems with delay. In: Proceedings of the 16th Mediterranean Conference on Control and Automation (MED08). Ajaccio, Corsica (2008)
38. Chen, Y., Xue, A., Lu, R., Zhou, S., Wang, J.: Delay-dependent robust H_∞ control of uncertain stochastic delayed system. In: Proceedings of the 17th World Congress The International Federation of Automatic Control, Seoul, Korea, July 6–11, vol. 1, pp. 4916–4921 (2008)
39. Chen, Y., Zheng, W.X., Xue, A.: A new result on stability analysis for stochastic neutral systems. Automatica **46**, 2100–2104 (2010)
40. Gershon, E., Shaked, U.: Robust H_∞ output-feedback control of retarded state-multiplicative stochastic systems. Int. J. Robust Nonlinear Control **21**, 1283–1296 (2010)
41. Gershon, E., Shaked, U., Berman, N.: H_∞ preview tracking control of retarded state-multiplicative stochastic systems. In: Systol10, Nice, France (2010)
42. Shaked, U., Gershon, E.: Robust H_∞ control of stochastic linear switched systems with dwell time. Int. J. Robust Nonlinear Control **24**(11), 1664–1676 (2013)
43. Gershon, E., Shaked, U.: H_∞ preview tracking control of retarded state-multiplicative stochastic systems. Int. J. Robust Nonlinear Control **24**(15), 2119–2135 (2014)
44. Gershon, E.: Robust reduced-order H_∞ output-feedback control of retarded state-multiplicative linear systems. IEEE Trans. Autom. Control **58**(11), 2898–2904 (2013)
45. Florchinger, P., Verriest, E.I.: Stabilization of nonlinear stochastic systems with delay feedback. In: Proceeding of the 32nd IEEE Conference Decision and Control, San Antonio, TX, vol. 1, pp. 859–860 (1993)
46. Gao, H., Lam, J., Wang, C.: Robust H_∞ filtering for discrete stochastic time-delay systems with nonlinear disturbances. Nonlinear Dyn. Syst. Theory **4**(3), 285–301 (2004)
47. Kolmanovskii, V., Richard, J.P.: Stability of some linear systems with delays. IEEE Trans. Autom. Control **44**, 984–989 (1999)
48. Moon, Y.S., Park, P., Kwon, W.H.: Delay-dependent robust stabilization of uncertain state-delayed systems. Int. J. Control **74**, 1447–1455 (2001)
49. Niculescu, S.I .: Delay Effects on Stability: A Robust Control Approach. Lecture Notes in Control and Information Sciences, vol. 269. Springer, London (2001)
50. Fridman, E.: New Lyapunov-Krasovskii functionals for stability of linear retarded and neutral type systems. Syst. Control. Lett. **43**, 309–319 (2001)

51. Fridman, E., Shaked, U.: Parameter dependent stability and stabilization of time-delay systems. IEEE Trans. Autom. Control **48**(5), 861–866 (2003)
52. Fridman, E., Shaked, U.: Input-output approach to stability and L_2-gain analysis of systems with time-varying delays. Syst. Control. Lett **55**, 1041–1053 (2006)
53. Kojima, A., Ishijima, S.: Formulas on preview and delayed H_∞ control. IEEE Trans. Autom. Control **51**, 1920–1937 (2006)
54. Fridman, E., Orlov, Y.: Exponential stability of linear distributed parameter systems with time-varying delays. Automatica **45**(2), 194–201 (2009)
55. Kao, C.Y., Lincoln, B.: Simple stability criteria for systems with time-varying delays. Automatica **40**, 1429–1434 (2004)

Chapter 10
Robust Vertex-Dependent Control and Filtering of Stochastic Systems

Abstract Linear discrete- and continuous-time systems with multiplicative noise and polytopic-type parameter uncertainties are considered. The problems of H_∞ state-feedback control and filtering of these systems are addressed. These problems are solved by applying a vertex-dependent Lyapunov function, based on the Finsler's lemma, which considerably reduces the overdesign associated with the classical "quadratic" design that applies a single Lyapunov function for the whole parameters range. In both settings, a cost function is defined which is the expected value of the standard H_∞ performance index with respect to the stochastic multiplicative parameters. The results are demonstrated via two numerical examples.

10.1 Introduction

We address the problems of robust H_∞ state-feedback control and filtering of continuous- and discrete-time, state-multiplicative linear stochastic systems with large parameter uncertainty of the polytopic type, via a new approach which significantly reduces the overdesign entailed with existing methods.

The field of stochastic state-multiplicative control systems has matured greatly over the last three decades. Numerous solution methods have been applied to both delay free and retarded systems. These include, among others, a game theoretical approach and and convex optimization techniques [1]. Following the research of the 60s and 70s where the main issues were stability and control of continuous-time state-multiplicative systems in the stochastic H_2 framework (see [2] and the references therein), research in the following decades has focused on the H_∞ control setting. Thus, the continuous-time stochastic state-multiplicative Bounded Real Lemma (BRL) was obtained in [3–5] and the discrete-time counterpart was derived in [6, 7]. The problems of H_∞ state-multiplicative state- and measurement feedback control was solved in [6–10] respectively. In [11], a discrete-time stochastic estimation for a guidance-motivated tracking problem was solved for the results of which were shown to achieve better results than those achieved by the Kalman filter. In [12], a practical continuous-time estimation problem was solved where a white-noise-modeled parameter uncertainty exists in the measurement of a radar altimeter.

© Springer Nature Switzerland AG 2019
E. Gershon and U. Shaked, *Advances in H∞ Control Theory*,
Lecture Notes in Control and Information Sciences 481,
https://doi.org/10.1007/978-3-030-16008-1_10

In the last decade, the continuous and discrete-time control and estimation problems were extended to systems with delay and the counterpart problems of the above ones were solved for the time delayed systems (see for example [13, 14]). In the H_∞ control context, the problems of state feedback and general-type filtering were solved for both: nominal and uncertain systems in [15].

In some practical situations where state-feedback control is applied, the systems are prone to relatively large parameter uncertainties which can not be compensated via simple control (the same applies to estimation processes). Hence, the relative importance of the robust design. The theory of robust stochastic H_∞ state-feedback control for uncertain systems has been concentrated on both the norm-bounded approach [16] and the polytopic-type description of the uncertain parameters (see [15] and the references therein). Modeling of uncertainties as norm bounded may lead to simple control and filtering solutions with considerable overdesign, whereas considering the uncertainty to be of the polytopic type a more realistic description of the system is obtained which enables the application of convex optimization method [17, 18]. To the best of our knowledge, the problems of state-feedback control and estimation of systems with multiplicative stochastic noise and polytopic type uncertain parameters were both solved only via the vertex-independent Lyapunov approach where a single Lyapunov function is uniformly assigned to all the vertices of the uncertainty polytope. This approach yields a solution, usually called the "quadratic solution", which is very conservative similar to what has been obtained in the control and estimation of deterministic uncertain systems. An adaptation of the less conservative Lyapunov vertex-dependent approach [18] that is applied to deterministic systems, also for systems with multiplicative noise is rather problematic. This is due to the fact that the use of the adjoint of the stochastic state-multiplicative system is not theoretically sound and, in fact, is not justified. We note that setting the delay to zero in the solution of the robust state feedback and filtering of retarded stochastic systems in [15], results in a LMI condition that is equivalent to the 'quadratic solution' for the delay-free systems, thereby it retains its conservatism (see Chaps. 7, 8 in [1]).

In the present chapter we treat both continuous-time and discrete-time systems. Starting with the discrete-time case, we first bring the solution of the discrete-time BRL where we develop a modified version of the Lyapunov vertex-dependent approach [18], which is applied to the original system rather than to its adjoint. This enables us to solve both the robust state-feedback and the estimation problems via simple Linear Matrix Inequalities (LMIs) conditions.

The discrete-time part of the chapter is organized as follows. Following the problem formulation in Sect. 10.2, we bring, in Sect. 10.3, the derivation of the less conservative vertex-dependent BRL condition. Based on the latter condition, we solve the robust state-feedback control in Sect. 10.4 and the robust estimation problem in Sect. 10.5. In Sect. 10.6 we bring two examples: one of a state-feedback control the second is an example of filtering. In both problems, we demonstrate the advent of our solutions comparing to the "quadratic" ones.

The continuous-time part of the chapter is organized similarly to the first part of the chapter. Following the problem formulation in Sect. 10.8, we bring the BRL result in Sect. 10.9. The state-feedback control problem is solved in Sect. 10.10 and the solution of the continuous-time filtering is brought in Sect. 10.11. This part of the chapter is concluded in Sect. 10.12.

10.2 Discrete-Time Case: Problem Formulation

We consider the following linear system:

$$
\begin{aligned}
x_{k+1} &= (A + D\nu_k)x_k + B_1 w_k + B_2 u_k, \quad x_l = 0, \ l \le 0, \\
y_k &= (C_2 + F\zeta_k)x_k + D_{21} w_k
\end{aligned}
\tag{10.1}
$$

with the objective vector

$$
z_k = C_1 x_k + D_{12} u_k,
\tag{10.2}
$$

where $x_k \in \mathcal{R}^n$ is the system state vector, $w_k \in \mathcal{R}^p$ is the exogenous disturbance signal, $u_k \in \mathcal{R}^\ell$ is the control input, $y_k \in \mathcal{R}^m$ is the measured output, and $z_k \in \mathcal{R}^r$ is the state combination (objective function signal) to be regulated. The variables ν_k and ζ_k are zero-mean real scalar white-noise sequences that satisfy

$$
E\{\nu_k \nu_j\} = \delta_{kj}, \ E\{\zeta_k \zeta_j\} = \delta_{kj}, \ E\{\zeta_k \nu_j\} = 0, \forall k, j \ge 0.
$$

We assume that the system matrices in (10.1a, b), (10.2) lie within the following polytope:

$$
\bar{\Omega} \triangleq \begin{bmatrix} A & B_1 & B_2 & C_1 & D_{12} & D_{21} \end{bmatrix},
\tag{10.3}
$$

which is described by the vertices:

$$
\bar{\Omega} = Co\{\bar{\Omega}_1, \bar{\Omega}_2, \dots, \bar{\Omega}_N\},
\tag{10.4}
$$

where

$$
\bar{\Omega}_i \triangleq \begin{bmatrix} A^{(i)} & B_1^{(i)} & B_2^{(i)} & C_1^{(i)} & D_{12}^{(i)} & D_{21}^{(i)} \end{bmatrix}
\tag{10.5}
$$

and where N is the number of vertices. In other words:

$$
\bar{\Omega} = \sum_{i=1}^{N} \bar{\Omega}_i f_i, \quad \sum_{i=1}^{N} f_i = 1, \quad f_i \ge 0.
\tag{10.6}
$$

We treat the following two problems:

(i) Robust H_∞ state-feedback control:

We consider the system of (10.1a) and (10.2) where the system matrices lie within the polytope $\bar{\Omega}$ of (10.3) and the following performance index:

$$J_E \triangleq ||z_k||^2_{\tilde{l}_2} - \gamma^2 ||w_k||^2_{\tilde{l}_2}. \tag{10.7}$$

Our objective is to find a state-feedback control law $u_k = Kx_k$ that achieves $J_E < 0$, for the worst-case disturbance $w_k \in \tilde{l}^2_{\mathcal{F}_k}([0, \infty); \mathcal{R}^q)$ and for a prescribed scalar $\gamma > 0$.

(ii) Robust H_∞ filtering:

We consider the system of (10.1a, b) and (10.2) where the system matrices lie within the polytope $\bar{\Omega}$ of (10.3) and where $B_2 = 0$ and $D_{12} = 0$. We consider the estimator of the following general form:

$$\begin{aligned} \hat{x}_{k+1} &= A_f \hat{x}_k + B_f y_k, \\ \hat{z}_k &= C_f \hat{x}_k. \end{aligned} \tag{10.8}$$

We denote

$$e_k = x_k - \hat{x}_k, \quad \text{and} \quad \bar{z}_k = z_k - \hat{z}_k, \tag{10.9}$$

and consider the following cost function:

$$J_F \triangleq ||\bar{z}_k||^2_{\tilde{l}_2} - \gamma^2 [||w_k||^2_{\tilde{l}_2}. \tag{10.10}$$

Given $\gamma > 0$, we seek an estimate $C_f \hat{x}_k$ of $C_1 x_k$ over the infinite time horizon $[0, \infty)$ such that J_F given by (10.10) is negative for all nonzero $w_k \in \tilde{l}^2_{\mathcal{F}_k}([0, \infty); \mathcal{R}^q)$.

10.3   Discrete-Time Case: The Bounded Real Lemma

Based on the stability result for discrete-time state-multiplicative linear systems, the following result was obtained in [1] given the following index of performance:

$$J_B \triangleq ||z_k||^2_{\tilde{l}_2} - \gamma^2 ||w_k||^2_{\tilde{l}_2}$$

(for detailed treatment see [1], Chap. 7).

Theorem 10.1 *Consider the system (10.1a) and (10.2) with $B_2 = 0$ and $D_{12} = 0$. The system is exponentially stable in the mean square sense and, for a pre-scribed scalar $\gamma > 0$, the requirement of $J_B < 0$ is achieved for all nonzero $w \in$*

$\tilde{l}^2_{\mathcal{F}_k}([0, \infty); \mathcal{R}^p)$, *if there exists $n \times n$ matrices $Q > 0$, that satisfies the following inequality:*

$$\Gamma_S \triangleq \begin{bmatrix} -Q & 0 & A^T & D^T & C_1^T \\ * & -\gamma^2 I_p & B_1^T & 0 & 0 \\ * & * & -Q^{-1} & 0 & 0 \\ * & * & * & -Q^{-1} & 0 \\ * & * & * & * & -I_r \end{bmatrix} < 0. \qquad (10.11)$$

Denoting $P = Q^{-1}$ and multiplying (10.11) by $diag[P, I, I, I, I]$ the following condition readily follows [1]:

$$\Gamma_P \triangleq \begin{bmatrix} -P & 0 & PA^T & PD^T & PC_1^T \\ * & -\gamma^2 I_p & B_1^T & 0 & 0 \\ * & * & -P & 0 & 0 \\ * & * & * & -P & 0 \\ * & * & * & * & -I_r \end{bmatrix} < 0. \qquad (10.12)$$

In the uncertain case, two approaches are possible, the first of which is the quadratic solution (see [1]). A new, possibly less conservative condition is obtained by applying the following vertex-dependent Lyapunov function approach:

Applying Schur complement to the LMI of (10.12), we obtain

$$\Psi + \Phi P \Phi^T < 0,$$

where

$$\Psi \triangleq \begin{bmatrix} -\gamma^2 I_p & B_1^T & 0 & 0 \\ * & -P & 0 & 0 \\ * & * & -P & 0 \\ * & * & * & -I_r \end{bmatrix}, \quad \Phi \triangleq \begin{bmatrix} 0 \\ A \\ D \\ C_1 \end{bmatrix}. \qquad (10.13)$$

The structure of (10.13) has been used in [18] to reduce the conservatism entailed by applying the Lyapunov function uniformly over the uncertainty polytope. The method of [18] was used to obtain vertex-dependent criteria. Unfortunately, the standard application of the latter method is not readily applicable to the stochastic case where taking the adjoint of the system, instead of the original one, is theoretically unjustified. We therefore derive below a vertex-dependent method that is more suitable and readily applicable to the uncertain stochastic case.

We start with (10.13) and obtain the following lemma.

Lemma 10.1 *Inequality (10.13a) is satisfied iff there exist matrices: $0 < P \in \mathcal{R}^{n \times n}$, $G \in \mathcal{R}^{n \times (2n+p+r)}$ and $H \in \mathcal{R}^{n \times n}$ that satisfy the following inequality:*

$$\Omega \triangleq \begin{bmatrix} \Psi + G^T \Phi^T + \Phi G & -G^T + \Phi H \\ * & -H - H^T + P \end{bmatrix} < 0. \qquad (10.14)$$

Proof Substituting $G = 0$ and $H = P$ in (10.14), inequality (10.13a) is obtained. To show that (10.14) leads to (10.13a) we consider

$$\begin{bmatrix} I & \Phi \\ 0 & I \end{bmatrix} \Omega \begin{bmatrix} I & 0 \\ \phi^T & I \end{bmatrix} = \begin{bmatrix} \Psi_{\Phi(1,1)} & \Psi_{\Phi(1,2)} \\ * & \Psi_{\Phi(2,2)} \end{bmatrix},$$

where

$$\Psi_{\Phi(1,1)} = \Psi + \Phi P \Phi^T, \quad \Psi_{\Phi(1,2)} = -G^T - \Phi H^T + \Phi P, \quad \Psi_{\Phi(2,2)} = -H - H^T + P.$$

Inequality (10.13a) thus follows from the fact that it is the left side (1,1) matrix block in the latter product.   ∎

In the uncertain case, we assume that the system parameters encounter uncertainty that is described in (10.3)–(10.6). Choosing then $V(x,t) = x^T(t) P^j(t) x(t)$, $j = 1, 2, \ldots N$, we consider

$$\Psi^j + \Phi^j P^j \Phi^{j,T} < 0, \quad \forall j = 1, 2, \ldots N \tag{10.15}$$

where Ψ^j and Φ^j are obtained from (10.13b, c) by assigning for each vertex j the appropriate matrices. We readily find the following vertex-dependent condition for the stochastic uncertain discrete-time case.

Corollary 10.1 *Inequality (10.15) is satisfied iff there exist matrices: $0 < P^j \in \mathcal{R}^{n \times n}$, $\forall j = 1, 2, \ldots N$, $G \in \mathcal{R}^{n \times (2n+p+r)}$ and $H \in \mathcal{R}^{n \times n}$ that satisfy the following inequality:*

$$\Omega \triangleq \begin{bmatrix} \Psi^j + G^T \Phi^{j,T} + \Phi^j G & * \\ -G + H^T \Phi^{j,T} & -H - H^T + P^j \end{bmatrix} < 0. \tag{10.16}$$

We note that condition (10.16) is suitable for both the robust state-feedback control problem and the robust estimation problem.

10.4   Discrete-Time Case: State-Feedback Control

In this section we consider the problem of finding the following state-feedback control law

$$u_k = K x_k \tag{10.17}$$

that stabilizes the system and achieves a prescribed level of attenuation. In the nominal case, the following result is obtained.

Theorem 10.2 [1], *Consider the system (10.1a) and (10.2). For a prescribed scalar $\gamma > 0$, there exists a state-feedback gain that achieves negative J_E for all nonzero*

$w_k \in \tilde{l}^2_{\mathcal{F}_k}([0, \infty); \mathcal{R}^p)$, *if there exist* $n \times n$ *matrix* $P > 0$, $l \times n$ *matrix* Y *that satisfy the following LMI condition:*

$$
\begin{bmatrix}
-P & 0 & PA^T + Y^T B_2^T & PD^T & PC_1^T + Y^T D_{12}^T \\
* & -\gamma^2 I_p & B_1^T & 0 & 0 \\
* & * & -P & 0 & 0 \\
* & * & * & -P & 0 \\
* & * & * & * & -I_r
\end{bmatrix} < 0.
\tag{10.18}
$$

In the latter case, the state-feedback gain is given by

$$
K = YP^{-1}.
\tag{10.19}
$$

Remark 10.1 We note that the above can be extended to include a white-noise sequence that multiplies the control input matrix B_2. In this case, an additional term, say $G\mu_k$, is added to the dynamics of (10.1a). Assuming, for simplicity, that the latter white-noise sequence is not correlated with ν_k the following LMI condition is then obtained

$$
\begin{bmatrix}
\Upsilon & Y^T G^T \\
* & -P
\end{bmatrix} < 0,
$$

where Υ is the left side of (10.18).

The above result provides a condition for H_∞ state-feedback control. Letting γ go to infinity, a corresponding condition for the H_2 state-feedback control (where we seek to minimize $E\{z_k^T z_k\}$ when w_k is assumed to be a zero-mean standard white-noise sequence) is readily found in the following corollary.

Corollary 10.2 *Consider the system (10.1a) and (10.2). There exists a state-feedback gain that minimizes the H_2-norm of the of the closed-loop system if there exist $n \times n$ matrix $P > 0$, $l \times n$ matrix Y, and $l \times l$ matrix H that satisfy the following LMI conditions:*

$$
\begin{bmatrix}
-P & PA^T + Y^T B_2^T & PD^T & PC_1^T + Y^T D_{12}^T \\
 & -P & 0 & 0 \\
* & & -P & 0 \\
* & & * & -I_r
\end{bmatrix} < 0 \quad
\begin{bmatrix}
H & B_1^T \\
B_1 & P
\end{bmatrix} > 0, \; and \; Trace\{H\} \to \min.
$$

$$
\tag{10.20}
$$

In the latter case, the state-feedback gain is given by

$$
K = YP^{-1}.
\tag{10.21}
$$

In the uncertain case, we consider the system of (10.1a) and (10.2) where the system matrices lie within the polytope $\bar{\Omega}$ of (10.3). We apply the control law of (10.17), where A is replaced by $(A^j + B_2^j K)$ and C_1 is replaced by $C_1^j + D_{12}^j K$ and *we readily obtain the well known vertex-independent condition (also known as the quadratic solution, see [1]). A less conservative, vertex-dependent result stems from Corollary 10.1. Thus, the following inequality follows:*

$$\Omega \overset{\Delta}{=} \begin{bmatrix} \Psi^j + \Phi_G^{j,T} + \Phi_G^j & -G^T + \Phi_H^j \\ * & -H - H^T + P^j \end{bmatrix} < 0, \quad where \quad \Psi^j \overset{\Delta}{=} \begin{bmatrix} -\gamma^2 I_p & B_1^{j,T} & 0 & 0 \\ * & -P^j & 0 & 0 \\ * & * & -P^j & 0 \\ * & * & * & -I_r \end{bmatrix},$$

$$\forall j = 1, 2, \ldots N$$

$$(10.22)$$

$$\Phi_G^{j,T} \overset{\Delta}{=} G^T \Phi^{j,T} \quad and \quad \Phi_H^j = \Phi^j H.$$

Taking $G = [0 \ \ G_K \ \ 0]$, where G_K is a $n \times n$ matrix, and taking $H = \alpha G_K$, the following result is obtained.

Corollary 10.3 *Consider the system (10.1a) and (10.2) where the system matrices lie within the polytope $\bar{\Omega}$ of (10.3). For a prescribed scalar $\gamma > 0$, and positive tuning scalar $\alpha > 0$, there exists a state-feedback gain that achieves negative J_E for all nonzero $w \in \tilde{l}^2_{\mathcal{F}_k}([0, \infty); \mathcal{R}^p)$, if there exist $n \times n$ matrices $P^j > 0$, $j = 1, 2, \ldots N$, $n \times n$ matrix G_K and a $l \times n$ matrix Y that satisfy the following LMI:*

$$\Gamma_F^j \overset{\Delta}{=} \begin{bmatrix} -\gamma^2 I_p & B_1^{j,T} & 0 & 0 & 0 \\ * & \Gamma_F^j(2,2) & G_K^T D^{j,T} & G_K^T C_1^T + Y^T D_{12}^T & -G_K^T + \alpha[A^j G_K + B_2^j Y] \\ * & * & -P^j & 0 & \alpha D^j G_K \\ * & * & * & -I_r & \alpha[C_1^j G_K + D_{12}^j Y] \\ * & * & * & * & -\alpha[G_K + G_K^T] + P^j \end{bmatrix} < 0,$$

$$(10.23)$$

where $\Gamma_F^j(2,2) = -P^j + A^j G_K + B_2^j Y + G_K^T A^{j,T} + Y^T B_2^{j,T}$.
In the latter case the state-feedback gain is given by

$$K = Y G_K^{-1}. \qquad (10.24)$$

We note that G_K is required to be nonsingular. This, however, follows from the fact that $\Gamma_F^j(5, 5)$ in (10.23) must be negative definite in order for (10.23) to be feasible.

The corresponding conditions for H_2 state-feedback control are given in the following corollary.

Corollary 10.4 *Consider the system (10.1a) and (10.2) where the system matrices lie within the polytope $\bar{\Omega}$ of (10.3). For a positive tuning scalar $\alpha > 0$, there exists a state-feedback gain which guarantees that the H_2-norm of the of the closed-loop system is less than a prescribed positive scalar $\bar{g}$ if there exist: $n \times n$ matrices $P^j > 0$, $j = 1, 2, \ldots N$, $n \times n$ matrix G_K, $l \times n$ matrix Y, and $l \times l$ matrix $\bar{H}$ that satisfy the following LMIs:*

$$\begin{bmatrix} \Gamma_F^j(2,2) & G_K^T D^{j,T} & G_K^T C_1^T + Y^T D_{12}^T & -G_K^T + \alpha[A^j G_K + B_2^j Y] \\ * & -P^j & 0 & \alpha D^j G_K \\ * & * & -I_r & \alpha[C_1^j G_K + D_{12}^j Y] \\ * & * & * & -\alpha[G_K + G_K^T] + P^j \end{bmatrix} < 0,$$

$$\begin{bmatrix} \bar{H} & B_1^T \\ B_1 & G_K \end{bmatrix} > 0, \ and \ Trace\{\bar{H}\} \leq \bar{g}^2, \tag{10.25}$$

where $\Gamma_F^j(2,2)$ is defined in (10.23).
In the latter case, the state-feedback gain is given by

$$K = Y G_K^{-1}. \tag{10.26}$$

10.5 Discrete-Time Case: Robust Filtering

In this section we address the discrete-time filtering problem of uncertain systems with state-multiplicative noise. We consider the system of (10.1a, b) and (10.2) with $B_2 = 0$ and $D_{12} = 0$ and the general-type filter of (10.8). Denoting $\xi_k^T \triangleq [x_k^T \ \hat{x}_k^T]$, we obtain the following augmented system:

$$\begin{aligned} \xi_{k+1} &= \tilde{A}\xi_k + \tilde{B}w_k + \tilde{D}\xi_k \nu_k + \tilde{F}\xi_k \zeta_k, \\ \tilde{z}_k &= \tilde{C}\xi_k, \quad \xi_l = 0, \ l \leq 0, \end{aligned} \tag{10.27}$$

where

$$\tilde{A} = \begin{bmatrix} A & 0 \\ B_f C_2 & A_f \end{bmatrix}, \ \tilde{B} = \begin{bmatrix} B_1 \\ B_f D_{21} \end{bmatrix}, \ \tilde{F} = \begin{bmatrix} 0 & 0 \\ B_f F & 0 \end{bmatrix}, \ \tilde{D} = \begin{bmatrix} D & 0 \\ 0 & 0 \end{bmatrix}, \ \tilde{C}^T = \begin{bmatrix} C_1^T \\ -C_f^T \end{bmatrix}. \tag{10.28}$$

Using the result of Theorem 10.1 the following inequality condition was obtained in [1].

Theorem 10.3 *Consider the system of (10.1a, b) and (10.2) with $B_2 = 0$ and $D_{12} = 0$ and the general-type filter of (10.8). For a prescribed scalar $\gamma > 0$, the following hold:*
(i) A necessary and sufficient condition for J_F of (10.10) to be negative for all nonzero $w_k \in \tilde{l}_2([0, \infty); \mathcal{R}^q)$, is that there exist $R = R^T \in \mathcal{R}^{n \times n}$, $W = W^T \in \mathcal{R}^{n \times n}$, $Z \in \mathcal{R}^{n \times m}$, $S \in \mathcal{R}^{n \times n}$ and $T \in \mathcal{R}^{r \times n}$, such that

$$\Sigma(R, W, Z, S, T, \gamma^2) < 0, \tag{10.29}$$

where $\Sigma(R, W, Z, S, T, \gamma^2) \triangleq$

$$\begin{bmatrix} -R & 0 & 0 & 0 & -RA & 0 & -RB_1 & 0 \\ * & -W & 0 & 0 & -WA - ZC_2 - S & S & -WB_1 - ZD_{21} & 0 \\ * & * & -R & 0 & -RD & 0 & 0 & 0 \\ * & * & * & -W & -WD & 0 & 0 & 0 \\ * & * & * & * & -R & 0 & 0 & T^T - C_1^T \\ * & * & * & * & * & -W & 0 & -T^T \\ * & * & * & * & * & * & -\gamma^2 I_p & 0 \\ * & * & * & * & * & * & * & -I_r \end{bmatrix}.$$

(ii) If (10.29) is satisfied, a mean square stabilizing filter in the form of (10.8) that achieves the negative J_F is given by

$$A_f = -W^{-1}S, \quad B_f = -W^{-1}Z \quad and \quad C_f = T. \tag{10.30}$$

The corresponding H_2 result that minimizes the variance of $\tilde{z}_k$, where ξ_k, ν_k and ζ_k are uncorrelated standard white sequences is obtained by multiplying (10.29), from both sides by

$$diag\{\gamma^{-1}I, \ \gamma^{-1}I, \ \gamma^{-1}I, ; \gamma^{-1}I, \ \gamma^{-1}I, \ \gamma^{-1}I, \ \gamma^{-1}I_p, \ \gamma I_r\}$$

and letting γ go to infinity. We thus obtain the following.

Corollary 10.5 *Consider the system of (10.1a, b) and (10.2) with $B_2 = 0$ and $D_{12} = 0$ and the general-type filter of (10.8). The variance of the estimation error $E\{\|z_k\|^2\}$ is less than a prescribed positive scalar h if there exist $R = R^T \in \mathcal{R}^{n \times n}$, $W = W^T \in \mathcal{R}^{n \times n}$, $Z \in \mathcal{R}^{n \times m}$, $S \in \mathcal{R}^{n \times n}$ $T \in \mathcal{R}^{r \times n}$, and $\bar{H} \in \mathcal{R}^{r \times r}$ such that*

$$\begin{bmatrix} -R & 0 & 0 & 0 & -RA & 0 & -RB_1 \\ * & -W & 0 & 0 & -WA - ZC_2 - S & S & -WB_1 - ZD_{21} \\ * & * & -R & 0 & -RD & 0 & 0 \\ * & * & * & -W & -WD & 0 & 0 \\ * & * & * & * & -R & 0 & 0 \\ * & * & * & * & * & -W & 0 \\ * & * & * & * & * & * & -I_p \end{bmatrix} < 0 \tag{10.31}$$

$$\begin{bmatrix} \bar{H} & C_1 - T & T \\ * & R & 0 \\ * & * & W \end{bmatrix} > 0, \quad and \quad Trace\{\bar{H}\} \leq h.$$

If (10.31) is satisfied, the matrix parameters of the H_2 stabilizing filter (in the form of (10.8)) are given by (10.30).

In the uncertain case, again, a vertex-independent condition is readily obtained (see [1]) using the vertices of (10.3). Applying the result of Corollary 10.1, a less conservative result can be derived similarly to the method used in Sect. 10.4. The LMI of (10.29) can then be written as

$$\tilde{\Psi}^j + \tilde{\Phi}^j R^j \tilde{\Phi}^{j,T} < 0,$$

where $\tilde{\Psi}^j (R, W, Z, S, T, \gamma^2) \triangleq$

$$
\begin{bmatrix}
-W & 0 & 0 & -WA^j - ZC_2^j - S & S & -WB_1^j - ZD_{21}^j & 0 \\
* & -R^j & 0 & -R^j D & 0 & 0 & 0 \\
* & * & -W & -WD & 0 & 0 & 0 \\
* & * & * & -R^j & 0 & 0 & T^T - C_1^{j,T} \\
* & * & * & * & -W & 0 & -T^T \\
* & * & * & * & * & -\gamma^2 I_p & 0 \\
* & * & * & * & * & * & -I_r
\end{bmatrix}
$$

and

$$\tilde{\Phi}^{j,T} \triangleq \begin{bmatrix} 0 & 0 & 0 & 0 & -A^j & 0 & -B_1^j & 0 \end{bmatrix}. \tag{10.32}$$

Applying the result of Corollary 10.1, we arrive at the following result.

Corollary 10.6 *Consider the system of (10.1a, b) and (10.2) with $B_2 = 0$ and $D_{12} = 0$ where the system matrices lie within the polytope $\bar{\Omega}$ of (10.3). For a prescribed scalar $\gamma > 0$,, there exists a filter of the structure (10.8) that achieves $J_F < 0$, where J_F is given in (10.10), for all nonzero $w \in \tilde{l}^2([0, \infty); \mathcal{R}^p)$, if there exist matrices $R^j = R^{j,T} \in \mathcal{R}^{n \times n}$, $j = 1, 2, \ldots N$, $W = W^T \in \mathcal{R}^{n \times n}$, $Z \in \mathcal{R}^{n \times m}$, $S \in \mathcal{R}^{n \times n}$, $T \in \mathcal{R}^{r \times n}$, $\tilde{G} \in \mathcal{R}^{n \times (5n+p+r)}$, and $\tilde{H} \in \mathcal{R}^{n \times n}$, such that the following set of LMIs is satisfied:*

$$\Omega \triangleq \begin{bmatrix} \tilde{\Psi}^j + \tilde{\Phi}_G^{j,T} + \tilde{\Phi}_G^j & -\tilde{G}^T + \tilde{\Phi}_H^j \\ * & -\tilde{H} - \tilde{H}^T + R^j \end{bmatrix} < 0,$$
$$\forall j = 1, 2, \ldots N, \tag{10.33}$$

where

$$\tilde{\Phi}_G^j \triangleq \tilde{\Phi}^j \tilde{G}, \quad and \quad \tilde{\Phi}_H^j \triangleq \tilde{\Phi}^j \tilde{H},$$

and where $\tilde{\Psi}^j$ and $\tilde{\Phi}^j$ are defined in (10.32).

In the latter case, the mean square stabilizing filter in the form of (10.8) that achieves the negative J_F is given by (10.30).

The corresponding robust H_2 filter that ensures an estimation error variance that is less than a prescribed positive scalar $\bar{h}$ over the entire uncertainty polytope is obtained by solving the following LMIs for $R^j = R^{j,T} \in \mathcal{R}^{n \times n}$, $j = 1, 2, \ldots N$, $W = W^T \in \mathcal{R}^{n \times n}$, $Z \in \mathcal{R}^{n \times m}$, $S \in \mathcal{R}^{n \times n}$, $T \in \mathcal{R}^{r \times n}$, $\tilde{G} \in \mathcal{R}^{n \times (5n+p+r)}$, $\tilde{H} \in \mathcal{R}^{n \times n}$, and $H \in \mathcal{R}^{r \times r}$:

$$\Omega_2 \triangleq \begin{bmatrix} \tilde{\Psi}_2^j + \tilde{G}^T \tilde{\Phi}_2^{j,T} + \tilde{\Phi}_2^j \tilde{G} & -\tilde{G}^T + \tilde{\Phi}_2^j \tilde{H} \\ * & -\tilde{H} - \tilde{H}^T + R^j \end{bmatrix} < 0, \quad \begin{bmatrix} H & C_1 - T & T \\ * & R^j & 0 \\ * & * & W \end{bmatrix} > 0, \text{ and } Trace\{H\} \le \bar{h},$$

$$\forall j = 1, 2, \ldots N$$

(10.34)

where

$$\tilde{\Psi}_2^j (R, W, Z, S, T,) = \begin{bmatrix} -W & 0 & 0 & -WA^j - ZC_2^j - S & S & -WB_1^j - ZD_{21}^j \\ * & -R^j & 0 & -R^j D & 0 & 0 \\ * & * & -W & -WD & 0 & 0 \\ * & * & * & -R^j & 0 & 0 \\ * & * & * & * & -W & 0 \\ * & * & * & * & * & -I_p \end{bmatrix}$$

and

$$\tilde{\Phi}_2^{j,T} \triangleq \begin{bmatrix} 0 & 0 & 0 & 0 & -A^j & 0 & -B_1^j \end{bmatrix}.$$

If a solution to the latter LMIs is obtained, the matrix parameters of the H_2 stabilizing filter are given by (10.30).

10.6   Discrete-Time Case: Examples

Example 10.1 **State-feedback control** We consider the system (10.1a) and (10.2) where

$$x_{k+1} = \begin{bmatrix} 0.1 & 0.6 \\ -1.0 & -2.0 - a \end{bmatrix} x_k + \begin{bmatrix} -0.0225 \\ 0.045 \end{bmatrix} w_k + \begin{bmatrix} 0.04 \\ 0.05 \end{bmatrix} u_k, + \begin{bmatrix} 0 & 0.63 \\ 0 & 0 \end{bmatrix} x_k \nu_k,$$

$$z_k = \begin{bmatrix} 2 & 0 \\ 0 & 0 \end{bmatrix} x_k + \begin{bmatrix} 0 \\ 0.1 \end{bmatrix} u_k,$$

and where $a \in [-0.9 \ 1]$. Solving for the nominal case (i.e., $a = 0$) by applying the result of Theorem 10.2, a near minimum attenuation level of $\gamma = 1.99$ was obtained. For the uncertain case, solving the problem by applying the "quadratic" result, an attenuation level of $\gamma = 11.8$ is obtained. The controller gain matrix is $K = [19.62 \ 40.05]$. Applying the result of Corollary 10.3 which yields the vertex-dependent Lyapunov solution, an attenuation level of $\gamma = 8.68$ is obtained for $\alpha = 1000$ where the controller gain matrix is $K = [19.04 \ 39.74]$.

Example 10.2 **Filtering** We consider the system (10.1a, b) and (10.2) where

$$x_{k+1} = \begin{bmatrix} 0.1 & 0.6 \\ -1.0 + b & -0.5 \end{bmatrix} x_k + \begin{bmatrix} -0.225 \\ 0.45 \end{bmatrix} w_k + \begin{bmatrix} 0 & 0.63 \\ 0 & 0 \end{bmatrix} x_k \nu_k,$$

$$y_k = [0 \ 1] x_k + 0.01 w_k, \quad z_k = \begin{bmatrix} 2 & 0 \\ 0 & 0 \end{bmatrix} x_k, \quad B_2 = 0, \quad D_{12} = 0,$$

and where $b \in [-0.15 \ \ 0.15]$. Solving for the nominal case (i.e., $b = 0$), by applying the result of Theorem 10.3, a near minimum attenuation level of $\gamma = 1.44$ is obtained. For the uncertain case, solving the problem by applying the "quadratic" result, an attenuation level of $\gamma = 21.15$ was obtained. The filter parameters are

$$A_f = \begin{bmatrix} 0.105 & 0.009 \\ -1.165 & -0.105 \end{bmatrix}, \quad C_f = \begin{bmatrix} -0.500 & 0.400 \\ 0 & 0 \end{bmatrix}, \quad B_f = \begin{bmatrix} 0.593 \\ -0.445 \end{bmatrix}.$$

Applying the result of Corollary 10.4, which brings the vertex-dependent Lyapunov solution method, an attenuation level of $\gamma = 12.35$ is obtained where the filter parameters are

$$A_f = \begin{bmatrix} 0.105 & 0.009 \\ -1.165 & -0.105 \end{bmatrix}, \quad C_f = \begin{bmatrix} -0.437 & 0.436 \\ 0 & 0 \end{bmatrix}, \quad B_f = \begin{bmatrix} 0.596 \\ -0.404 \end{bmatrix}.$$

Clearly, a significant improvement is achieved by the new approach.

10.7 Discrete-Time Case: Conclusions

In this part the theory of robust linear H_∞ state-feedback control and filtering is developed, via a new approach, for discrete-time uncertain systems with multiplicative noise that is encountered in both the dynamic and the measurement matrices in the state space model of the system.

Sufficient conditions are derived for the BRL of uncertain polytopic-type systems applying a vertex-dependent Lyapunov function. Based on this BRL, the robust state-feedback control and filtering problems are formulated and solved by applying the new approach. In the state-feedback case, an inherent overdesign is admitted to our solution by assigning a special structure to the matrix G. In the robust filtering case, however, the two decision variables that stem from the vertex-dependent approach (i.e., $\tilde{G}$ and $\tilde{H}$) are independent one of the another.

The robust state-feedback control and the robust filtering examples demonstrate the considerable advantage of the vertex-dependent approach. Extension of the theory presented here to cases where the various multiplicative white-noise sequences are correlated can be readily accounted for in the various LMI conditions for both the nominal and the uncertain cases. This would lead, however, to cumbersome matrix conditions.

10.8 Continuous-Time Case: Problem Formulation

We consider the following linear system:

$$\begin{aligned} dx(t) &= Ax(t)dt + Dx(t)d\nu(t) + B_1 w(t)dt + B_2 u(t)dt, \\ x(\tau) &= 0, \ \tau \le 0, \\ y(t) &= (C_2 + F\zeta(t))x(t) + D_{21}w(t), \end{aligned} \tag{10.35}$$

with the objective vector

$$z(t) = C_1 x(t) + D_{12} u(t),\tag{10.36}$$

where $x(t) \in \mathcal{R}^n$ is the system state vector, $w(t) \in \mathcal{R}^p$ is the exogenous disturbance signal, $u(t) \in \mathcal{R}^\ell$ is the control input, $y_k \in \mathcal{R}^m$ is the measured output and $z(t) \in \mathcal{R}^r$ is the state combination (objective function signal) to be regulated. The variables $\nu(t)$ and $\zeta(t)$ are zero-mean real scalar Wiener processes that satisfy

$$\mathcal{E}\{d\nu(t)\} = 0, \ \ \mathcal{E}\{d\zeta(t)\} = 0, \ \ \mathcal{E}\{d\nu(t)^2\} = dt, \ \ \mathcal{E}\{d\zeta(t)^2\} = dt,$$

$$\mathcal{E}\{d\nu(t)d\zeta(t)\} = 0.$$

We assume that the system matrices in (10.35a, b), (10.36) lie within the following polytope:

$$\bar{\Omega} \triangleq \begin{bmatrix} A & B_1 & B_2 & C_1 & D_{12} & D_{21} \end{bmatrix},\tag{10.37}$$

which is described by the vertices:

$$\bar{\Omega} = Co\{\bar{\Omega}_1, \bar{\Omega}_2, \dots, \bar{\Omega}_N\},\tag{10.38}$$

where

$$\bar{\Omega}_i \triangleq \begin{bmatrix} A^{(i)} & B_1^{(i)} & B_2^{(i)} & C_1^{(i)} & D_{12}^{(i)} & D_{21}^{(i)} \end{bmatrix}\tag{10.39}$$

and where N is the number of vertices. In other words,

$$\bar{\Omega} = \sum_{i=1}^N \bar{\Omega}_i f_i, \quad \sum_{i=1}^N f_i = 1 \ , f_i \geq 0.\tag{10.40}$$

We treat the following two problems:

(i) Robust H_∞ state-feedback control:

We consider the system of (10.35a) and (10.36) where the relevant system matrices lie within the polytope $\bar{\Omega}$ of (10.37) and the following performance index:

$$J_E \triangleq \|z(t)\|_{\tilde{L}_2}^2 - \gamma^2 \|w(t)\|_{\tilde{L}_2}^2.\tag{10.41}$$

Our objective is to find a state-feedback control law $u(t) = Kx(t)$ that achieves $J_E < 0$, for the worst-case disturbance $w(t) \in \tilde{L}^2([0, \infty); \mathcal{R}^p)$ and for a prescribed scalar $\gamma > 0$.

(ii) Robust H_∞ filtering:

We consider the system of (10.35a, b) and (10.36) where the system matrices lie within the polytope $\bar{\Omega}$ of (10.37) and where $B_2 = 0$ and $D_{12} = 0$. We consider an estimator of the following general form:

$$d\hat{x}(t) = A_f \hat{x}(t)dt + B_f y(t),$$
$$\hat{z}(t) = C_f \hat{x}(t). \qquad (10.42)$$

We denote

$$e(t) = x(t) - \hat{x}(t), \quad \text{and} \quad \bar{z}(t) = z(t) - \hat{z}(t), \qquad (10.43)$$

and consider the following cost function:

$$J_F \triangleq ||\bar{z}(t)||^2_{\tilde{L}_2} - \gamma^2[||w(t)||^2_{\tilde{L}_2}. \qquad (10.44)$$

Given $\gamma > 0$, we seek an estimate $C_f \hat{x}(t)$ of $C_1 x(t)$ over the infinite time horizon $[0, \infty)$ such that J_F given by (10.44) is negative for all nonzero $w(t) \in \tilde{L}^2([0, \infty); \mathcal{R}^p)$.

10.9 Continuous-Time Case: The Bounded Real Lemma

Based on the stability result for continuous-time state-multiplicative linear systems, the following result was obtained in [1] given the following index of performance:

$$J_B \triangleq ||z(t)||^2_{\tilde{L}_2} - \gamma^2||w(t)||^2_{\tilde{L}_2}.$$

(for detailed treatment see [1], Chap. 2).

Theorem 10.4 *Consider the system (10.35a) and (10.36) with $B_2 = 0$ and $D_{12} = 0$. The system is exponentially stable in the mean square sense and for a prescribed scalar $\gamma > 0$,, the requirement of $J_B < 0$ is achieved for all nonzero $w(t) \in \tilde{L}^2([0, \infty); \mathcal{R}^p)$, if there exists $n \times n$ matrices $Q > 0$ that satisfies the following inequality:*

$$\Gamma_Q \triangleq \begin{bmatrix} A^T Q + QA & C_1^T & QB_1 & D^T Q \\ * & -\gamma^2 I_r & 0 & 0 \\ * & * & -I_p & 0 \\ * & * & * & -Q \end{bmatrix} < 0. \qquad (10.45)$$

Denoting $P = Q^{-1}$ and multiplying (10.45) by $diag[P, I, I, P]$ from the left and the right, the following condition readily follows [1]:

$$\Gamma_P \triangleq \begin{bmatrix} PA^T + AP & PC_1^T & B_1 & PD^T \\ * & -\gamma^2 I_r & 0 & 0 \\ * & * & -I_p & 0 \\ * & * & * & -P \end{bmatrix} < 0. \qquad (10.46)$$

In the uncertain case, two approaches are possible, the first of which is the quadratic solution (see [1], Chap. 2). A new, possibly less conservative condition is obtained by applying the following modified vertex-dependent Lyapunov function approach [18].

We consider the following matrix inequality:

$$\Psi \overset{\Delta}{=} \begin{bmatrix} \Psi_{11} & P - G^T + AH & G^T C_1^T & B_1 \\ * & -H - H^T & H^T C_1^T & 0 \\ * & * & -\gamma^2 I_r & 0 \\ * & * & * & -I_p \end{bmatrix} < 0, \qquad (10.47)$$

where $\Psi_{11} = AG + G^T A^T + \Psi_{11}(P)$ *and* $\Psi_{11}(P) = P D^T Q^{-1} D P.$

We obtain the following lemma.

Lemma 10.2 *Inequality (10.46) is satisfied iff there exist matrices:* $0 < P \in \mathcal{R}^{n \times n}$, $G \in \mathcal{R}^{n \times n}$ *and* $H \in \mathcal{R}^{n \times n}$ *that satisfy (10.47).*

Proof Substituting $G = P$ and $H = \epsilon I_n$, $\epsilon \to 0$ in (10.47), inequality (10.46) is obtained. To show that (10.47) leads to (10.46), we multiply (10.47) from the left by Υ_P and from the right by Υ_P^T where

$$\Upsilon_P = \begin{bmatrix} I_n & A & 0 & 0 \\ 0 & I_n & 0 & 0 \\ 0 & C_1 & I_r & 0 \\ 0 & 0 & 0 & I_p \end{bmatrix}.$$

We obtain

$$\begin{bmatrix} \Psi_{11} & P - G^T + AH & PC_1^T & B_1 \\ * & -H - H^T & -HC_1^T & 0 \\ * & * & -\gamma^2 I_r & 0 \\ * & * & * & -I_p \end{bmatrix} < 0,$$

where $\Psi_{11} = AP + PA^T + \Psi_{11}(P)$.

Inequality (10.46) thus follows from the fact that it is the left side (1,1) matrix block in the latter product following a Schur's complement with respect to the third and fourth columns and rows. ∎

In the uncertain case, we assume that the system parameters encounter uncertainty that is described in (10.37)–(10.40). We readily find the following vertex-dependent condition for the stochastic uncertain continuous-time case.

Corollary 10.7 *Consider the system (10.35a) and (10.36) with $B_2 = 0$ and $D_{12} = 0$ where the system matrices lie within the polytope $\bar{\Omega}$ of (10.37). The system is exponentially stable in the mean square sense and, for a prescribed scalar $\gamma > 0$, the requirement of $J_B < 0$ is achieved for all nonzero $w(t) \in \tilde{L}^2([0, \infty); \mathcal{R}^p)$, if there exist matrices $0 < P^j \in \mathcal{R}^{n \times n}$, $\forall j = 1, 2, \ldots N$, $G \in \mathcal{R}^{n \times n}$ and $H \in \mathcal{R}^{n \times n}$ that satisfies the following set of LMIs:*

$$
\Gamma_j \stackrel{\Delta}{=}
\begin{bmatrix}
\Psi_{11,j} & P^j - G^T + A^j H & G^T C_1^{j,T} & B_1^j & P D^{j,T} \\
* & -H - H^T & H^T C_1^{j,T} & 0 & 0 \\
* & * & -\gamma^2 I_r & 0 & 0 \\
* & * & * & -I_p & 0 \\
* & * & * & * & -P^j
\end{bmatrix}
< 0, \qquad (10.48)
$$

$\forall j = 1, 2, \ldots N$, where $\Psi_{11,j} = A^j G + G^T A^{j,T}$.

We note that condition (10.48) is suitable for both the robust state-feedback control problem and the robust estimation problem where a slight modification is needed in the latter case.

10.10 Continuous-Time Case: State-Feedback Control

In this section, we consider the problem of finding the following state-feedback control law

$$
u(t) = K x(t) \qquad (10.49)
$$

that stabilizes the system and achieves a prescribed level of attenuation. In the nominal case, the following result is obtained.

Theorem 10.5 [1], *Consider the system (10.35a) and (10.36). For a prescribed scalar $\gamma > 0$, there exists a state-feedback gain that achieves negative J_E for all nonzero $w(t) \in \tilde{L}^2([0, \infty); \mathcal{R}^p)$, if there exist $n \times n$ matrix $P > 0$, and $l \times n$ matrix Y that satisfy the following LMI condition:*

$$
\begin{bmatrix}
\Gamma_{1,1} & P C_1^T + Y^T D_{12}^T & B_1 & P D^T \\
* & -\gamma^2 I_r & 0 & 0 \\
* & * & -I_p & 0 \\
* & * & * & -P
\end{bmatrix}
< 0, \qquad (10.50)
$$

where $\Gamma_{1,1} = P A^T + Y^T B_2^T + A P + B_2 Y$.

In the latter case, the state-feedback gain is given by

$$
K = Y P^{-1}. \qquad (10.51)
$$

Remark 10.2 The above solution can be extended to include a white-noise sequence that multiplies the control input matrix B_2. In this case, an additional term, say $G\mu(t)$, is added to the dynamics of (10.35a). Assuming, for simplicity, that the latter zero-mean white-noise sequence is not correlated with $\nu(t)$ the following LMI condition is then obtained

$$\begin{bmatrix} \Upsilon & (GY)^T \\ * & -P \end{bmatrix} < 0,$$

where Υ is the left side of (10.50).

The above results provide conditions for H_∞ state-feedback control. Multiplying (10.50), from both sides, by $diag\{\gamma^{-1}I, \ \gamma^{-1}I, \ \gamma I, \ \gamma^{-1}I\}$, denoting $\tilde{P} = \gamma^{-2}P$ and $\tilde{Y} = \gamma^{-2}Y$, and letting γ tend to infinity, a corresponding condition for the H_2 state-feedback control (where we seek to minimize $\|z(t)\|_{\tilde{L}_2}^2$ when $w(t)$ is assumed to be a standard white-noise signal) is readily obtained in the following corollary.

Corollary 10.8 *Consider the system (10.35a) and (10.36). There exists a state-feedback gain that makes the H_2 norm of the closed-loop system less than a prescribe positive scalar h if there exist $n \times n$ matrix $\tilde{P} > 0$, $l \times n$ matrix $\tilde{Y}$, and $l \times l$ matrix H that satisfy the following LMI conditions:*

$$\begin{bmatrix} \tilde{\Gamma}_{1,1} & \tilde{P}C_1^T + \tilde{Y}^T D_{12}^T & \tilde{P}D^T \\ * & -I_r & 0 \\ * & * & -\tilde{P} \end{bmatrix} < 0, \ \begin{bmatrix} H & B_1^T \\ B_1 & \tilde{P} \end{bmatrix} > 0, \ and \ Trace\{H\} \leq h^2,$$

(10.52)

where $\tilde{\Gamma}_{1,1} = \tilde{P}A^T + \tilde{Y}^T B_2^T + A\tilde{P} + B_2\tilde{Y}$.

In the latter case, the state-feedback gain is given by

$$K = \tilde{Y}\tilde{P}^{-1}.$$

(10.53)

In the uncertain case, we consider the system of (10.35a) and (10.36) where the system matrices lie within the polytope $\bar{\Omega}$ of (10.37). We apply the control law of (10.49), where A is replaced by $(A^j + B_2^j K)$ and C_1 is replaced by $C_1^j + D_{12}^j K$ and we readily obtain the well-known vertex-independent condition (also known as the quadratic solution, see [1]). A less conservative, vertex-dependent result stems from Corollary 10.7. Thus, we obtain the following result.

Corollary 10.9 *Consider the system (10.35a) and (10.36) where the system matrices lie within the polytope $\bar{\Omega}$ of (10.37). For a prescribed scalar $\gamma > 0$, there exists a state-feedback gain that achieves negative J_E for all nonzero $w(t) \in \tilde{L}^2([0, \infty); \mathcal{R}^p)$, if there exist $n \times n$ matrices $P^j > 0$, $j = 1, 2, \ldots N$, $l \times n$ matrix G_K and $n \times n$ matrices H and G that satisfy the following set of LMIs:*

$$
\begin{bmatrix}
\Gamma_F^j(1,1) & \Gamma_F^j(1,2) & \Gamma_F^j(1,3) & B_1^j & P^j D^T \\
* & \Gamma_F^j(2,2) & H^T C_1^{j,T} & 0 & 0 \\
* & * & -\gamma^2 I_r & 0 & 0 \\
* & * & * & -I_p & 0 \\
* & * & * & * & -P^j
\end{bmatrix} < 0,
\qquad (10.54)
$$

$\forall j = 1, 2, \ldots N$ where

$$
\begin{aligned}
\Gamma_F^j(1,1) &= G^T A^{j,T} + G_K^T B_2^{j,T} + A^j G + B_2^j G_K, \\
\Gamma_F^j(1,2) &= P^j - G^T + A^j H, \\
\Gamma_F^j(1,3) &= G^T C_1^{j,T} + G_K^T D_{12}^{j,T}, \\
\Gamma_F^j(2,2) &= -H - H^T.
\end{aligned}
$$

In the latter case, the state-feedback gain is given by

$$
K = G_K G^{-1}. \qquad (10.55)
$$

The corresponding H_2 result is obtained similarly to the way it was derived above for the nominal system. We obtain the following.

Corollary 10.10 *Consider the system (10.35a) and (10.36) where the system matrices lie within the polytope $\bar{\Omega}$ of (10.37). For a prescribed scalar $\bar{h}$, there exists a state-feedback gain that achieves $\|z\|_{\tilde{L}_2}^2 < \bar{h}^2$ where $w(t)$ is a standard white-noise signal, if there exist $n \times n$ matrices $\tilde{P}^j > 0$, $j = 1, 2, \ldots N$, $l \times n$ matrix $\tilde{G}_K$, $n \times n$ matrices $\tilde{H}$ and $\tilde{G}$ and $r \times r$ matrix $\bar{H}$ that satisfy the following set of LMIs for all $j = 1, 2, \ldots N$:*

$$
\begin{bmatrix}
\tilde{\Gamma}_F^j(1,1) & \tilde{\Gamma}_F^j(1,2) & \tilde{\Gamma}_F^j(1,3) & \tilde{P}^j D^T \\
* & \tilde{\Gamma}_F^j(2,2) & \tilde{H}^T C_1^{j,T} & 0 \\
* & * & -I_r & 0 \\
* & * & * & -I_p
\end{bmatrix} < 0, \quad
\begin{bmatrix}
\bar{H} & B_1^{jT} \\
* & G
\end{bmatrix} > 0, \text{ and } Trace\{\bar{H}\} \le \bar{h}^2,
$$

$$ \qquad (10.56) $$

where

$$
\begin{aligned}
\tilde{\Gamma}_F^j(1,1) &= \tilde{G}^T A^{j,T} + \tilde{G}_K^T B_2^{j,T} + A^j \tilde{G} + B_2^j \tilde{G}_K, \\
\tilde{\Gamma}_F^j(1,2) &= \tilde{P}^j - \tilde{G}^T + A^j \tilde{H}, \\
\tilde{\Gamma}_F^j(1,3) &= \tilde{G}^T C_1^{j,T} + \tilde{G}_K^T D_{12}^{j,T}, \\
\tilde{\Gamma}_F^j(2,2) &= -\tilde{H} - \tilde{H}^T.
\end{aligned}
$$

In the latter case, the state-feedback gain is given by

$$
K = \tilde{G}_K \tilde{G}^{-1}. \qquad (10.57)
$$

10.11   Continuous-Time Case: Robust Filtering

In this section we address the continuous-time filtering problem of uncertain systems with state-multiplicative noise. We consider the system of (10.35a, b) and (10.36) with $B_2 = 0$ and $D_{12} = 0$ and the general-type filter of (10.42). Denoting $\xi^T(t) \triangleq [x^T(t) \ \hat{x}^T(t)]$, we obtain the following augmented system:

$$
\begin{aligned}
&d\xi(t) = \tilde{A}\xi(t)dt + \tilde{B}w(t)dt + \tilde{D}\xi(t)d\nu(t) + \tilde{F}d\xi(t)\zeta(t), \\
&\tilde{z}(t) = \tilde{C}\xi(t), \quad \xi(\tau) = 0, \ \tau \le 0,
\end{aligned}
\tag{10.58}
$$

where

$$
\tilde{A} = \begin{bmatrix} A & 0 \\ B_f C_2 & A_f \end{bmatrix}, \ \tilde{B} = \begin{bmatrix} B_1 \\ B_f D_{21} \end{bmatrix}, \ \tilde{F} = \begin{bmatrix} 0 & 0 \\ B_f F & 0 \end{bmatrix}, \ \tilde{D} = \begin{bmatrix} D & 0 \\ 0 & 0 \end{bmatrix}, \ \text{and}
$$

$$
\tilde{C}^T = \begin{bmatrix} C_1^T \\ -C_f^T \end{bmatrix}.
$$

$$
\tag{10.59}
$$

Using the result of Theorem 10.4, the following inequality condition has been obtained in [1].

Theorem 10.6 *Consider the system of (10.35a, b) and (10.36) with $B_2 = 0$ and $D_{12} = 0$ and the general-type filter of (10.42). For a prescribed scalar $\gamma > 0$, the following hold:*
(i) A necessary and sufficient condition for J_F of (10.44) to be negative for all nonzero $w(t) \in \tilde{L}^2([0, \infty); \mathcal{R}^p)$, is that there exist $0 < R \in \mathcal{R}^{n \times n}$, $0 < W \in \mathcal{R}^{n \times n}$, $Z \in \mathcal{R}^{n \times m}$, $S \in \mathcal{R}^{n \times n}$ and $T \in \mathcal{R}^{r \times n}$, such that

$$
\Sigma(R, W, Z, S, T, \gamma^2) < 0,
\tag{10.60}
$$

where by $\Sigma(R, W, Z, S, T, \gamma^2)$ we denote

$$
\begin{bmatrix}
RA + A^T R & A^T W^T + C_2^T Z^T + S^T & C_1^T - T^T & D^T R^T \\
WA + ZC_2 + S & S - S^T & T^T & 0 \\
C_1 - T & T & -I_r & 0 \\
RD & 0 & 0 & -R \\
WD + ZF & 0 & 0 & 0 \\
ZF & 0 & 0 & 0 \\
B_1^T R & B_1^T W + D_{21}^T Z^T & 0 & 0
\end{bmatrix}
$$

$$
\left[
\begin{array}{ccc}
D^T W^T + F^T Z^T & F^T Z^T & RB_1 \\
0 & 0 & W^T B_1 + Z D_{21} \\
0 & 0 & 0 \\
0 & 0 & 0 \\
-W & 0 & 0 \\
0 & -W & 0 \\
0 & 0 & -\gamma^2 I_p
\end{array}
\right].
\tag{10.61}
$$

(ii) If (10.61) is satisfied, a mean square stabilizing filter in the form of (10.42) that achieves $J_F < 0$ is given by

$$
A_f = -W^{-1}S, \quad B_f = -W^{-1}Z \quad \text{and} \quad C_f = T.
\tag{10.62}
$$

The corresponding H_2 result that minimizes $\|\tilde{z}\|^2_{\bar{L}_2}$, when ξ, ν and ζ are uncorrelated standard white signals is obtained by multiplying (10.60), from both sides by $\operatorname{diag}\{\gamma^{-1}I, \ \gamma^{-1}I, \ \gamma I, \ \gamma^{-1}I, \ \gamma^{-1}I, \ \gamma^{-1}I, \ \gamma^{-1}I_p, \ \gamma^{-1}I_r\}$ and letting γ go to infinity. The following is obtained.

Corollary 10.11 *Consider the system of (10.35a, b) and (10.36) with $B_2 = 0$ and $D_{12} = 0$ and the general-type filter of (10.42). For a prescribed scalar $h > 0$, the estimation error variance $\|z(t)\|^2_{\bar{L}_2}$ will be less than h, where $w(t) \in \mathcal{R}^p$ is a standard white-noise signal, if there exist $0 < R \in \mathcal{R}^{n \times n}$, $0 < W \in \mathcal{R}^{n \times n}$, $Z \in \mathcal{R}^{n \times m}$, $S \in \mathcal{R}^{n \times n}$, $T \in \mathcal{R}^{r \times n}$, and $H \in \mathcal{R}^{r \times r}$ such that*

$$
\left[
\begin{array}{cccccc}
RA + A^T R & A^T W^T + C_2^T Z^T + S^T & D^T R^T & D^T W^T + F^T Z^T & F^T Z^T & RB_1 \\
WA + ZC_2 + S & S - S^T & 0 & D^T W^T + F^T Z^T & F^T Z^T & RB_1 \\
RD & 0 & -R & 0 & 0 & W^T B_1 + Z D_{21} \\
WD + ZF & 0 & 0 & -W & 0 & 0 \\
ZF & 0 & 0 & 0 & -W & 0 \\
B_1^T R & B_1^T W + D_{21}^T Z^T & 0 & 0 & 0 & I_p
\end{array}
\right] < 0.
$$

$$
\left[
\begin{array}{ccc}
H & C_1 - T & T \\
* & R & 0 \\
* & * & W
\end{array}
\right] > 0, \quad \text{and} \quad Trace\{H\} \leq h.
\tag{10.63}
$$

If (10.63) is satisfied, a H_2 filter in the form of (10.42) that achieves $\|z(t)\|^2_{\bar{L}_2} < h$ is given by

$$
A_f = -W^{-1}S, \quad B_f = -W^{-1}Z \quad \text{and} \quad C_f = T.
\tag{10.64}
$$

Similarly to the less conservative condition obtained in Corollary 10.9 for the state-feedback case, a less conservative condition is obtained for the filtering case by applying the vertex-dependent approach. We thus obtain the following result.

Corollary 10.12 *Consider the system of (10.35a, b) and (10.36) with $B_2 = 0$ and $D_{12} = 0$. For a prescribed scalar $\gamma > 0$, there exists a filter of the structure (10.42) that achieves $J_F < 0$, where J_F is given in (10.44), for all nonzero*

$w(t) \in \tilde{L}^2([0, \infty); \mathcal{R}^p)$, *if there exist* $0 < R \in \mathcal{R}^{n \times n}$, $0 < W \in \mathcal{R}^{n \times n}$, $Z \in \mathcal{R}^{n \times m}$, $S \in \mathcal{R}^{n \times n}$, $T \in \mathcal{R}^{r \times n}$, $\tilde{G} \in \mathcal{R}^{n \times n}$ *and* $\tilde{H} \in \mathcal{R}^{n \times n}$, *such that the following LMI condition is satisfied:*

$$
\bar{\Phi} = \begin{bmatrix}
\tilde{G}^T A + A^T \tilde{G} & R - \tilde{G}^T + A^T \tilde{H} & \tilde{G}^T B_1 & \bar{\Phi}_{1,4} \\
R - \tilde{G} + \tilde{H}^T A & -\tilde{H} - \tilde{H}^T & \tilde{H}^T B_1 & 0 \\
B_1^T \tilde{G} & B_1^T \tilde{H} & -\gamma^2 I & \bar{\Phi}_{3,4} \\
WA + ZC_2 + S & 0 & \bar{\Phi}_{3,4}^T & S - S^T \\
C_1 - T & 0 & 0 & T \\
RD & 0 & 0 & 0 \\
WD + ZF & 0 & 0 & 0 \\
ZF & 0 & 0 & 0
\end{bmatrix}
$$

$$
\begin{bmatrix}
C_1^T - T^T & D^T R & D^T W + F^T Z^T & F^T Z^T \\
0 & 0 & 0 & 0 \\
0 & 0 & 0 & 0 \\
T^T & 0 & 0 & 0 \\
-I & 0 & 0 & 0 \\
0 & -R & 0 & 0 \\
0 & 0 & -W & 0 \\
0 & 0 & 0 & -W
\end{bmatrix} < 0, \qquad (10.65)
$$

where
$$
\bar{\Phi}_{1,4} = A^T W + Z^T C_2^T + S^T,
$$
$$
\bar{\Phi}_{3,4} = B_1^T W + D_{21}^T Z^T.
$$

Proof Substituting $\tilde{G} = R$ and $\tilde{H} = \varepsilon I$, $\varepsilon > 0$ in (10.65), applying the Schur's complements formula on the resulting second row- and column blocks of (10.65), letting $\varepsilon \to 0$ and moving the 2nd row and column blocks in the resulting matrix to be the last row and column, the LMI of (10.61) is recovered. To show that (10.65) leads to (10.61), we define

$$
\bar{\Gamma} = \begin{bmatrix}
I & A^T & 0 & 0 & \dots & 0 \\
0 & I & 0 & 0 & \dots & 0 \\
0 & B_1^T & I & 0 & \dots & 0 \\
\vdots & \vdots & \vdots & \vdots & \vdots & 0 \\
0 & 0 & 0 & 0 & \dots & I
\end{bmatrix}.
$$

Multiplying $\bar{\Psi} = \bar{\Gamma} \bar{\Phi} \bar{\Gamma}^T$ and moving the second row and column blocks in $\bar{\Psi}$ to be the last row and column, we find that if (10.65) is satisfied by $R > 0$, $\tilde{G}$, $\tilde{H}$ then for the same R, (10.61) is also satisfied. ∎

In the uncertain case, again two possible approaches can be applied. The first of which is the traditional "quadratic" solution where a single Lyapunov matrix function is applied to all the uncertainty interval. The second approach is obtained by applying the result of Corollary 10.12 to the uncertain case. We thus arrive at the following result for the uncertain polytopic case.

Corollary 10.13 *Consider the system of (10.35a, b) and (10.36) with $B_2 = 0$ and $D_{12} = 0$ where the system matrices lie within the polytope $\bar{\Omega}$ of (10.37). For a prescribed scalar $\gamma > 0$, there exists a filter of the structure (10.42) that achieves $J_F < 0$, where J_F is given in (10.44), for all nonzero $w(t) \in \tilde{L}^2([0, \infty); \mathcal{R}^p)$, if there exist $0 < R^j \in \mathcal{R}^{n \times n}, \forall j = 1, 2, \ldots N, 0 < W \in \mathcal{R}^{n \times n}, Z \in \mathcal{R}^{n \times m}, S \in \mathcal{R}^{n \times n}, T \in \mathcal{R}^{r \times n}, G \in \mathcal{R}^{n \times n}$, and $H \in \mathcal{R}^{n \times n}$, such that the following set of LMIs is satisfied:*

$$
\begin{bmatrix}
G^T A^j + A^{j,T} G & R^j - G^T + A^{j,T} H & G^T B_1^j & \Phi_{1,4}^j \\
R^j - G + H^T A^j & -H - H^T & H^T B_1^j & 0 \\
B_1^{j,T} G & B_1^{j,T} H & -\gamma^2 I & \Phi_{3,4}^j \\
W A^j + Z C_2^j + S & 0 & W B_1^j + Z D_{21}^j & S - S^T \\
C_1^j - T & 0 & 0 & T \\
R^j D & 0 & 0 & 0 \\
W D + Z F & 0 & 0 & 0 \\
Z F & 0 & 0 & 0
\end{bmatrix}
$$

$$
\begin{bmatrix}
C_1^{j,T} - T^T & D^T R^j & D^T W + F^T Z^T & F^T Z^T \\
0 & 0 & 0 & 0 \\
0 & 0 & 0 & 0 \\
T^T & 0 & 0 & 0 \\
-I & 0 & 0 & 0 \\
0 & -R & 0 & 0 \\
0 & 0 & -W & 0 \\
0 & 0 & 0 & -W
\end{bmatrix} < 0, \qquad (10.66)
$$

$\forall j = 1, 2, \ldots N$, *where*

$$
\begin{aligned}
\Phi_{1,4}^j &= A^{j,T} W + Z^T C_2^T + S^T, \\
\Phi_{3,4}^j &= B_1^{j,T} W + D_{21}^{j,T} Z^T.
\end{aligned}
$$

In the latter case, the mean square stabilizing filter in the form of (10.42) that achieves the negative J_F is given by (10.64).

The corresponding robust H_2 filter that ensures an estimation error variance that is less than a prescribed positive scalar $\bar{h}$ over the entire uncertainty polytope is obtained next.

Corollary 10.14 *Consider the system of (10.35a, b) and (10.36) with $B_2 = 0$ and $D_{12} = 0$ where the system matrices lie within the polytope $\bar{\Omega}$ of (10.37). For a prescribed scalar $\bar{h} > 0$, there exists a filter of the structure (10.42) that achieves $\|z(t)\|_{\tilde{L}_2}^2 < \bar{h}$, where $w(t) \in VL\mathcal{R}^p)$ is a standard white-noise signal if there exist $0 < \tilde{R}^j \in \mathcal{R}^{n \times n}, \forall j = 1, 2, \ldots N, 0 < \tilde{W} \in \mathcal{R}^{n \times n}, \tilde{Z} \in \mathcal{R}^{n \times m}, \tilde{S} \in$*

$\mathcal{R}^{n \times n}$, $\tilde{T} \in \mathcal{R}^{r \times n}$, $\tilde{G} \in \mathcal{R}^{n \times n}$, $\tilde{H} \in \mathcal{R}^{n \times n}$ and $\bar{H} \in \mathcal{R}^{r \times r}$ such that the following set of LMIs is satisfied:

$$
\begin{bmatrix}
\tilde{G}^T A^j + A^{j,T} \tilde{G} & \tilde{R}^j - \tilde{G}^T + A^{j,T} \tilde{H} & \tilde{G}^T B_1^j & \Phi_{1,4}^j & D^T \tilde{R}^j & D^T \tilde{W} + F^T \tilde{Z}^T & F^T \tilde{Z}^T \\
\tilde{R}^j - \tilde{G} + \tilde{H}^T A^j & -\tilde{H} - \tilde{H}^T & \tilde{H}^T B_1^j & 0 & 0 & 0 & 0 \\
B_1^{j,T} \tilde{G} & B_1^{j,T} \tilde{H} & -I & \Phi_{3,4}^j & 0 & 0 & 0 \\
\Phi_{1,4}^{jT} & 0 & \Phi_{3,4}^{jT} & \tilde{S} - \tilde{S}^T & 0 & 0 & 0 \\
\tilde{R}^j D & 0 & 0 & 0 & -\tilde{R} & 0 & 0 \\
\tilde{W} D + \tilde{Z} F & 0 & 0 & 0 & 0 & -\tilde{W} & 0 \\
\tilde{Z} F & 0 & 0 & 0 & 0 & 0 & -\tilde{W}
\end{bmatrix} < 0
$$

$$
\begin{bmatrix}
H \ C_1 - T & T \\
* & R^j & 0 \\
* & * & \tilde{W}
\end{bmatrix} > 0, \ and \ Trace\{\bar{H}\} \leq \bar{h}.
$$

(10.67)

$\forall j = 1, 2, \ldots N$, where

$$
\Phi_{1,4}^j = A^{j,T} \tilde{W} + \tilde{Z}^T C_2^T + \tilde{S}^T,
$$
$$
\Phi_{3,4}^j = B_1^{j,T} \tilde{W} + D_{21}^{j,T} \tilde{Z}^T.
$$

In the latter case, the H_2 filter in the form of (10.42) that achieves $\|z(t)\|_{\tilde{L}_2}^2 < \bar{h}$ is given by (10.64).

10.12   Continuous-Time Case: Conclusions

In this part, the theory of robust linear H_∞ state-feedback control and filtering of state-multiplicative noisy systems is developed and extended for continuous-time uncertain systems with multiplicative noise that is encountered in both the dynamic and the measurement matrices in the state space model of the system.

Sufficient conditions are derived for the BRL of uncertain polytopic-type systems by applying a vertex-dependent Lyapunov function. Based on this BRL, the robust state-feedback control and filtering problems are formulated and solved by applying the quadratic solution method and the new approach. In both, the robust state-feedback control case and the filtering case, the two decision variables that stem from the vertex-dependent approach (i.e., G and H) are independent one of the another and are not assigned any special structure. This is strikingly different from the discrete-time counterpart case, where special structures must be assigned to the latter decision variables in order to obtain an LMI solution. Extension of the theory that is presented here to cases where the various multiplicative white-noise sequences are correlated, can be readily accounted for in LMI conditions for both the nominal and the uncertain cases. This would lead, however, to cumbersome matrix conditions.

References

1. Gershon, E., Shaked, U., Yaesh, I.: H_∞ Control and Estimation of State-Multiplicative Linear Systems. Lecture Notes in Control and Information Sciences, LNCIS, vol. 318. Springer (2005)
2. Willems, J.L., Willems, J.C.: Feedback stabilizability for stochastic systems with state and control dependent noise. Automatica **12**, 277–283 (1976)
3. Dragan, V., Morozan, T., Halanay, A.: Optimal stabilizing compensator for linear systems with state-dependent noise. Stoch. Anal. Appl. **10**, 557–572 (1992)
4. Dragan, V., Halanay, A., Stoica, A.: A small gain theorem for linear stochastic system. Syst. Control Lett. **30**, 243–251 (1997)
5. Hinriechsen, D., Pritchard, A.J.: Stochasic H_∞. SIAM J. Control Optim. **36**(5), 1504–1538 (1998)
6. El Bouhtouri, A., Hinriechsen, D., Pritchard, A.J.: H_∞ type control for discrete-time stochasic systems. Int. J. Robust Nonlinear Control **9**, 923–948 (1999)
7. Dragan, V., Stoica, A.: A γ Attenuation Problem for Discrete-Time Time-Varying Stochastic Systems with Multiplicative Noise. Reprint Series of the Institute of Mathematics of the Romanian Academy, vol. 10 (1998)
8. Costa, O.L.V., Kubrusly, C.S.: State-feedback H_∞-control for discrete-time infinite-dimensional stochastic bilinear systems. J. Math. Syst. Estim. Control **6**, 1–32 (1996)
9. Ugrinovsky, V.A.: Robust H_∞ control in the presence of of stochastic uncertainty. Int. J. Control **71**, 219–237 (1998)
10. Gershon, E., Shaked, U., Yaesh, I.: H_∞ control and filtering of discrete-time stochastic systems with multiplicative noise. Automatica **37**, 409–417 (2001)
11. Gershon, E., Shaked, U., Yaesh, I.: Robust H_∞ filtering of stationary discrete-time linear systems with stochastic parameter uncertainties. Syst. Control Lett. **45**, 257–269 (2002)
12. Gershon, E., Limebeer, D.J.N., Shaked, U., Yaesh, I.: Robust H_∞ filtering of stationary continuous-time linear systems with stochastic uncertainties. IEEE Trans. Autom. Control **46**(11), 1788–1793 (2001)
13. Verriest, E.I., Florchinger, P.: Stability of stochastic systems with uncertain time delays. Syst. Control Lett. **24**(1), 41–47 (1995)
14. Yue, D., Tian, E., Zhang, Y.: Stability analysis of discrete systems with stochastic delay and its applications. Int. J. Innov. Comput. Inf. Control **5**(8), 2391–2403 (2009)
15. Gershon, E., Shaked, U.: Advanced Topics in Control and Estimation of State-Multiplicative Noisy Systems. Lecture Notes in Control and Information Sciences, LNCIS, vol. 439. Springer (2013)
16. Xie, L., Fu, M., deSouza, C.E.: H_∞ control and quadratic stabilization of systems with parameter uncertainty via output feedback. IEEE Trans. Autom. Control **37**, 1253–1246 (1992)
17. Boyd, S., El Ghaoui, L., Feron, E., Balakrishnan, V.: Linear Matrix Inequality in Systems and Control Theory. SIAM Frontier Series (1994)
18. de Oliveira, M.C., Skelton, R.E.: Stability test for constrained linear systems. In: Reza Moheimani, S.O. (ed.) Perspectives in Robust Control. Lecture Notes in Control and Information Sciences 268. Springer, London (2001)

Chapter 11
Robust Output-Feedback Control of Stochastic Discrete-Time Systems

Abstract Linear discrete-time systems with stochastic and deterministic polytopic-type uncertainties in their state-space model are considered. A dynamic output-feedback controller is obtained via a new approach, which allows a derivation of a controller in spite of parameter uncertainty. In the proposed approach, the system is described via a difference equation and an augmented system is then used to obtain the output-feedback controller parameters. The controller is obtained without assuming a specific structure to the quadratic Lyapunov function. It is the first time that an output-feedback controller is obtained for robust state-multiplicative systems. The controller minimizes the stochastic l_2-gain of the closed-loop system, where the cost function is defined to be the expected value of the standard H_∞ performance index with respect to the stochastic uncertainty. An example is given that demonstrates the merit of the theory.

11.1 Introduction

The field of stochastic H_∞ control and estimation of state-multiplicative systems has been developed following the advance in the stochastic H_2 control theory that was developed in the early 60s and 70s [1] It has emerged, as a natural extension, in parallel with the progress made to the field of deterministic H_∞ control theory, drawing on the advent of the latter field to cope with MIMO control problems [2] and with the robust control and estimation problems [3]. This field is recognized now as a central subfield within the theory of stochastic control [4–6] (see also [1] for a comprehensive review). Numerous solution methods have been applied to various control and estimation patterns (see [1] and the references therein) and over the last decade, the latter field have been expanded to include stochastic systems with time delay of various kinds (i.e., constant time delay, slow and fast varying delay) in both the nominal and the uncertain cases (see [7], see also [8, 9] for the deterministic case).

The problem of H_∞ dynamic output feedback has always been a central issue in modern control theory. In the stochastic discrete-time case, this problem was tackled by various research groups [10–13]. The solution in [10] includes the finite- and the

© Springer Nature Switzerland AG 2019
E. Gershon and U. Shaked, *Advances in H$_\infty$ Control Theory*,
Lecture Notes in Control and Information Sciences 481,
https://doi.org/10.1007/978-3-030-16008-1_11

infinite-time horizon problems without transients. One drawback of [10] is the fact that in the infinite-time horizon case, an infinite number of Linear Matrix Inequality (LMI) sets should be solved. We also note that the fact that in [10], the measurement coupling matrix cannot accommodate any uncertainty is a practical restriction, for example, in cases where the measurements include state derivatives (e.g., acceleration control of an aircraft or missile). The treatment of [11] includes the derivation of the stochastic Bounded Real Lemma (BRL) and concerns only the stationary case where two coupled nonlinear inequalities were obtained.

In [12], a solution to the dynamic output-feedback problem was obtained using the adjoint system. A modified Riccati recursion has been obtained there which guarantees a given H_∞ estimation level, while minimizing an upper-bound on the covariance of the estimation error. While the recursion solution obtained in [12] for the finite-horizon case is rather easy to implement, the theoretical justification for using the adjoint in stochastic systems, particularly in the H_∞ control field, is somewhat debatable.

In [13], the solution of the output-feedback control problem has been obtained by transforming the problem to one of filtering. In the finite-horizon case, the solution has been obtained there via DLMIs [Difference Linear Matrix Inequalities] whereas in the stationary case the solution has been obtained by assigning a special structure to the Lyapunov matrix solution, thereby imposing an overdesign on the solution. The robust case could not be treated in [13] owing to the fact that the LMI obtained there is not affine in the system matrices. The problem of finding an optimal dynamic output feedback for state-multiplicative systems has thus been constrained in the past by the solution method used and conservative solutions have then been obtained.

In the present chapter, we solve the problem of discrete-time dynamic output-feedback control via a new approach that enables us to solve the problem of measurement control of uncertain systems without the restrictions that have been indicated in the above mentioned works. Applying a linear fractional approach, we apply the method of [14] and transform the problem of finding a dynamic controller to one of finding a state-feedback controller gain matrix. We are thus able to treat two control design cases which are fundamental in control engineering, on one hand, and on the other hand, we are also able to cope with the robust version of the latter two control patterns, without preassigning structure to the relevant Lyapunov matrix function solution. In the uncertain case, we bring a vertex-dependent solution (in contrast to the traditional "quadratic" solution).

The chapter is organized as follows: starting with the observable canonical system representation, we first introduce in Sect. 11.2 the linear fractional intermediate description of the system. In Sect. 11.3, we consider the stabilization and the l_2-gain of the nominal closed-loop system. Based on the result of Sect. 11.3, the solution of the robust dynamic output-feedback control problem is given in Sect. 11.4 for the case where the system encounters polytopic-type parameter uncertainty. Two different control solution methods are applied, which are the quadratic- and the vertex-dependent approaches. In Sect. 11.5, two examples are given that demonstrate the applicability and tractability of our method to uncertain systems.

11.2 Problem Formulation

We consider the r-inputs m-outputs system that is described by the following difference equation (given in the observability canonical form [15]):

$$y_{k+l} + \sum_{i=0}^{l-1}(\bar{D}_i + F_i\nu_k)y_{k+i} = \sum_{i=0}^{l-1}(\bar{N}_{l-1-i} + F_{2l-1-i}\nu_k)u_{k+i}, \qquad (11.1)$$

where $y_k \in \mathcal{R}^m$ is the system output, $u_k \in \mathcal{R}^r$ is the control input, $\{\nu_k\}$ is a standard zero-mean real scalar white-noise sequence with $E\{\nu_k\nu_j\} = \delta_{kj}$ where $\bar{D}_i \in R^{m \times m}$, $\bar{N}_i \in R^{m \times r}$, $F_i \in R^{m \times r}$, $i = 1, \ldots, 2l - 1$.

We assume that the system parameters are not precisely known and that they reside in the following polytope:

$$\Omega = Co\{\Omega_1, \Omega_2, \ldots, \Omega_{\bar{L}}\}, \qquad (11.2)$$

where

$$\Omega_i = \begin{bmatrix} \bar{N}_{0,\,i} & \cdots & \bar{N}_{l-1,\,i} \\ \bar{D}_{0,\,i} & \cdots & \bar{D}_{l\ \ 1,\,i} \\ F_{0,\,i} & \cdots & F_{2l-2,\,i} \end{bmatrix}, \quad i = 1, \ldots, \bar{L}, \qquad (11.3)$$

and where $\bar{L}$ is the number of vertices.

We seek an output-feedback dynamical controller of the type

$$u_{k+l-1} + \sum_{i=0}^{l-2}\bar{A}_i u_{k+i} = \sum_{i=0}^{l-1}\bar{B}_{l-1-i}y_{k+i} \qquad (11.4)$$

that over the polytope, Ω stabilizes the system exponentially in the mean square sense and minimizes the induced l_2-norm of the resulting closed-loop.

The above system is described (for clarity of representation we take $F_i = 0$, $i = l, \ldots, 2l - 1$) in the block diagram of Fig. 11.1 where δ denotes the one unit time-shift operator.

We represent the above system by the augmented observability canonical form [15] as follows: we define the following augmented state vector in $\Re^n$, $n \overset{\Delta}{=} l(m + r) - r$,

$$\xi_k = \begin{bmatrix} y_{k+l-1}^T & y_{k+l-2}^T & \cdots & y_k^T & u_{k+l-2}^T & \cdots & u_k^T \end{bmatrix}^T, \qquad (11.5)$$

and obtain

$$\xi_{k+1} = \tilde{A}\xi_k + \nu_k\tilde{F}\xi_k + (\tilde{B} + \nu_k\hat{B})K\xi_k, \qquad (11.6)$$

where

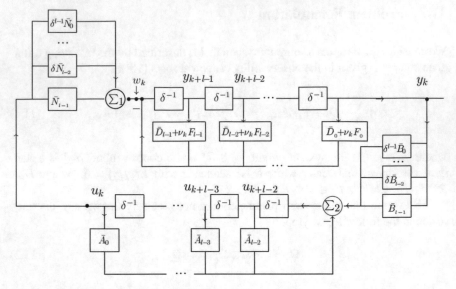

Fig. 11.1 The closed-loop fractional description

$$
\tilde{A} =
\begin{bmatrix}
-\bar{D}_{l-1} & \ldots & & -\bar{D}_0 & \bar{N}_1 & \ldots & & \bar{N}_{l-1} \\
I_m & 0 & . & . & 0 & 0 & \ldots & & 0 \\
0 & I_m & . & . & 0 & 0 & \ldots & & 0 \\
. & & . & . & . & . & . & . & . \\
0 & & . & I_m & 0 & 0 & \ldots & & 0 \\
0 & & . & . & 0 & 0 & \ldots & & 0 \\
0 & & . & . & 0 & I_r & 0 & . & . & 0 \\
0 & & . & . & 0 & 0 & I_r & . & . & 0 \\
. & & . & & . & . & . & . & . & . \\
0 & & . & . & 0 & 0 & . & . & I_r & 0
\end{bmatrix},
$$

$$
\tilde{F} =
\begin{bmatrix}
-F_{l-1} & \ldots & & -F_0 & F_{l+1} & \ldots & & F_{2l-1} \\
0 & 0 & . & . & 0 & 0 & \ldots & & 0 \\
0 & 0 & . & . & 0 & 0 & \ldots & & 0 \\
. & & . & . & . & . & . & . & . \\
0 & & . & . & 0 & 0 & 0 & \ldots & & 0 \\
0 & & . & . & 0 & 0 & \ldots & & 0 \\
0 & & . & . & 0 & 0 & 0 & . & . & 0 \\
0 & & . & . & 0 & 0 & 0 & . & . & 0 \\
. & & . & . & . & . & . & . & . & . \\
0 & & . & . & 0 & 0 & . & 0 & 0
\end{bmatrix},
$$

$$\hat{B} = \begin{bmatrix} \bar{N}_0^T & 0 & \dots & 0 & I_r & 0 & \dots & 0 \end{bmatrix}^T, \quad \hat{B} = \begin{bmatrix} F_l^T & 0 & \dots & 0 & 0 \end{bmatrix}^T$$

and

$$K = [K_1 \, K_2, \dots, K_l \, K_{l+1}, \dots, K_{2l-1}] = \begin{bmatrix} \bar{B}_0 \bar{B}_1 \dots \bar{B}_{l-1} -\bar{A}_{l-2} -\bar{A}_{l-3} \dots -\bar{A}_0 \end{bmatrix}.$$
(11.7)

We note that the above state-space model is non-minimal, for both the nominal and the uncertain cases. It is already observable but may be uncontrollable due to the additional roots added to the augmented system. A "state feedback" from the state vector ξ_k in this model to the "control input" u_{k+l-1} will produce the components A_i and B_i of the controller (11.4) that relates y_k with u_k.

In the stochastic H_∞ control setting, we obtain the following state-space representation:

$$\begin{aligned} \xi_{k+1} &= \tilde{A}\xi_k + \tilde{B}_w w_k + \tilde{B}K\xi_k + \nu_k(\tilde{F} + \hat{B}K)\xi_k, \\ z_k &= \tilde{C}_1\xi_k + \nu_k \tilde{W}\xi_k + \tilde{D}_{11}w_k, \end{aligned}$$
(11.8)

where $\tilde{W}$ and $\hat{B}$ depend on the way by which ν_k affects $\bar{N}_i$, $i = 1, 2, N$. In our model, the disturbance acts on y_{k+l} with

$$\tilde{B}_w = [I_m \, 0 \, 0 \, \dots 0]^T \text{ and } \tilde{C}_1 = [0 \dots 0 \, \bar{N}_1 \dots \bar{N}_{l-1}] + \bar{N}_0 K.$$
(11.9)

The matrix $\tilde{D}_{11}$ depends on the control problem to be solved. It describes the way that the disturbance affects the objective function.

In the following, we describe two problems in which the disturbance effect is considered differently.

Problem 11.1 Sensitivity Minimization

The sensitivity matrix function is the operator that relates the input $\{w_k\}$ in Fig. 11.1 with the signal that is measured at the bullet that immediately proceeds the disturbance entry point. The minimization of the H_∞ norm of the sensitivity function is thus achieved by taking $\tilde{D}_{11} = 0$ in (11.8b).

Problem 11.2 Complementary Sensitivity Minimization

The complementary sensitivity is the operator that relates the input $\{w_k\}$ in Fig. 11.1 to the signal that precedes the entry point. In our case, we seek to reduce the l_2-gain of the latter transference. We thus require that $\tilde{D}_{11} = 0$. We note that in our system the measurement signal is not contaminated by sensor noise. However, the complementary sensitivity function is also the transference (noting that the phase is reversed) from the traditional entry point of the measurement noise to the output measurement signal. By minimizing the induced l_2 norm of this function one also minimizes the effect of measurement noise if it exists.

11.3   The H_∞ Controller for Nominal Systems

In order to solve the last two minimization problems for the augmented system, we first bring the stochastic BRL result that was obtained in (see [1], p. 104). This result refers to the following nth-order system

$$
\begin{aligned}
x_{k+1} &= (A + F\nu_k)x_k + B_1 w_k, \quad x_i = 0, \ i \le 0, \\
z_k &= (C_1 + W\nu_k)x_k + D_{11}w_k,
\end{aligned}
\tag{11.10}
$$

with the index of performance:

$$
J_E \overset{\Delta}{=} \|z_k\|_{\tilde{l}_2}^2 - \gamma^2 \|w_k\|_{\tilde{l}_2}^2,
\tag{11.11}
$$

where $w_k \in R^p$, $z_k \in R^r$. The following has been obtained in [1, 10].

Lemma 11.1 *Consider the system (11.10a, b) and the above J_E. The system is exponentially stable in the mean square sense and for a prescribed scalar $\gamma > 0$,, the requirement of $J_E < 0$ is achieved for all nonzero $w_k \in \tilde{l}_{\mathcal{F}_k}^2([0, \infty); \mathcal{R}^p)$, iff there exists $n \times n$ matrix $P > 0$, that satisfies the following inequality:*

$$
\Gamma_P \overset{\Delta}{=}
\begin{bmatrix}
-P & 0 & PA^T & PF^T & PC_1^T & PW^T \\
* & -\gamma^2 I_p & B_1^T & 0 & D_{11}^T & 0 \\
* & * & -P & 0 & 0 & 0 \\
* & * & * & -P & 0 & 0 \\
* & * & * & * & -I_r & 0 \\
* & * & * & * & * & -I_r
\end{bmatrix} < 0.
\tag{11.12}
$$

Based on Lemma 11.1, the "state-feedback controller" can be readily derived for the augmented system by applying the latter result to the system of (11.8a, b). We thus obtain the following result.

Theorem 11.1 *Consider the augmented system (11.8a, b) and (11.11). For a prescribed scalar $\gamma > 0$, there exists a state-feedback gain that achieves negative J_E for all nonzero $w_k \in \tilde{l}_{\mathcal{F}_k}^2([0, \infty); \mathcal{R}^p)$, if there exist $n_1 \times n_1$ matrix $P > 0$, where $n_1 = l(m + r) - r$, and a matrix Y that satisfy the following LMI condition:*

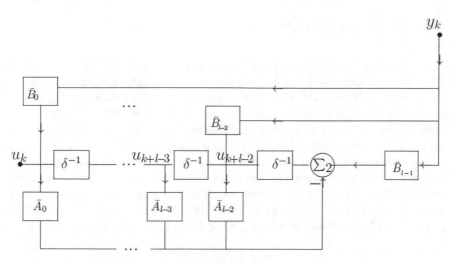

Fig. 11.2 The structure of the controller

$$
\begin{bmatrix}
-P & 0 & P\tilde{A}^T + Y^T\tilde{B}^T & P\tilde{F}^T + Y^T\hat{B}^T & P\tilde{C}_1^T & P\tilde{W}^T \\
* & -\gamma^2 I_p & \tilde{B}_w^T & 0 & \tilde{D}_{11}^T & 0 \\
* & * & -P & 0 & 0 & 0 \\
* & * & * & -P & 0 & 0 \\
* & * & * & * & -I_r & 0 \\
* & * & * & * & * & -I_r
\end{bmatrix} < 0. \qquad (11.13)
$$

In the latter case, the "state-feedback gain" is given by

$$
K = \begin{bmatrix} \bar{B}_0 \ \bar{B}_1 \dots \bar{B}_{l-1} -\bar{A}_{l-2} -\bar{A}_{l-3} \dots -\bar{A}_0 \end{bmatrix} = YP^{-1}, \qquad (11.14)
$$

from where the system matrices of the actual output-feedback controller of Fig. 11.2 are obtained.

Proof We replace A, B_1, C_1, W, and D_{11} of (11.10a, b) by $\tilde{A} + \hat{B}K$, $\tilde{B}_w$ and $\tilde{C}_1$, $\tilde{W}$, $\tilde{D}_{11}$ of (11.8a, b), respectively, and we denote KP by Y. Substituting in (12), the LMI of (13) is readily obtained. ∎

Remark 11.1 To the best of our knowledge, the result of Theorem 11.1 is the first to produce a dynamic output-feedback controller for the stochastic control problem **without** assuming special structure for the Lyapunov function or solving nonlinear inequalities. This is true for the case where the system parameters are all known and as will be shown below, it is certainly the only solution that derives an optimal dynamic output-feedback controller in the case where the stochastic system encounters a polytopic-type uncertainty.

11.4   The H_∞ Controller for Uncertain Systems

The main advantage of our approach is the ability it provides to obtain a solution to
the output-feedback control problem in the case of polytopic uncertain parameters.
Two possible solution methods can be considered. The first produces a "quadratic"
solution where a single Lyapunov function is considered over the whole uncertainty
interval. This is readily obtained by applying the result of Theorem 11.1 to all the
vertices of the uncertainty polytope applying the same matrices P and Y. The second
solution method is based on the fact that a less conservative solution may be obtained
by considering vertex-dependent Lyapunov functions over the whole uncertainty
polytope. In order to apply the latter approach, we first present the vertex-dependent
solution of the BRL of Lemma 11.1. We consider the system of (11.10a, b) and the
LMI of (11.12). Applying Schur's complements to the latter LMI, we obtain

$$\Psi + \Phi P \Phi^T \le 0,$$

where

$$\Psi \triangleq \begin{bmatrix} -\gamma^2 I_p & B_1^T & 0 & 0 & 0 \\ * & -P & 0 & 0 & 0 \\ * & * & -P & 0 & 0 \\ * & * & * & -I_r & 0 \\ * & * & * & * & -I_r \end{bmatrix}, \quad \Phi \triangleq \begin{bmatrix} 0 \\ A \\ F \\ C_1 \\ W \end{bmatrix}. \tag{11.15}$$

The structure of (11.15) can be used in order to reduce the conservatism entailed by
applying the Lyapunov function uniformly over the uncertainty polytope. However,
the standard application of the latter method [16] is not readily applicable to the
stochastic case where taking the adjoint of the system, instead of the original one, is
theoretically unjustified. We therefore bring below a vertex-dependent method that
is more suitable and readily applicable to the uncertain stochastic case.

We start with (11.15) and obtain the following lemma.

Lemma 11.2 *Inequality (11.15a) is satisfied iff there exist matrices:* $0 < P \in \mathcal{R}^{n \times n}$,
$G \in \mathcal{R}^{n \times (2n+p+r)}$ *and* $H \in \mathcal{R}^{n \times n}$ *that satisfy the following inequality:*

$$\Omega \triangleq \begin{bmatrix} \Psi + G^T \Phi^T + \Phi G & -G^T + \Phi H \\ * & -H - H^T + P \end{bmatrix} < 0. \tag{11.16}$$

Proof Substituting $G = 0$ and $H = P$ in (11.16), inequality (11.15a) is obtained. To
show that (11.16) leads to (11.15a) we consider

$$\begin{bmatrix} I & \Phi \\ 0 & I \end{bmatrix} \Omega \begin{bmatrix} I & 0 \\ \phi^T & I \end{bmatrix} = \begin{bmatrix} \Psi_{\Phi(1,1)} & \Psi_{\Phi(1,2)} \\ * & \Psi_{\Phi(2,2)} \end{bmatrix},$$

where

$$\Psi_{\Phi(1,1)} = \Psi + \Phi P \Phi^T, \;\; \Psi_{\Phi(1,2)} = -G^T - \Phi H^T + \Phi P, \;\; \Psi_{\Phi(2,2)} = -H - H^T + P.$$

Inequality (11.15a) thus follows from the fact that it is the left side $(1,1)$ matrix block in the latter product.   ∎

In the uncertain case, we assume that the system parameters encounter polytopic-type uncertainty with $\bar{L}$ vertices. Choosing then

$$V^{(j)}(x, t) = x^T(t)P^{(j)}x(t), \;\; j = 1, 2, \dots \bar{L},$$

we consider

$$\Psi^{(j)} + \Phi^{(j)}P^{(j)}\Phi^{(j)T} \le 0, \;\; \forall j = 1, 2, \dots \bar{L}, \tag{11.17}$$

where $\Psi^{(j)}$ and $\Phi^{(j)}$ are obtained from (11.15b, c) by assigning for each vertex j the appropriate matrices. We readily find the following vertex-dependent condition for the stochastic uncertain discrete-time case.

Corollary 11.1 *Inequality (11.17) is satisfied iff there exist matrices:* $0 < P^{(j)} \in \mathcal{R}^{n \times n}, \; \forall j = 1, 2, \dots \bar{L}, \; G \in \mathcal{R}^{n \times (2n+p+r)}$ *and* $H \in \mathcal{R}^{n \times n}$ *that satisfy the following inequality*

$$\Omega^{(j)} \triangleq \begin{bmatrix} \Psi^{(j)} + G^T\Phi^{(j)T} + \Phi^{(j)}G & * \\ -G + H^T\Phi^{(j)T} & -H - H^I + P^{(j)} \end{bmatrix} < 0, \;\; \forall j = 1, 2, \dots \bar{L}. \tag{11.18}$$

Applying the above result, the solution of the vertex-dependent robust BRL problem can be readily derived for the augmented system. Considering the augmented system (11.8a, b) where $K = 0$ and (11.11), where the system matrices components lie within the polytope of (11.2), we obtain the following condition:

$$\hat{\Omega}^{(j)} \triangleq \begin{bmatrix} \hat{\Psi}^{(j)} + G^T\Phi^{(j)T} + \Phi^{(j)}G & -G^T + \Phi^{(j)}H \\ * & -H - H^T + P^{(j)} \end{bmatrix} < 0,$$

$$\hat{\psi}^{(j)} \triangleq \begin{bmatrix} -\gamma^2 I_p & \tilde{B}_w^T & 0 & \tilde{D}_{11}^T & 0 \\ ** & -P^{(j)} & 0 & 0 & 0 \\ ** & * & -P^{(j)} & 0 & 0 \\ ** & * & * & -I_r & 0 \\ ** & * & * & * & -I_r \end{bmatrix}, \tag{11.19}$$

$$\Phi^{(j),T} = [0 \;\; \tilde{A}^{(j)T} \;\; \tilde{F}^T \;\; \tilde{C}_1^{(j)T} \;\; \tilde{W}^T], \;\; \forall j = 1, 2, \dots \bar{L}.$$

Taking $G = [0 \;\; \alpha H \;\; [0 \; 0 \; 0]]$ where H is a $n \times n$ matrix, the following result is obtained.

Theorem 11.2 *Consider the augmented system (11.8a, b) and (11.11) where the original system matrices components* $\tilde{N}_j$, $\bar{D}_j$, $j = 1, \dots l - 1$, *lie within the polytope of (11.2). For a prescribed scalar* $\gamma > 0$, *and positive tuning scalar* $\alpha > 0$, *there exists a dynamic output-feedback controller that achieves negative* J_E *for all nonzero* $w \in \tilde{l}^2_{\mathcal{F}_k}([0, \infty); \mathcal{R}^p)$, *if there exist* $n_1 \times n_1$ *matrices* $P^{(j)} > 0$, $j = 1, 2, \dots \bar{L}$, *where* $n_1 = l(m + r) - r$, *a matrix* H *and a matrix* Y *that satisfy the following set of LMIs* $\Gamma_{OF}^{(j)} \triangleq$

$$
\begin{bmatrix}
-\gamma^2 I_p & \tilde{B}_w^T & 0 & \tilde{D}_{11}^T & 0 & 0 \\
* & \Gamma_{OF}^{(j)}(2,2) & \Gamma_{OF}^{(j)}(2,3) & \alpha H^T \tilde{C}_1^{(j)T} & \alpha H^T \tilde{W}^T & -\alpha H^T + \tilde{A}^{(j)} H + \tilde{B}^{(j)} Y \\
* & * & -P^{(j)} & 0 & 0 & \tilde{F}^{(j)} H + \hat{B}^{(j)} Y \\
* & * & * & -I_r & 0 & \tilde{C}_1^{(j)} H \\
* & * & * & * & -I_r & \tilde{W} H \\
* & * & * & * & * & -H - H^T + P^{(j)}
\end{bmatrix} < 0,
$$

$$(11.20)$$

$j = 1, 2, \dots \bar{L}$, *where* $\Gamma_{OF}^{(j)}(2,2) = -P^{(j)} + \alpha \tilde{A}^{(j)} H + \alpha \tilde{B}^{(j)} Y + \alpha H^T \tilde{A}^{(j)T} + \alpha Y^T \tilde{B}^{(j)T}$ *and* $\Gamma_{OF}^{(j)}(2,3) = \alpha H^T \tilde{F}^{(j)T} + Y^T \hat{B}^{(j)T}$.

In the latter case, the "state-feedback" gain is given by

$$
K = \begin{bmatrix} \bar{B}_0 \bar{B}_1 \dots \bar{B}_{l-1} -\bar{A}_{l-2} -\bar{A}_{l-3} \dots -\bar{A}_0 \end{bmatrix} = YH^{-1}, \qquad (11.21)
$$

from where the system matrices of the actual dynamic output-feedback controller of Fig. 11.2 are obtained.

Proof We replace $\tilde{A}^{(j)}$ and $\tilde{F}^{(j)}$ of (11.19c) by $\tilde{A}^{(j)} + \tilde{B}^{(j)} K$, $\tilde{F}^{(j)} + \hat{B}^{(j)} K$, respectively and we denote KH by Y. Substituting in (11.19), the LMI of (11.20) is readily obtained. We note that H in (11.20) is required to be nonsingular. This, however, follows from the fact that $\Gamma_{OF}^{(j)}(6, 6)$ in (11.20) must be negative definite in order for (11.20) to be feasible. ∎

11.5 Examples

Example 11.1 **Output-Feedback Control**

We consider the following uncertain system:

$$
y_{k+2} - 1.6 y_{k+1} - (1.2 + a) y_k - 0.16 \nu y_{k+1} + 0.08 \nu y_k = u_{k+1} - 1.2 u_k, \quad (11.22)
$$

where a is an uncertain parameter that lies in the interval $a \in [-0.1 \ 0.1]$. It is required to find a dynamic output- feedback controller that will stabilize the system in the mean square sense and that minimizing the induced l_2 norm of the system closed-loop complementary sensitivity.

The system is readily transformed to the augmented model of (11.8a, b):

$$\xi_{k+1} = \begin{bmatrix} 1.6 & 1.2+a & -1.2 \\ 1 & 0 & 0 \\ 0 & 0 & 0 \end{bmatrix} \xi_k + \begin{bmatrix} 1 \\ 0 \\ 0 \end{bmatrix} w_k + \begin{bmatrix} 1 \\ 0 \\ 1 \end{bmatrix} K\xi_k + \begin{bmatrix} 0.16 & -0.08 & 0 \\ 0 & 0 & 0 \\ 0 & 0 & 0 \end{bmatrix} \xi_k \nu_k,$$

where $z_k = [0 \ \ 0 \ -1.2]\xi_k + K\xi_k$ describes the signal that is fedback to the summation Σ_1 in Fig. 11.1 due to the disturbance w that acts on the bullet that proceeds this summation and where, because $N_0 = 1$ in the configuration of (11.9b), $N_0 K = K$ is the "state-feedback gain" in the augmented system that produces the dynamic controller parameters according to (11.14). Solving for the nominal case (i.e., $a = 0$) by applying the result of Theorem 11.1, a near minimum attenuation level of $\gamma = 3.712$ was obtained for the controller whose transfer function is $G_c(z) = -\frac{2.88z+1.641}{z-1.641}$.

In the uncertain case, solving the problem by applying the "quadratic" result, an attenuation level of $\gamma = 5.77$ was obtained. The transfer function of the resulting controller is $G_c(z) = -\frac{2.8z+1.661}{z+1.661}$. Applying next Theorem 11.2 which yields the vertex-dependent Lyapunov solution, an attenuation level of $\gamma = 5.7318$ was obtained for $\alpha = 0.0025$ with the controller $G_c(z) = -\frac{2.8z+1.657}{z+1.659}$.

Example 11.2 **Robot Manipulator System**

We consider an example that is given in [17] that concerns the control of one joint of a real-life Space Robot Manipulator (SRM) with the following state-space description:

$$\dot{x}(t) = A_p x(t) + B_p u(t) = \begin{bmatrix} 0 & 1 & 0 & 0 \\ 0 & 0 & a_1 & 0 \\ 0 & 0 & 0 & 1 \\ 0 & -a_2 & -a_3 & -a_4 \end{bmatrix} x(t) + \begin{bmatrix} 0 \\ 1 \\ 0 \\ -1 \end{bmatrix} \delta u(t), \qquad (11.23)$$

where

$$a_1 = \frac{c}{N^2 I_m}, \quad a_2 = \frac{\beta}{I_{son}}, \quad a_3 = c(\frac{1}{N^2 I_m} + \frac{1}{I_{son}}), \quad a_4 = a_2, \text{ and } \delta = \frac{K_t}{N I_m},$$

and where the plant parameters appear in Table 11.1.

Table 11.1 Plant parameters

Parameter	Sym.	Value
Gearbox ratio	N	−260.6
Motor torque constant	K_t	0.6
The damping coefficient	β	0.4
Inertia of the input axis	I_m	0.0011
Inertia of the output system	I_{son}	400
Spring constant	c	130,000
Motor current	i_c	variable

Since

$$(sI - A_p)^{-1} = [s(s^3 + a_4 s^2 + a_3 s + a_1 a_2)]^{-1} *$$

$$\begin{bmatrix} s^3 + a_4 s^2 + a_3 s + a_1 a_2 & s^2 + a_4 s + a_3 & a_1 s + a_1 a_4 & a_1 \\ 0 & s^3 + a_4 s^2 + a_3 s & a_1 s^2 + a_1 a_4 s & a_1 s \\ 0 & -a_2 s & s^3 + a_4 s^2 & s^2 \\ 0 & -a_2 s^2 & -a_3 s^2 - a_1 a_2 s & s^3 \end{bmatrix}, \quad (11.24)$$

we readily find that the transference from u to y_1 and y_2 is given by the following transfer function matrix.

$$G(s) = \begin{bmatrix} \frac{c}{I_{son}} \\ N s^3 + \frac{N\beta}{I_{son}} s^2 + \frac{cN}{I_{son}} s \end{bmatrix} \frac{\delta}{s(s^3 + a_2 s^2 + (a_1 + \frac{c}{I_{son}})s + a_1 a_2)}. \quad (11.25)$$

Applying Euler's forward discretization, using sampling interval of $T = 0.02\,\text{s}$, one should replace s in the above by $\frac{z-1}{T}$. We then obtain

$$G(z) = \begin{bmatrix} G_1(z) \\ G_2(z) \end{bmatrix} = \quad (11.26)$$

$$\begin{bmatrix} \frac{c}{I_{son}} \\ \frac{N}{T^3}(z^3 - 3z^2 + 3z - 1) + \frac{N\beta}{I_{son}T^2}(z^2 - 2z + 1) + \frac{cN}{TI_{son}}(z - 1) \end{bmatrix} \frac{\delta}{\Delta(z)},$$

where

$$\Delta(z) = \frac{z-1}{T} \Big[\frac{1}{T^3}(z^3 - 3z^2 + 3z - 1) + \frac{a_2}{T^2}(z^2 - 2z + 1) + \frac{a_1 + \frac{c}{I_{son}}}{T}(z - 1) + a_1 a_2 \Big].$$

The objective is to best attenuate the disturbance that acts on y_1. We thus seek first a large scalar output-feedback gain K_2 that stabilizes the loop from y_2 to the input u. Applying discrete-time root loci it is found that any $K_2 > 400$ will stabilize that loop. For $K_2 = 1000$, we find that the resulting transference from u to y_1 can be described by the ratio of the numerators of G_1 and G_2, namely,

$$G_k(z) = \frac{\frac{c}{I_{son}}}{\frac{N}{T^3}(z^3 - 3z^2 + 3z - 1) + \frac{N\beta}{I_{son}T^2}(z^2 - 2z + 1) + \frac{cN}{TI_{son}}(z - 1)}.$$

We thus obtain the following corresponding augmented discrete-time description.

$$y_{k+3} + \bar{D}_2 y_{k+2} + \bar{D}_1 y_{K+1} + \bar{D}_0 y_k = \bar{N}_2 u_k, \tag{11.27}$$

where $\bar{N}_2 = \frac{T^3 c}{NI_{son}}$, $\bar{D}_2 = \frac{\beta T}{I_{son}} - 3$, $\bar{D}_1 = 3 + \frac{T}{I_{son}}(cT - 2\beta)$, $\bar{D}_0 = -1 + \frac{\beta T}{I_{son}} - \frac{cT^2}{I_{son}}$.

The case where the parameters c and β are both deterministically unknown can be treated by applying the method of [14] in order to minimize the effect of the disturbance that acts on the shaft angular position ($z(t) = y_1(t)$), namely to minimize the H_∞-norm of the sensitivity transfer function matrix $S(t)$. Using the latter method we define

$$\tilde{A}^{(j)} = \begin{bmatrix} -\bar{D}_2^{(j)} & -\bar{D}_1^{(j)} & -\bar{D}_0^{(j)} & 0 & \bar{N}_2^{(j)} \\ 1 & 0 & 0 & 0 & 0 \\ 0 & 1 & 0 & 0 & 0 \\ 0 & 0 & 0 & 0 & 0 \\ 0 & 0 & 0 & 1 & 0 \end{bmatrix}, B_u = \begin{bmatrix} 0 \\ 0 \\ 0 \\ 1 \\ 0 \end{bmatrix}, B_w = \begin{bmatrix} 1 \\ 0 \\ 0 \\ 0 \\ 0 \end{bmatrix}, \tag{11.28}$$

$$\tilde{C}_1^{(j)} = \begin{bmatrix} 0 & 0 & 0 & 0 & \bar{N}_2^{(j)} \end{bmatrix}, \text{ and } D_{11} = 1, \ j = 1, 2.$$

On the other hand, allowing stochastic uncertainty in the damping coefficient β, we denote $\beta = \bar{\beta} + f\nu_k$ where $\bar{\beta}$ is the average of β (namely $\beta = 0.4$), f is the standard deviation of β (we take 0.04 to be 2f), and ν_k is a standard white sequence and considering the interval $c \in [117,000 - 143,000]$ we obtain

$$y_{k+3} + (\bar{D}_2 + \frac{T}{I_{son}}f\nu_k)y_{k+2} - (\bar{D}_1 + \frac{2T}{I_{son}}f\nu_k)y_{k+1} + (\bar{D}_0 + \frac{T}{I_{son}}f\nu_k)y_k = \bar{N}_2 u_k. \tag{11.29}$$

In this case (since $\bar{N}_0 = 0$) we obtain

$$\tilde{F} = \begin{bmatrix} -f\frac{T}{I_{son}} & 2f\frac{T}{I_{son}} & -f\frac{T}{I_{son}} & 0 & 0 \\ 1 & 0 & 0 & 0 & 0 \\ 0 & 1 & 0 & 0 & 0 \\ 0 & 0 & 0 & 0 & 0 \\ 0 & 0 & 0 & 1 & 0 \end{bmatrix}.$$

We note that in (11.29) only $\bar{D}_0$, $\bar{D}_1$ and $\bar{N}_2$ are vertex dependent (because only they depend on c). Solving for the nominal case, we apply the result of Theorem 11.1 and obtain $\gamma = 1.522$. In the uncertain case, solving the problem by applying the quadratic approach, we obtain $\gamma = 1.768$ which is 16% higher than the nominal case. On the other hand, applying the vertex-dependent approach of Theorem 11.2, $\gamma = 1.541$ is obtained for $\alpha = 41$. The latter is only about 1% higher than the one achieved for the nominal case.

11.6   Conclusions

The problem of dynamic H_∞ output-feedback control of discrete-time linear systems with multiplicative stochastic uncertainties is solved. Solutions are obtained for both nominal and uncertain polytopic-type systems. In the latter case, a less conservative approach is presented which applies a vertex-dependent Lyapunov function. The main merit of this chapter is that, for the first time, the problem of the dynamic output-feedback control for stochastic systems is solved without preassigning any structure to the Lyapunov function. This is true for both nominal and uncertain systems. The solutions in this chapter are obtained via simple LMIs that are readily implementable and are easily solved. The simple system in Example 11.1 is chosen in order to demonstrate the tractability and applicability of our solution method. In Example 11.2, a real-life control engineering system is solved where a space robot manipulator system is considered.

References

1. Gershon, E., Shaked, U., Yaesh, I.: H_∞ Control and Estimation of State-Multiplicative Linear Systems. Lecture Notes in Control and Information Sciences, LNCIS, vol. 318. Springer (2005)
2. Green, M., Limebeer, D.J.N.: Linear Robust Control. Prentice Hall, N.J. (1995)
3. Scherer, C., Weiland, S.: Linear Matrix Inequalities in Control. e-book at http://www.st.ewi.tudelft.nl/~roos/courses/WI4218/lmi052.pdf (2006)
4. Dragan, V., Morozan, T.: Mixed input-output optimization for time-varying Ito systems with state dependent noise. Dyn. Contin. Discret. Impuls. Syst. **3**, 317–333 (1997)
5. Hinriechsen, D., Pritchard, A.J.: Stochastic H_∞. SIAM J. Control Optim. **36**(5), 1504–1538 (1998)
6. Chen, B.S., Zhang, W.: Stochastic H_2/H_∞ control with state-dependent noise. IEEE Trans. Autom. Control **49**(1), 45–57 (2004)
7. Gershon, E., Shaked, U.: Advanced Topics in Control and Estimation of State-Multiplicative Noisy Systems. Lecture Notes in Control and Information Sciences, LNCIS, vol. 439. Springer (2013)
8. Li, X., Gao, H.: A new model model transformation of discrete-time systems with time-varying delay and its application to stability analysis. IEEE Trans. Autom. Control **56**(9), 2072–2178 (2011)
9. Li, X., Gao, H.: A unified approach to the stability of generalized static neural networks with linear fractional uncertainties and delays. IEEE Trans. Syst. Cybern. **41**(5), 1275–1286 (2011)
10. Dragan, V., Stoica, A.: A γ Attenuation Problem for Discrete-Time Time-Varying Stochastic Systems with Multiplicative Noise. Reprint Series of the Institute of Mathematics of the Romanian Academy, vol. 10 (1998)
11. Bouhtouri, A., Hinriechsen, D., Pritchard, A.J.: H_∞-type control for discrete-time stochasic systems. Int. J. Robust Nonlinear Control **9**, 923–948 (1999)
12. Gershon, E., Shaked, U., Yaesh, I.: H_∞ control and filtering of discrete-time stochastic systems with multiplicative noise. Automatica **37**, 409–417 (2001)
13. Gershon, E., Shaked, U.: H_∞ output-feedback of discrete-time systems with state-multiplicative noise. Automatica **44**(2), 574–579 (2008)
14. Suplin, V., Shaked, U.: Robust H_∞ output-feedback control of linear discrete-time systems. Syst. Control Lett. **54**, 799–808 (2005)
15. Kailath, T.: Linear Systems. Prentice-Hall Inc, Englewood Cliffs, N.J. (1980)

16. Oliveira, M.C., Skelton, R.E.: Stability Test for Constrained Linear Systems. In: S. O. Reza Moheimani (ed.) Perspectives in Robust Control. Lecture Notes in Control and Information Sciences, vol. 268. Springer, London (2001)
17. Kanev, S., Scherer, C., Verhaegen, M., De Schutter, B.: Robust output-feedback controller design via local BMI optimization. Automatica **40**, 1115–1127 (2004)

Chapter 12
Predictor-Based Control of Systems with State-Multiplicative Noise

Abstract Linear, continuous-time systems with state-multiplicative noise and time-delayed input are considered. The problems of H_∞ state-feedback and output-feedback control are solved for these systems when uncertainty in their deterministic parameters is encountered. Cascaded sub-predictors of the Luenberger-type are applied that considerably increase the size of the input time delay that can be solved for. Two examples are given. The first is a practical control engineering design and the second is an illustrative example that compares several solution methods for the stochastic state-feedback control.

12.1 Introduction

H_∞ analysis and design of linear control systems with stochastic state-multiplicative uncertainties have matured over the last three decades (see for example, [1, 2]). Such systems are encountered in many areas of control engineering including altitude and tracking control, to name a few (see [2] for a comprehensive study). H_∞ stability analysis and control of these systems have been extended, in the last decade, to include time-delay systems of various types (i.e., constant time-delay, slow and fast varying delay) and they have become a central issue in the theory of stochastic state-multiplicative systems (see [3, 4] for continuous-time systems and [5, 6] for the discrete-time counterpart). Many of the results that have been derived for the stability and control of deterministic retarded systems, since the 90s ([7–17], see [18] for a comprehensive study), have also been applied to the stochastic case, mainly for continuous-time systems (see [4] and the references therein).

In the continuous-time stochastic setting, the predominant tool for the solution of the traditional control and estimation stochastic problems is the Lyapunov–Krasovskii (L-K) approach. For example, in [19, 20], the (L-K) approach is applied to systems with constant delays, and stability criteria are derived for cases with norm-bounded uncertainties. The H_∞ state-feedback control problem for systems with time-varying delay is treated also in [3, 21]) where the latter work treats also the H_∞ estimation of time-delay systems. An alternative, relatively simpler, approach for studying stochastic systems, is the *input–output* approach (see [4] for a comprehensive study). The *input–output* approach is based on the representation of the

© Springer Nature Switzerland AG 2019
E. Gershon and U. Shaked, *Advances in H∞ Control Theory*,
Lecture Notes in Control and Information Sciences 481,
https://doi.org/10.1007/978-3-030-16008-1_12

system's delay action by delay-free linear operators which allows one to replace the underlying system with an equivalent one that possesses norm-bounded uncertainty, and may therefore be treated by the theory of such uncertain non-retarded systems with state-multiplicative noise [4]. Albeit its simplicity and the convenience of its application to various control and estimation problems, the input–output approach is quite conservative since it inherently entails overdesign [4]. We note that in [22], a robust delay-dependent state-feedback solution is obtained for systems with both state and input delay.

The predictor-based control that transforms the problem of stabilizing a system with constant input delay to one that seeks a stabilizing controller for a corresponding delay-free system is an efficient classical control design method [23, 24]. The reduction of the problem of controlling systems with input time-delay to one of finding a controller to a delay-free counterpart system can be achieved by using the model reduction approach which is based on a change of the state variable [25, 26]. This reduction approach has been extended to linear systems with norm-bounded and delay uncertainties [27] as well as to sampled-data control [28]. However, this approach has not been studied yet in the stochastic framework, in general, and it has not been used in cases where the systems encounter parameter uncertainty of a stochastic type, in particular.

In the present chapter, we consider continuous-time linear systems with stochastic uncertainty in their dynamics and a control input that is either constantly delayed or is subject to a delay that is fast varying in time. The stochastic uncertainty is modeled as white multiplicative noise in the state-space description of the system. We extend the reduction approach to stochastic systems which is based on a transformed predictor-type state-space description [18]. We also derive a modified criterion for bounding the L_2-gain of the considered system and present a design method for minimizing this bound by applying a predictor state-feedback control. The theory developed is demonstrated by two examples where the second example is taken from the field of process control. A predictor-based state-feedback controller is derived there for a given input time-delay that is caused by a transport process.

12.2 Problem Formulation and Preliminaries

We consider the following system:

$$
\begin{aligned}
dx(t) &= [Ax(t) + B_1w(t)]dt + Gx(t)d\beta(t) + B_2u(t - \tau_0)dt, \\
x(t) &= 0, \quad t \in [-\tau_0, 0) \\
\bar{z}(t) &= C_1x(t) + D_{12}u(t),
\end{aligned}
\tag{12.1}
$$

where $x(t) \in R^n$ is the state vector, $w(t) \in \tilde{L}^2_{\mathcal{F}_t}([0, \infty); \mathcal{R}^q)$ is an exogenous disturbance, $\bar{z}(t) \in R^m$ is the objective vector, $u(t) \in R^\ell$ is the control input signal, and τ_0 is a known time-delay. For the sake of clarity we take $C_1^T D_{12} = 0$.

The zero-mean real scalar Wiener process $\beta(t)$ satisfies $\mathcal{E}\{\beta(t)\beta(s)\} = min(t, s)$. We treat the following problem.

Given the system of (12.1a–c), we seek a state-feedback control law $u(t) = Kz(t)$ that stabilizes the system and guarantees a prescribed bound on the L_2-gain of the resulting closed-loop.

In order to solve the above problem, we introduce the following modified state vector [26]:

$$z(t) = e^{A\tau_0}x(t) + \int_{t-\tau_0}^{t} e^{A(t-s)}B_2u(s)ds \tag{12.2}$$

and obtain that

$$\begin{aligned}
dz(t) &= e^{A\tau_0}dx(t) + A[z(t) - e^{A\tau_0}x(t)]dt + B_2udt - e^{A\tau_0}B_2u(t-\tau_0)dt \\
&= Az(t)dt + e^{A\tau_0}Gx(t)d\beta(t) + e^{A\tau_0}B_1w(t)dt + B_2u(t)dt \\
&= Az(t)dt + \tilde{G}[z - \int_{t-\tau_0}^{t} e^{A(t-s)}B_2u(s)ds]d\beta(t) + e^{A\tau_0}B_1w(t)dt + B_2u(t)dt \\
&= [Adt + \tilde{G}d\beta(t)]z - \tilde{G}[\int_{t-\tau_0}^{t} e^{A(t-s)}B_2u(s)ds]d\beta(t) + e^{A\tau_0}B_1w(t)dt + B_2u(t)dt,
\end{aligned} \tag{12.3}$$

where $\tilde{G} \triangleq e^{A\tau_0}Ge^{-A\tau_0}$.

We seek first a state-feedback gain matrix K such that

$$u(t) = Kz(t) \tag{12.4}$$

stabilizes the system and then

$$\begin{aligned}
dz(t) &= [(A + B_2K)dt + \tilde{G}d\beta(t)]z(t) \\
&- \tilde{G}[\int_{t-\tau_0}^{t} e^{A(t-s)}B_2Kz(s)ds]d\beta(t) + e^{A\tau_0}B_1w(t)dt.
\end{aligned} \tag{12.5}$$

We use the *Itô* lemma [29]. By this lemma, taking $w(t) \equiv 0$, if one has

$$dz(t) = f(z(t), t)dt + g(z(t), t)d\beta(t),$$

then for a scalar function $V(z, t)$,

$$dV = \frac{\partial V}{\partial t}dt + \frac{\partial V}{\partial z}dz + \frac{1}{2}g^T V_{z,z}gdt. \tag{12.6}$$

In our case $f = (A + B_2K)z$ and $g = \tilde{G}[z - \int_{t-\tau_0}^{t} e^{A(t-s)}B_2Kz(s)ds]$.

We consider the following Lyapunov function:

$$V(t) = V_1(t, z) + V_2(t), \tag{12.7}$$

where

$$\begin{aligned}
V_1 &= z^T Qz(t), \quad Q > 0 \in R^{n\times n} \\
V_2 &= \int_{-\tau_0}^{0} \int_{t+\theta}^{t} z^T(s)K^T B_2^T e^{-A^T\theta} Re^{-A\theta}B_2Kz(s)ds\, d\theta, \quad R > 0 \in R^{n\times n}.
\end{aligned} \tag{12.8}$$

For $w(t) \equiv 0$, we obtain, using (12.6),

$$
\begin{aligned}
\{(LV)(t)\} &= 2\mathcal{E}\{< Qz(t),\ (A + B_2K)z(t) >\} \\
&+ \{Tr\{Q\tilde{G}[z - \int_{t-\tau_0}^{t} e^{A(t-s)} B_2 K z(s)ds][\tilde{G}(z - \int_{t-\tau_0}^{t} e^{A(t-s)} B_2 K z(s)ds)]^T\} \\
&+ z^T(t)\left[\int_{-\tau_0}^{0} K^T B_2^T e^{-A^T\theta} R e^{-A\theta} B_2 K d\theta\right] z(t) - \Gamma(\tau_0)\},
\end{aligned}
$$

$$(12.9)$$

where $(LV)(t)$ is the infinitesimal generator [29] of the system with respect to V and

$$
\Gamma(\tau) \overset{\Delta}{=} \int_{-\tau}^{0} z^T(t + \theta) K^T B_2^T e^{-A^T\theta} R e^{-A\theta} B_2 K z(t + \theta) d\theta.
$$

Using Jensen inequality [30], we obtain

$$
- \Gamma(\tau_0) \le -\frac{1}{\tau_0} \int_{-\tau_0}^{0} z^T(t + \theta) K^T B_2^T e^{-A^T\theta} d\theta R \int_{-\tau_0}^{0} e^{-A\theta} B_2 K z(t + \theta) d\theta.
$$

$$(12.10)$$

Thus, choosing

$$
R = rI, \qquad (12.11)
$$

where r is a positive scalar, we obtain

$$
\begin{aligned}
\{(LV)(t)\} &\le 2 < Qz(t),\ (A + B_2K)z(t) > + \\
&[\tilde{G}(z - \int_{t-\tau_0}^{t} e^{A(t-s)} B_2 K z(s)ds)]^T Q[\tilde{G}(z - \int_{t-\tau_0}^{t} e^{A(t-s)} B_2 K z(s)ds)] \\
&+ z^T(t)\left[\int_{-\tau_0}^{0} K^T B_2^T e^{-A^T\theta} R e^{-A\theta} B_2 K d\theta\right] z(t) \\
&- \frac{1}{\tau_0} \int_{-\tau_0}^{0} z(t + \theta) K^T B_2^T e^{-A^T\theta} d\theta R \int_{-\tau_0}^{0} e^{-A\theta} B_2 K z(t + \theta) d\theta \\
&= \left[z^T(t)\ \int_{-\tau_0}^{0} z(t + \theta) K^T B_2^T e^{-A^T\theta} d\theta\right] \Psi(Q, r, K) \begin{bmatrix} z(t) \\ \int_{-\tau_0}^{0} e^{-A\theta} B_2 K z(t + \theta) d\theta \end{bmatrix},
\end{aligned}
$$

$$(12.12)$$

where $\Psi(Q, r, K) =$

$$
\begin{bmatrix} Q(A+B_2K)+(A+B_2K)^T Q+\tilde{G}^T Q\tilde{G}+r\left[\int_{-\tau_0}^{0} K^T B_2^T e^{-A^T\theta} e^{-A\theta} B_2 K d\theta\right] & -\tilde{G}^T Q \\ * & \tilde{G}^T Q\tilde{G}-\frac{r}{\tau_0}I \end{bmatrix}.
$$

$$(12.13)$$

We thus obtain the following result.

Theorem 12.1 *The system (12.1a,b) with $w(t) \equiv 0$ is stabilized by the feedback control (12.4) if there exist matrices $P > 0$ and Y, and a tuning parameter $r > 0$ that satisfy the following inequality:*

$$
\begin{bmatrix} AP+B_2Y+(AP+B_2Y)^T & 0 & P\tilde{G}^T & rY^T B_2^T \Phi_A \\ * & -\frac{r}{\tau_0}I -\tilde{G}^T & 0 & 0 \\ * & * & -P & 0 \\ * & * & * & -r\Phi_A \end{bmatrix} < 0. \qquad (12.14)
$$

The controller gain is then given by $K = YP^{-1}$.

Proof It follows from (12.12), (12.13) that stability is guaranteed if

$$\Psi(Q, r, K) < 0. \tag{12.15}$$

Noting that $\Psi(Q, r, K) =$

$$diag\{Q(A+B_2K)+(A+B_2K)^T Q+r\left[\int_{-\tau_0}^0 K^T B_2^T e^{-A^T\theta}e^{-A\theta}B_2K d\theta\right], \ -\frac{r}{\tau_0}I\}$$
$$+\begin{bmatrix}\tilde{G}^T \\ -\tilde{G}^T\end{bmatrix} Q\begin{bmatrix}\tilde{G} & -\tilde{G}\end{bmatrix},$$

we obtain, applying Schur's complements, that the requirement of (12.15) becomes

$$\begin{bmatrix} Q(A+B_2K)+(A+B_2K)^T Q+r\left[\int_{-\tau_0}^0 K^T B_2^T e^{-A^T\theta}Qe^{-A\theta}B_2K d\theta\right] & 0 & \tilde{G}^T\tilde{Q} \\ * & -\frac{r}{\tau_0}I_n-\tilde{G}^T Q \\ * & & -Q \end{bmatrix} < 0.$$

Defining $P = Q^{-1}$ and $Y = KP$, and multiplying the latter inequality, from both sides, by $diag\{P, \ I, \ P\}$, we obtain the following condition for stability:

$$\begin{bmatrix} (AP+B_2Y)+(AP+B_2Y)^T+r\left[\int_{-\tau_0}^0 Y^T B_2^T e^{-A^T\theta}e^{-A\theta}B_2Y d\theta\right] & 0 & P\tilde{G}^T \\ * & -\frac{r}{\tau_0}I & -\tilde{G}^T \\ * & * & -P \end{bmatrix} < 0.$$
$$\tag{12.16}$$

Denoting $\Phi_A = \int_{-\tau_0}^0 e^{-(A^T+A)\theta}d\theta$ the result of (12.14) is obtained.   □

Remark 12.1 The condition of Theorem 12.1 is sufficient only. The conservatism of this condition stems from the use of Jensen inequality in (12.10) and requiring R to be a scalar matrix. The latter restriction can be eliminated by choosing a full matrix R and solving the resulting inequality as a BMI.

The above stabilization method has a little meaning in the case where the system is open-loop stable. In the case where a disturbance attenuation is the objective, the above should be extended to establish conditions for achieving L_2-gain bound on the closed-loop system.

12.3   The L_2-gain Bound

In the case where $w(t)$ is not zero, we consider the following performance index

$$J \stackrel{\Delta}{=} \int_0^\infty J_1(s)ds, \quad \text{where} \quad J_1(t) \stackrel{\Delta}{=} \mathcal{E}\{||C_1x(t)||^2 + ||D_{12}u(t)|| - \gamma^2||w(t)||^2\},$$
$$\tag{12.17}$$

and where γ is a prescribed positive scalar. We obtain the following result.

Theorem 12.2 *The L_2-gain of the system (12.1a–c) with the feedback law (12.4) is less than a prescribed scalar $\gamma > 0$ if there exist matrices $P > 0$ and Y, and a tuning parameter $r > 0$ that satisfy the following inequality.*

$$
\begin{bmatrix}
AP+B_2Y+PA^T+Y^TB_2^T & 0 & Y^T\hat{\Phi}_A(r) & e^{A\tau_0}B_1 & P\tilde{G}^T & P\tilde{C}_1^T \\
* & -\frac{r}{\tau_0}I & 0 & 0 & -\tilde{G}^T & -\tilde{C}_1^T \\
* & * & -\hat{\Phi}_A(r) & 0 & 0 & 0 \\
* & * & * & -\gamma^2 I & 0 & 0 \\
* & * & * & * & -P & 0 \\
* & * & * & * & * & -I
\end{bmatrix} < 0, \qquad (12.18)
$$

where $\hat{\Phi}_A(r) = rB_2^T\Phi_AB_2 + \bar{R}$, $\bar{R} = D_{12}^TD_{12}$ and where Φ_A is defined following (12.16). The controller gain that achieves the L_2-gain bound is given by $u(t) = Kz(t)$ where $K = YP^{-1}$.

Proof It follows from (12.9) that

$$
\begin{aligned}
\{(LV)(t)\} + J_1(t) &= 2\mathcal{E}\{< Qz(t), \ (A + B_2K)z(t) + e^{A\tau_0}B_1w(t) >\} + J_1(t) \\
&+ \{Tr\{Q\tilde{G}[z - \int_{t-\tau_0}^t e^{A(t-s)}B_2Kz(s)ds][\tilde{G}(z - \int_{t-\tau_0}^t e^{A(t-s)}B_2Kz(s)ds)]^T\} \\
&+ z^T(t)\left[\int_{-\tau_0}^0 K^TB_2^Te^{-A^T\theta}Re^{-A\theta}B_2Kd\theta\right]z(t) - \Gamma(\tau_0)\}.
\end{aligned}
$$

$$(12.19)$$

Requiring then that for all $t \geq 0$

$$
\{(LV)(t)\} + J_1(t) \leq 0 \quad \forall w(t) \in \tilde{L}_{\mathcal{F}_t}^2([0, \infty); \mathcal{R}^q),
$$

the stability and the desired disturbance attenuation are achieved.

Applying the steps that lead to (12.12) and choosing, again, $R = rI$ the following sufficient condition is obtained:

$$
\begin{aligned}
&\left[z^T(t) \ \int_{-\tau_0}^0 z(t+\theta)K^TB_2^Te^{-A^T\theta}d\theta\right] \Psi(Q, r, K) \begin{bmatrix} z(t) \\ \int_{-\tau_0}^0 e^{-A\theta}B_2Kz(t+\theta)d\theta \end{bmatrix} \\
&+ 2 < Qz(t), e^{A\tau_0}B_1w(t) > + J_1(t) - \gamma^2\|w\|^2 \leq 0.
\end{aligned}
$$

Denoting

$$
\xi(t) = col\{z(t), \ \int_{-\tau_0}^0 e^{-A\theta}B_2Kz(t+\theta)d\theta, \ w(t)\},
$$

the latter inequality becomes

$$
\xi^T(t)\Phi(Q, z(t), r, K)\xi(t) < 0, \qquad (12.20)
$$

where

$$\Phi \triangleq \begin{bmatrix} \Phi_1 & -\tilde{G}^T Q \tilde{G} - \tilde{C}_1^T \tilde{C}_1 & Q e^{A\tau_0} B_1 \\ * & \tilde{G}^T Q \tilde{G} - \frac{r}{\tau_0} I + \tilde{C}_1^T \tilde{C}_1 & 0 \\ * & * & -\gamma^2 I \end{bmatrix}, \qquad (12.21)$$

where $\Phi_1 = Q(A + B_2 K) + (A + B_2 K)^T Q + \tilde{C}_1^T \tilde{C}_1 + \tilde{G}^T Q \tilde{G} + K^T (r B_2^T \Phi_A B_2 + \bar{R}) K$ and where $\tilde{C}_1 \triangleq C_1 e^{-A\tau_0}$. Inequality (12.20) can then be written, applying Schur's complements, as

$$\begin{bmatrix} Q(A+B_2 K)+(A+B_2 K)^T Q + K^T (r B_2^T \Phi_A B_2 + \bar{R}) K & 0 & Q e^{A\tau_0} B_1 & \tilde{G}^T Q & \tilde{C}_1^T \\ * & -\frac{r}{\tau_0} I & 0 & -\tilde{G}^T Q & -\tilde{C}_1^T \\ * & * & -\gamma^2 I & 0 & 0 \\ * & * & * & -Q & 0 \\ * & * & * & * & -I \end{bmatrix} < 0.$$

$$(12.22)$$

Denoting $P = Q^{-1}$, $Y = KP$ and multiplying (12.22), from both sides, by $diag\{P, I, I, P, I\}$ we obtain the result of Theorem 12.2. $\qquad\square$

The condition of Theorem 12.2, if satisfied, guarantees a L_2-gain bound of γ. The definition of the L_2-gain is based, however, on the performance index J_E of (12.17). Since the control u is delayed by τ_0 seconds, the controller cannot affect the system in the time interval $[0, \tau_0]$ and a more useful criterion would be to achieve

$$J_z \triangleq \mathcal{E}\{\int_{\tau_0}^{\infty} (||C_1 x(t)||^2 + u(t)^T \bar{R} u(t)) dt$$
$$-\gamma^2 \int_0^{\infty} ||w(t)||^2 dt\} < 0, \ \forall \ w(t) \in \tilde{L}_{\mathcal{F}_t}^2([0, \infty); \mathcal{R}^q). \qquad (12.23)$$

It follows from the definition (12.2) that $z(t)$ is the predictor value of $x(t)$ which is based on the control input $u(t)$ that is delayed by τ_0 seconds, where the effect of $w(t)$ on $x(\tau)$ in the interval $\tau \in [t - \tau_0, \ t]$ is discarded. Thus, adopting the predictor-type performance index

$$J_P \triangleq \mathcal{E}\{\int_{\tau_0}^{\infty} (||C_1 z(t)||^2 + u(t)^T \bar{R} u(t)) dt - \gamma^2 [\lim_{T \to \infty} \int_0^{T-\tau_0} ||w(t)||^2] dt\},$$

$$(12.24)$$

the contribution of $w(t)$ at the interval $[T - \tau_0, \ T]$ for T that tends to infinity is ignored. The latter is neglectable when T tends to infinity and thus J_z and J_P have the same physical meaning. We thus obtain the following from the arguments that led to (12.22).

Corollary 12.1 *J_P will be negative for all $w(t) \in \tilde{L}_{\mathcal{F}_t}^2([0, \infty); \mathcal{R}^q)$, using the control input (12.4), if there exist matrices $P > 0$ and Y, and a tuning parameter $r > 0$ that satisfy the following inequality.*

$$\begin{bmatrix} AP+B_2Y+PA^T+Y^TB_2^T & 0 & Y^T\hat{\Phi}_A(r) & e^{A\tau_0}B_1 & P\tilde{G}^T & PC_1^T \\ * & -\frac{r}{\tau_0}I & 0 & 0 & -\tilde{G}^T & 0 \\ * & * & -\hat{\Phi}_A(r) & 0 & 0 & 0 \\ * & * & * & -\gamma^2 I & 0 & 0 \\ * & * & * & * & -P & 0 \\ * & * & * & * & * & -I \end{bmatrix} < 0. \qquad (12.25)$$

12.4   Time-Varying Delay

In the above, the delay τ_0 was assumed to be a known constant τ_0. The results can be readily extended, however, also to the case of time-varying delay

$$\tau(t) = \tau_0 + \eta(t), \quad \text{where } |\eta(t)| \le \mu \le \tau_0. \qquad (12.26)$$

We first obtain the following stability condition.

Theorem 12.3 *The system (12.1a,b) with the delay (12.26) is stabilized by the feedback control (12.4) if there exist matrices $P > 0$ and Y, and tuning parameters $r, \varepsilon > 0$ that satisfy the following (12.27). The controller gain is then given by $K = YP^{-1}$.*

$$\begin{bmatrix} (AP+B_2Y)+(AP+B_2Y)^T & 0 & P\tilde{G}^T & -e^{A\tau_0}B_2Y & \varepsilon\bar{\tau}_0(PA^T+Y^TB_2^T) & rY^TB_2^T\Phi_A \\ * & -\frac{r}{\tau_0}I & -\tilde{G}^T & 0 & 0 & 0 \\ * & * & -P & 0 & 0 & 0 \\ * & * & * & -\varepsilon P & -\varepsilon\bar{\tau}_0Y^TB_2^Te^{A^T\tau_0} & 0 \\ * & * & * & * & -\varepsilon P & 0 \\ * & * & * & * & * & -r\Phi_A \end{bmatrix} < 0, \qquad (12.27)$$

where Φ_A is defined following (12.16).

Proof Applying the definition of $z(t)$ as in (12.2) the differential change $dz(t)$ of (12.2) will now have an additional term $e^{A\tau_0}B_2[u(t-\tau(t)) - u(t-\tau_0)]dt$ and for $u(t)$ defined in (12.4) the following is obtained.

$$dz(t) = [(A+B_2K)dt + \tilde{G}d\beta(t)]z(t)$$
$$- \tilde{G}[\int_{t-\tau_0}^t e^{A(t-s)}B_2Kz(s)ds]d\beta(t) + e^{A\tau_0}B_2K[z(t-\tau(t)) - z(t-\tau_0)]dt$$
$$+ e^{A\tau_0}B_1w(t)dt. \qquad (12.28)$$

In order to avoid the use of the derivative of z (since only $dz(t)$ is allowed), we apply the input–output approach for delayed systems (see [4], Chap. 2). Thus, Eq. (12.28) can be written as

$$dz(t) = [(A+B_2K)dt + \tilde{G}d\beta(t)]z(t)-$$
$$\tilde{G}[\int_{t-\tau_0}^t e^{A(t-s)}B_2Kz(s)ds]d\beta(t) - e^{A\tau_0}B_2K(\int_{t-\tau}^{t-\tau_0}\bar{y}(s)ds)dt \qquad (12.29)$$
$$- \Gamma_\beta dt + e^{A\tau_0}B_1w(t)dt,$$

where

$$\bar{y}(t) = (A + B_2 K)z(t) - e^{A\tau_0} B_2 K w_2(t) - \Gamma_\beta + e^{A\tau_0} B_1 w(t),$$

and where we define

$$w_2(t) = (\Delta_2 \bar{y})(t), \quad (\Delta_2 g)(t) \overset{\Delta}{=} \int_{t-\tau(t)}^{t-\tau_0} g(s)ds$$

and denote

$$\Gamma_\beta = \int_{t-\tau}^{t-\tau_0} \tilde{G} \left[z(t) - \int_{t-\tau_0}^{t} e^{A(t-s)} B_2 K z(s)ds \right] d\beta(t).$$

It is well known (see lemma 2 in [31]) that

$$\bar{y}^T(t)\Delta_2^T \Delta_2 \bar{y}(t) \le \bar{\tau}_0^2 \|\bar{y}(t)\|^2, \quad \text{where} \quad \bar{\tau}_0 = \frac{7}{4}\tau_0.$$

Replacing then (12.12) by

$$\{(LV)(t)\} \le 2 < Qz(t), \ (A + B_2 K)z(t) - e^{A\tau_0} B_2 K(\Delta_2 \bar{y})(t) >$$
$$+ [\tilde{G}(z - \int_{t-\tau_0}^{t} e^{A(t-s)} B_2 K z(s)ds)]^T Q[\tilde{G}(z - \int_{t-\tau_0}^{t} e^{A(t-s)} B_2 K z(s)ds)]$$
$$+ z^T(t) \left[\int_{-\tau_0}^{0} K^T B_2^T e^{-A^T \theta} R e^{-A\theta} B_2 K d\theta \right] z(t) -$$
$$\frac{1}{\tau_0} \int_{-\tau_0}^{0} z(t+\theta) K^T B_2^T e^{-A^T \theta} d\theta R \int_{-\tau_0}^{0} e^{-A\theta} B_2 K z(t+\theta)d\theta$$
$$= \left[z^T(t) \int_{-\tau_0}^{0} z(t+\theta) K^T B_2^T e^{-A^T \theta} d\theta \right] \Psi(Q, r, K) \begin{bmatrix} z(t) \\ \int_{-\tau_0}^{0} e^{-A\theta} B_2 K z(t+\theta)d\theta \end{bmatrix} -$$
$$2z^T(t) Q \Delta_2 \bar{y}(t)$$

and using the fact that

$$-w_2^T(t) R_2 w_2(t) + \bar{\tau}_0^2 \bar{y}^T(t) R_2 \bar{y}(t) \ge 0,$$

where R_2 is a positive definite matrix, the condition for stability becomes

$$\bar{\xi}^T(t)\bar{\Phi}(Q, z(t), r, K)\bar{\xi}(t) + \bar{\tau}_0^2 \bar{y}^T(t) R_2 \bar{y}(t) < 0, \qquad (12.30)$$

where

$$\bar{\xi}(t) = col\{z(t), \ \int_{-\tau_0}^{0} e^{-A\theta} B_2 K z(t+\theta)d\theta, \ w_2(t)\},$$

and where

$$\bar{\Phi} \triangleq \begin{bmatrix} Q(A+B_2K)+(A+B_2K)^TQ+\tilde{G}^TQ\tilde{G}+rK^TB_2^T\Phi_AB_2K & -\tilde{G}^TQ\tilde{G} & -Qe^{A\tau_0}B_2K \\ * & \tilde{G}^TQ\tilde{G}-\frac{r}{\tau_0}I & 0 \\ * & * & -R_2 \end{bmatrix}.$$

$$(12.31)$$

We thus require that

$$\begin{bmatrix} \bar{\Phi}_1 & -\tilde{G}^TQ\tilde{G} & -Qe^{A\tau_0}B_2K & \bar{\tau}_0(A+B_2K)^TR_2 \\ *\tilde{G}^TQ\tilde{G}-\frac{r}{\tau_0}I & 0 & 0 \\ * & * & -R_2 & -\bar{\tau}_0K^TB_2^Te^{A^T\tau_0}R_2 \\ * & * & * & -R_2 \end{bmatrix} < 0, \qquad (12.32)$$

where $\bar{\Phi}_1 = Q(A + B_2K) + (A + B_2K)^TQ+\tilde{G}^TQ\tilde{G} + rK^TB_2^T\Phi_AB_2K$. Following the steps that led to (12.14) the following requirement is then obtained:

$$\begin{bmatrix} (AP+B_2Y)+(AP+B_2Y)^T & 0 & P\tilde{G}^T & -e^{A\tau_0}B_2K & \bar{\tau}_0(PA^T+Y^TB_2^T) & rY^TB_2^T\Phi_A \\ * & -\frac{r}{\tau_0}I & -\tilde{G}^T & 0 & 0 & 0 \\ * & * & -P & 0 & 0 & 0 \\ * & * & * & -R_2 & -\bar{\tau}_0K^TB_2^Te^{A^T\tau_0} & 0 \\ * & * & * & * & -R_2^{-1} & 0 \\ * & * & * & * & * & -r\Phi_A \end{bmatrix} < 0.$$

$$(12.33)$$

Choosing $R_2 = \varepsilon Q$, where ε is a tuning positive scalar, and multiplying the fourth column and row block in (12.33) by P, we obtain the result of (12.27) in Theorem 12.3.   □

Adopting the performance index (12.24), the problem of securing a prescribed bound on the L_2-gain of the closed-loop system can be solved by applying similar arguments to those that led to the LMI condition of (12.25). The following result is obtained.

Theorem 12.4 *The L_2-gain of the system (12.1a–c) with a feedback law (12.4) is less than a prescribed scalar $\gamma > 0$, for the time-varying delay (12.26) if there exist matrices $P > 0$ and Y, and tuning parameters $r > 0$ and $\varepsilon > 0$ that satisfy the following inequality.*

$$\begin{bmatrix} \Psi_1 & 0 & P\tilde{G}^T & -e^{A\tau_0}B_2Y & \varepsilon\bar{\tau}_0(PA^T+Y^TB_2^T) & Y^T\hat{\Phi}_A(r) & e^{A\tau_0}B_1 & PC_1^T \\ * & -\frac{r}{\tau_0}I & -\tilde{G}^T & 0 & 0 & 0 & 0 & 0 \\ * & * & -P & 0 & 0 & 0 & 0 & 0 \\ * & * & * & -\varepsilon P & -\varepsilon\bar{\tau}_0Y^TB_2^Te^{A^T\tau_0} & 0 & 0 & 0 \\ * & * & * & * & -\varepsilon P & 0 & e^{A\tau_0}B_1 & 0 \\ * & * & * & * & * & -\hat{\Phi}_A(r) & 0 & 0 \\ * & * & * & * & * & * & -\gamma^2 I & 0 \\ * & * & * & * & * & * & * & -I \end{bmatrix} < 0, \qquad (12.34)$$

where $\bar{\Phi}_1 = Q(A + B_2K) + (A + B_2K)^TQ+\tilde{G}^TQ\tilde{G} + rK^TB_2^T\Phi_AB_2K$. The controller gain that achieves the L_2-gain bound is given by $u(t) = Kz(t)$ where $K = YP^{-1}$.

Remark 12.2 We note that a simple strategy is applied for the numerical solution of (12.34) which also applies to (12.27). The solution of (12.34) involves a search for two scalar variables: r and ε. One may start by taking arbitrary values for the latter two parameters and seek, using line-searching, values for these tuning parameters that lead to a stabilizing controller of minimum γ. Once such a controller is obtained, standard optimization techniques can be used, say MATLAB function "fminsearch", which seek the combination of the two scalar parameters that bring γ to a local minimum.

12.5 Examples

Example 12.1 **Stability and L_2-gain**
We consider the system of (1a–c) with

$$A = \begin{bmatrix} -1 & 1 \\ 0.3 & -0.08 \end{bmatrix}, \quad G = \begin{bmatrix} 0 & 0.08 \\ 0 & 0 \end{bmatrix}, \quad B_1 = \begin{bmatrix} 1 \\ -1 \end{bmatrix}, \quad B_2 = \begin{bmatrix} 0 \\ 1 \end{bmatrix}$$

and

$$C_1 = \begin{bmatrix} -0.5 & 0.4 \\ 0 & 0 \end{bmatrix}, \quad D_{12} = [0 \ 0.1]^T.$$

Applying Theorem 12.1 we find that the system can be stabilized for $\tau_0 \leq 2.93$ s. For $\tau_0 = 2.93$ s., a stabilizing feedback gain $K = [0.0104 \ -0.4684]$ is obtained for $r = 1$. An upper-bound of $\gamma = 2.01$ is guaranteed for the predictor-based L_2-gain of the closed-loop system using the condition of Corollary 12.1 for $\tau_0 = 2.8$ and $r = 0.1$. We note that one can apply the method of [22] for solving the L_2-gain of this example, however, there is no clear indication on how to optimize the obtained results there [see Theorem 2 and Remark 4 in [22]].

In [4], an alternative method for solving the problem has been introduced. It is based on augmenting the system dynamics to include the delayed input as a delayed state vector and it applies the input–output approach for delayed state systems. Using the latter method, a near maximum constant delay $\tau_0 = 1.5$ s. was reached for which an upper-bound of $\gamma = 277.73$ was found on the L_2-gain of the closed-loop system. For time-varying delay, stability is guaranteed using Theorem 12.3 for maximum value of $\tau_0 = \mu = 0.93$ s., namely for a fast varying delay from $[0, 1.86]$ s., using $r = 0.005$, $\varepsilon = 0.6$ and the feedback gain matrix $K = [-0.1260 \ -0.2640]$. A bound of $\gamma_{max} = 6.56$ is obtained for $\tau_0 = \mu = 0.9$ s., using (12.34) with $r = 0.005$, $\varepsilon = 0.6$ and $K = [-0.1378 \ -0.2698]$. The results are summarized in Table 12.1.

Example 12.2 **Process Control**
Time delay is an inherent physical phenomenon in process control dynamics, which is usually caused by transport of heat or mass from a certain point in the physical plant [say, from the source of a certain substance] to another point [the main reactor].

Table 12.1 Range of delay and L_2-gain bounds

Method	Max delay	γ
Theorem 1, Corollary 1	$\tau_0 = 2.93$ (constant)	2.01 (for $\tau_0 = 2.8$)
Reference [4]	$\tau_0 = 1.5$ (constant)	not feasible (for $\tau_0 = 2.8$)
Theorem 3, (12.27)	$\tau(t) \in [0, 1.86]$	6.56 (for $\tau(t) = [0, 1.80]$)

Since, compared to electrical or electronic dynamical systems, chemical reactions are relatively slow, typical delays are of tens of seconds. The Continuous Stirred Tank Reactor (CSTR) is one of the fundamental physical plants in process control engineering systems. We consider the following irreversible reaction scheme which is known as the van der Vusse reaction [32, 33]

$$\bar{A} \xrightarrow{k_1} \bar{B} \xrightarrow{k_2} \bar{C}, \quad 2\bar{A} \xrightarrow{k_3} \bar{D},$$

where $\bar{A}$ is the source material [Cyclopentadine] applied to the reactor, $\bar{B}$ is an intermediate substance [Cyclopentenol] , $\bar{C}$ [Cyclopentanediol] and $\bar{D}$ [Dicyclopentadiene] are the competing products. The parameters k_1, k_2 and k_3 are the reaction rate constants. The linearized dynamics, around a set point, of this reaction is described by the model of (12.1a–c) (excluding, in the meantime, the delay and the multiplicative noise) with the following matrices:

$$A = \begin{bmatrix} -\frac{F_s}{V} - k_1 - 2k_3\bar{C}_{As} & 0 \\ k_1 & -\frac{F_s}{V} - k_2 \end{bmatrix}, \quad B_1 = \begin{bmatrix} -\frac{F_s}{V} \\ 0 \end{bmatrix},$$

$$B_2 = \begin{bmatrix} \bar{C}_{Afs} - \bar{C}_{As} \\ -\bar{C}_{Bs} \end{bmatrix}, \quad C_1 = \begin{bmatrix} 0 & 1 \\ 0 & 0 \end{bmatrix}, \quad D_{12} = \begin{bmatrix} 0 \\ 0.1 \end{bmatrix}, \qquad (12.35)$$

where $\frac{F_s}{V} = 0.5714 \, \text{min}^{-1}$, $\bar{C}_{As} = 3 \, \text{gmoll}^{-1}$, $\bar{C}_{Afs} = 10 \, \text{gmoll}^{-1}$, $\bar{C}_{Bs} = 1.117$ gmoll^{-1} and where $k_1 = 0.83 \, \text{min}^{-1}$ and $k_3 = 0.166 \, \text{moll}^{-1} \text{min}^{-1}$. The rate k_2 depends on the fluctuating concentration of a catalyst. We model then k_2 as a white-noise process with average of $2.43 \, \text{min}^{-1}$ and standard deviation of $0.77 \, \text{min}^{-1}$. In our model, the state vector is $x(t) \triangleq \begin{bmatrix} \bar{C}_A - \bar{C}_{As} \\ \bar{C}_B - \bar{C}_{Bs} \end{bmatrix}$, where $\bar{C}_A$ and $\bar{C}_B$ are the concentrations of $\bar{A}$ and $\bar{B}$, respectively and $\bar{C}_{As}$ and $\bar{C}_{Bs}$ are the steady-state concentrations of the the latter substances . The input to the system $\frac{F}{V}$ is the flow of substance $\bar{A}$ into the reactor.

We consider the above van der Vusse reaction scheme in an isothermal CSTR [i.e., with constant temperature] subject to transport time delay of 0.5 min in the inflow of substance $\bar{A}$ [Cyclopentadine]. The statistical nature of k_2 leads to a state-multiplicative noisy system (k_2 appears in the dynamical matrix of (12.35a)). Substituting for the above-given system parameters, we obtain

$$A = \begin{bmatrix} -2.4 & 0 \\ 0.83 & -3.0 \end{bmatrix}, \quad G = \begin{bmatrix} 0 & 0 \\ 0 & -0.77 \end{bmatrix}, \quad B_1 = \begin{bmatrix} 0.57 \\ 0 \end{bmatrix}, \quad B_2 = \begin{bmatrix} 7 \\ -1.11 \end{bmatrix}, \quad C_1 = \begin{bmatrix} 0 & 1 \\ 0 & 0 \end{bmatrix}$$

and $D_{12} = \begin{bmatrix} 0 \\ 0.1 \end{bmatrix}$.

Applying Corollary 12.1 and solving (12.25), for $r = 1.1$, a controller is found for a delay of $\tau_0 = 0.5$ minutes with an upper-bound of $\gamma = 0.040$. The corresponding stabilizing control law is $\frac{F}{V} = -[0.0171(\bar{C}_A - \bar{C}_{As}) + 0.0094(\bar{C}_B - \bar{C}_{Bs})]$.

12.6 Conclusions

A predictor-type state-feedback control is applied to systems with stochastic multiplicative type uncertainty and time-delayed actuators. A new state vector is defined which, given the feedback gain matrix, predicts the value of the true state of the resulting closed-loop. A condition for closed-loop stability is derived for these systems which are used to find a stabilizing state-feedback control. Similar conditions are obtained that guarantee prescribed bounds on the L_2-gain of the resulting closed-loop systems.

Two types of delays in the actuators are treated. The first is a constant known delay and the second is time-varying delay that may be fast varying. Two examples were given. In Example 12.1, an open-loop unstable system is considered. It is shown here that a much larger time-delay can be handled compared to a corresponding result that is obtained by another existing method. It is also shown in the example that the fact that the delay can be fast varying, reduces the maximum range of the delay for which stabilization can be achieved. A practical control problem taken from the field of process control is treated in Example 12.2. The result that is obtained there is shown to be very practical and may thus be used in the future to develop a new methodology in process control.

The results of this chapter are based on the input–output approach where by applying the norm of the delay operator a delay-free stabilization problem is obtained and solved. A future work may be to properly apply the direct Lyapunov–Krasovskii approach in an attempt to reduce the encountered overdesign.

References

1. Dragan, V., Morozan, T.: Mixed input-output optimization for time-varying Ito systems with state dependent noise. Dyn. Contin., Discret. Impuls. Syst. **3**, 317–333 (1997)
2. Gershon, E., Shaked, U., Yaesh, I.: H_∞ Control and Estimation of State-Multiplicative Linear Systems. Lecture Notes in Control and Information Sciences, LNCIS, vol. 318. Springer, Berlin (2005)
3. Boukas, E.K., Liu, Z.K.: Deterministic and Stochastic Time Delay Systems. Birkhauser, Basel (2002)

4. Gershon, E., Shaked, U.: Advanced Topics in Control and Estimation of State-Multiplicative Noisy Systems. Lecture Notes in Control and Information Sciences, LNCIS, vol. 439. Springer, Berlin (2013)
5. Xu, S., Lam, J., Chen, T.: Robust H_∞ control for uncertain discrete stochastic time-delay systems. Syst. Control Lett. **51**, 203–215 (2004)
6. Yue, D., Tian, E., Zhang, Y.: Stability analysis of discrete systems with stochastic delay and its applications. Int. J. Innov. Comput., Inf. Control **5**(8), 2391–2403 (2009)
7. Kolmanovskii, V., Richard, J.P.: Stability of some linear systems with delays. IEEE Trans. Autom. Control **44**, 984–989 (1999)
8. Huang, Y.P., Zhou, K.: Robust stability of uncertain time-delay systems. IEEE Trans. Autom. Control **45**, 2169–2173 (2000)
9. Moon, Y.S., Park, P., Kwon, W.H., Lee, Y.S.: Delay-dependent robust stabilization of uncertain state-delayed systems. Int. J. Control **74**, 1447–1455 (2001)
10. Niculescu, S.I.: Delay effects on stability: a robust control approach. Lecture Notes in Control and Information Sciences, vol. 269. Springer, London (2001)
11. Fridman, E.: New Lyapunov-Krasovskii functionals for stability of linear retarded and neutral type systems. Syst. Control Lett. **43**, 309–319 (2001)
12. Kao, C.Y., Lincoln, B.: Simple stability criteria for systems with time-varying delays. Automatica **40**, 1429–1434 (2004)
13. Fridman, E., Shaked, U.: Input-output approach to stability and L_2−gain analysis of systems with time-varying delays. Syst. Control Lett. **55**, 1041–1053 (2006)
14. Gao, H., Chen, T.: New results on stability of discrete-time systems with time-varying state delay. IEEE Trans. Autom. Control **52**(2), 328–334 (2007)
15. Fridman, E., Orlov, Y.: Exponential stability of linear distributed parameter systems with time-varying delays. Automatica **45**(2), 194–201 (2009)
16. Li, X., Gao, H., Yu, X.: A unified approach to the stability of generalized static neural networks with linear fractional uncertainties and delay. IEEE Trans. Syst., Man Cybernatics **41**(5), 1275–1286 (2011)
17. Li, X., Gao, H.: A new model transformation of discrete-time systems with time-varying delay and its application to stability analysis. IEEE Trans. Autom. Control **56**(9), 2172–2178 (2011)
18. Fridman, E.: Introduction to Time Delay Systems: Analysis and Control. Birkhauser, Basel (2014)
19. Xu, S., Lam, J., Mao, X., Zou, Y.: A new LMI condition for delay-dependent robust stability of stochastic time-delay systems. Asian J. Control **7**, 419–423 (2005)
20. Chen, W.H., Guan, Z., Lu, X.: Delay-dependent exponential stability of uncertain stochastic systems with multiple delays: an LMI approach. Syst. Control Lett. **54**, 547–555 (2005)
21. Xu, S., Chen, T.: Robust H_∞ control for uncertain stochastic systems with state-delay. IEEE Trans. Autom. Control **47**, 2089–2094 (2002)
22. Li, H., Chen, B., Zhou, Q., Lin, C.: A delay-dependent approach to robust H_∞ control for uncertain stochastic systems with state and input delays. Circuits, Syst. Signal Process. **28**, 169–183 (2009)
23. Smith, O.: A controller to overcome dead time. ISA **6**(2), 28–33 (1959)
24. Manitius, A.Z., Olbrot, A.W.: Finite spectrum assignment problem for systems with delay. IEEE Trans. Autom. Control **24**, 541–553 (1979)
25. Kwon, W., Pearson, A.: Feedback stabilization of linear systems with delayed control. IEEE Trans. Autom. Control **25**(2), 266–269 (1980)
26. Artstein, Z.: Linear systems with delayed controls: a reduction. IEEE Trans. Autom. Control **27**(4), 869–879 (1982)
27. Yue, D., Han, Q.L.: Delayed feedback control of uncertain systems with time-varying input delay. Automatica **41**(2), 233–240 (2005)
28. Mazenc, F., Normand-Cyrot, D.: Reduction model approach for linear systems with sampled delayed inputs. IEEE Trans. Autom. Control **58**(5), 1263–1268 (2013)
29. Klebaner, F.C.: Introduction to Stochastic Calculus with Applications. Imperial College Press, London (2012)

30. Cvetkovski, Zdravko: Inequalities: Theorems, Techniques and Selected Problems. Springer, Berlin (2012)
31. Shustin, E., Fridman, E.: On delay-derivative-dependent stability of systems with fast-varying delays. Automatica **43**, 1649–1655 (2007)
32. Seborg, D.E., Edgar, T.F., Mellichamp, D.A.: Process Dynamics and Control, 2nd edn. Wiley and Sons, Hoboken (2004)
33. Vishnoi, V., Padhee, S., Kaur, G.: Controller performance evaluation for concentration control of isothermal continuous stirred tank reactor. Int. J. Sci. Res. Publ. **2**(6), 1–7 (2012)

Chapter 13
Static Output-Feedback Control of Retarded Stochastic Systems

Abstract Linear, state-delayed, discrete-time systems with stochastic uncertainties in their state-space model are considered. The problem of H_∞ static output-feedback control is solved, for the stationary case, via the input–output approach where the system is replaced by a non-retarded system with deterministic norm-bounded uncertainties. Based on the BRL result of the above systems, solutions are obtained for the nominal and the uncertain polytopic cases where, for the former, a single linear matrix inequality solution is obtained and where, in the latter case, quadratic, less conservative, vertex-dependent approaches are adopted. A numerical example is given that demonstrates the applicability and tractability of the theory.

13.1 Introduction

In this chapter, we address the problem of H_∞ static output-feedback control of state-delayed, discrete-time, state-multiplicative linear systems based on the Bounded Real Lemma (BRL) for these systems, which was developed in [1, 2]. In our systems, different white-noise sequences multiply both the delayed and the non-delayed states of the system. We treat first the nominal case and we then solve the problem for systems with polytopic-type uncertainties. A linear quadratic solution is obtained and a different, possibly less conservative, solution is suggested that applies the Finsler's lemma [3].

The problem of deterministic static output-feedback control has been a focus of many works in the past (see, for example, [4, 5]) and in recent times [6, 7] where a bound on an H_∞ performance index was guaranteed. The main advantage of the static output feedback is the simplicity of its implementation and the ability it provides for designing controllers of prescribed structure such as PI and PID.

The analysis and design of controllers for systems with stochastic multiplicative noise have matured greatly in the last two decades (see [8] and the references therein). Numerous solutions to various stochastic control and filtering problems including those that ensure a worst-case performance bound in the H_∞ sense have been derived for both delay-free [8–15] and retarded linear stochastic systems [16–21].

© Springer Nature Switzerland AG 2019
E. Gershon and U. Shaked, *Advances in H_∞ Control Theory*,
Lecture Notes in Control and Information Sciences 481,
https://doi.org/10.1007/978-3-030-16008-1_13

Delay-free systems with parameter uncertainties that are modeled as white-noise processes in a linear setting have been treated in [9–13], for the continuous-time case and in [14, 15] for the discrete-time case. Such models of uncertainties are encountered in many areas of applications such as nuclear fission and heat transfer, population models, and immunology. In control theory, such models are encountered in gain scheduling when the scheduling parameters are corrupted with measurement noise.

In the stochastic setting, the problem of zero-order control was addressed in [13] for continuous-time, delay-free systems where, by adding a simple component of large bandwidth to the measured output, tighter bounds on the H_2 and the H_∞ performance criteria were achieved when numerically compared to other design techniques [13]. The full-order dynamic output-feedback problem for stochastic **retarded systems** was solved in [18], for nominal systems and norm-bounded systems with uncertainties. Recently [19], an extended solution which includes full and reduced-order controllers has been applied to retarded continuous-time, state-multiplicative stochastic systems without applying the low-pass filter of [13].

The stochastic, discrete-time, state-delayed counterpart systems have been addressed to a lesser extent compared to the continuous-time case. The mean square exponential stability and the control and filtering of these systems were treated by several groups [20, 21]. In [20], the state-feedback control problem is solved for retarded norm-bounded uncertain systems in the restrictive case where the same multiplicative noise sequence multiplies both the states and the input of the system. The solution there is delay-dependent. The static output-feedback control solution for delay-free discrete-time stochastic state-multiplicative system was obtained in [15]. Unfortunately, when delay is encountered in the system states, a different approach must be applied to the problem solution using, in our case, the input–output approach [1].

To the best of our knowledge, the input–output approach has not been applied to the solution of the static output-feedback problem in the discrete-time stochastic setting. In [1, 2], this method was applied only to the state-feedback control and filtering problems. Based on a stability condition, a BRL condition was introduced there. In the continuous-time setting, the input–output approach has been shown in [18] to achieve better results, when compared to other solution methods. This approach provides simple solutions to the latter problems for polytopic-type uncertain systems.

In the present chapter, we apply the input–output approach, to the solution of the stochastic static output-feedback control problem. This approach is based on representing the system's delay action by delay-free linear operators which allow one to replace the underlying system with an equivalent system with norm-bounded parameter uncertainty. The problem can then be treated by the theory of norm-bounded uncertain, delay-free systems with state-multiplicative noise [8]. Similarly to the systems treated in [1, 2], our systems may encounter a time-varying delay where the uncertain stochastic parameters multiply both the delayed and the non-delayed states in the state-space model of the systems. We treat both the nominal case and the uncertain case where the system matrices encounter polytopic uncertainties. In the latter case, we bring the simple "quadratic" solution where the same decision variables are assigned to all the vertices of the polytope, and we also treat a possibly less conservative solution which is achieved by applying a vertex-dependent approach [22]. The latter approach is first applied to the stability issue of the systems under study, followed by the solution of BRL for the uncertain case.

This chapter is organized as follows: We first bring the previously obtained stability and BRL results [1] for both nominal and uncertain systems. These results, brought in Sects. 13.3 and 13.4, respectively, are required for the derivation of the static control problem in the nominal case and they are extended to the less conservative vertex-dependent solution in the robust uncertain case. Based on the BRL result, the solution of the static output-feedback control problem is given in Sect. 13.5.1 for the nominal case. The robust polytopic case is treated in Sect. 13.5.2, where a quadratic solution is suggested and where the vertex-dependent approach is applied. In Sect. 13.6, a control engineering example is brought that demonstrates the applicability and tractability of our method.

13.2 Problem Formulation

We consider the following linear retarded system:

$$x_{k+1} = (A_0 + Dv_k)x_k + (A_1 + F\mu_k)x_{k-\tau(k)} + B_1 w_k + B_2 u_k, \quad x_l = 0, \ l \le 0,$$
$$y_k = C_2 x_k + D_{21} n_k$$

$$(13.1)$$

with the objective vector

$$z_k = C_1 x_k + D_{12} u_k, \tag{13.2}$$

where $x_k \in \mathcal{R}^n$ is the system state vector, $w_k \in \mathcal{R}^q$ is the exogenous disturbance signal, $n_k \in \mathcal{R}^p$ is the measurement noise signal, $u_k \in \mathcal{R}^\ell$ is the control input, $y_k \in \mathcal{R}^m$ is the measured output, and $z_k \in \mathcal{R}^r$ is the state combination (objective function signal) to be regulated. The time delay is denoted by the integer τ_k, and it is assumed that $0 \le \tau_k \le h$, $\forall k$. The variables $\{v_k\}$, and $\{\mu_k\}$ are zero-mean real scalar white-noise sequences that satisfy

$$E\{v_k v_j\} = \delta_{kj}, \ E\{\mu_k \mu_j\} = \delta_{kj}, \ E\{\mu_k v_j\} = 0, \ \forall k, j \ge 0.$$

The matrices in (13.1a, b) and (13.2) are constant matrices of appropriate dimensions.

We seek a constant output-feedback controller

$$u_k = K y_k \tag{13.3}$$

that achieves a certain performance requirement. We treat the following two problems:

(i) H_∞ static output-feedback control:

We consider the system of (13.1a, b) and (13.2) and the following performance index:

$$J_{OF} \triangleq ||\bar{z}_k||_{\tilde{l}_2}^2 - \gamma^2 [||w_k||_{\tilde{l}_2}^2 + ||n_{k+1}||_{\tilde{l}_2}^2].   \tag{13.4}$$

The objective is to find a controller of the type of (13.3) such that J_{OF} is negative for all nonzero w_k, n_k where $w_k \in \tilde{l}_{\mathcal{F}_k}^2([0, \infty); \mathcal{R}^q)$, $n_k \in \tilde{l}_{\mathcal{F}_k}^2([0, \infty]; \mathcal{R}^p)$.

(ii) Robust H_∞ static output-feedback control:

In the robust stochastic H_∞ control problem, we assume that the system parameters lie within the following polytope:

$$\bar{\Omega} \triangleq \begin{bmatrix} A_0 \ A_1 \ B_1 \ B_2 \ C_1 \ D_{12} \ D_{21} \end{bmatrix},   \tag{13.5}$$

which is described by the vertices:

$$\bar{\Omega} = \mathcal{C}o\{\bar{\Omega}_1, \bar{\Omega}_2, \dots, \bar{\Omega}_N\},   \tag{13.6}$$

where

$$\bar{\Omega}_i \triangleq \begin{bmatrix} A_0^{(i)} \ A_1^{(i)} \ B_1^{(i)} \ B_2^{(i)} \ C_1^{(i)} \ D_{12}^{(i)} \ D_{21}^{(i)} \end{bmatrix},   \tag{13.7}$$

and where N is the number of vertices. In other words,

$$\bar{\Omega} = \sum_{i=1}^{N} \bar{\Omega}_i f_i, \quad \sum_{i=1}^{N} f_i = 1 \ , f_i \geq 0.   \tag{13.8}$$

Similarly to the nominal case, our objective is to find a controller of the type of (13.3) such that J_{OF} of (13.4) is negative for all nonzero w_k, n_k where $w_k \in \tilde{l}_{\mathcal{F}_k}^2([0, \infty); \mathcal{R}^q)$, $n_k \in \tilde{l}_{\mathcal{F}_k}^2([0, \infty]; \mathcal{R}^p)$.

13.3 Mean Square Exponential Stability

In order to solve the above two problems, we bring first the stability result for the retarded, stochastic, discrete-time system (for detailed treatment, see [1, 2]). We apply the following scalar operators that were needed, in [1, 2], in order to transform the delayed system to an equivalent norm-bounded uncertain system:

$$\Delta_1(g_k) = g_{k-h}, \quad \Delta_2(g_k) = \sum_{j=k-h}^{k-1} g_j.   \tag{13.9}$$

Denoting $\bar{y}_k = x_{k+1} - x_k$ and using the fact that $\Delta_2(\bar{y}_k) = x_k - x_{k-h}$, the following state-space description of the system is obtained:

$$x_{k+1} = (A_0 + Dv_k + M)x_k + (A_1 - M + F\mu_k)\Delta_1(x_k) - M\Delta_2(\bar{y}_k) + B_1 w_k + B_2 u_k,$$
$$x_l = 0, \quad l \le 0,$$

$$y_k = C_2 x_k + D_{21} n_k,$$
$$z_k = C_1 x_k + D_{12} u_k, \tag{13.10}$$

where the matrix M is a free decision variable.

Considering the system of (13.10a) with $B_1 = 0$ and $B_2 = 0$ and the Lyapunov function $V_k \triangleq x_k^T Q x_k$, we obtain the following theorem, which is brought here for the sake of completeness, for achieving $E\{V_{k+1}\} - V_k < 0, \forall k$.

Theorem 13.1 *[1, 2] The exponential stability in the mean square sense of the system (13.1a) with $B_1 = 0$ and $B_2 = 0$ is guaranteed if there exist $n \times n$ matrices $Q > 0, R_1 > 0, R_2 > 0,$ and M that satisfy the following inequality:*

$$\bar{\Gamma} \triangleq \begin{bmatrix} \bar{\Gamma}_{1,1} & (A_0+M)^T Q & 0 & 0 & \bar{\Gamma}_{1,5} \\ * & -Q & Q(A_1 - M) & QM & 0 \\ * & * & \bar{\Gamma}_{3,3} & 0 & \bar{\Gamma}_{3,5} \\ * & * & * & -R_2 & -hM^T R_2 \\ * & * & * & * & -R_2 \end{bmatrix} < 0,$$

where

$$\begin{aligned} \bar{\Gamma}_{1,1} &= Q + D^T(Q + h^2 R_2)D + R_1, \\ \bar{\Gamma}_{1,5} &= h(A_0^T + M^T)R_2 - R_2 h, \\ \bar{\Gamma}_{3,3} &= -R_1 + F^T(Q + h^2 R_2)F, \\ \bar{\Gamma}_{3,5} &= h(A_1^T - M^T)R_2. \end{aligned} \tag{13.11}$$

We note that inequality (13.11a) is bilinear in the decision variables because of the terms QM and $R_2 M$. In order to remain in the linear domain, we can define $Q_M = QM$ and choose $R_2 = \epsilon Q$ where ϵ is a positive tuning scalar. The resulting Linear Matrix Inequality (LMI) can be found in [1, 2].

In the polytopic uncertain case, we obtain the following two results, the first of which is the quadratic solution which appears in [1, 2]. A new result is obtained by applying a vertex-dependent Lyapunov function [22]. Using Schur's complement, (13.11a) can be written as

$$\Psi + \Phi Q \Phi^T < 0, \tag{13.12}$$

with

$$\Psi \triangleq \begin{bmatrix} \Psi_{1,1} & 0 & 0 & h(A_0^T + M^T)R_2 - R_2 h \\ * & \Psi_{2,2} & 0 & h(A_1^T - M^T)R_2 \\ * & * & -R_2 & -hM^T R_2 \\ * & * & * & -R_2 \end{bmatrix}, \quad \Phi \triangleq \begin{bmatrix} A_0^T + M^T \\ A_1^T - M^T \\ M^T \\ 0 \end{bmatrix}, \tag{13.13}$$
$$\Psi_{1,1} = -Q + D^T(Q + h^2 R_2)D + R_1$$

$$\text{and } \Psi_{2,2} = -R_1 + F^T(Q + h^2 R_2)F.$$

The following result is thus obtained.

Lemma 13.1 *Inequality (13.12) is satisfied iff there exist matrices:* $0 < Q \in \mathcal{R}^{n \times n}$, $G \in \mathcal{R}^{n \times 4n}$, $M \in \mathcal{R}^{n \times n}$, *and* $H \in \mathcal{R}^{n \times n}$ *that satisfy the following inequality:*

$$\Omega \triangleq \begin{bmatrix} \Psi + G^T \Phi^T + \Phi G & -G^T + \Phi H \\ -G + H^T \Phi^T & -H - H^T + Q \end{bmatrix} < 0. \tag{13.14}$$

Proof Substituting $G = 0$ and $H = Q$ in (13.14), inequality (13.12) is obtained. To show that (13.14) leads to (13.12), we consider

$$\begin{bmatrix} I & \Phi \\ 0 & I \end{bmatrix} \Omega \begin{bmatrix} I & 0 \\ \Phi^T & I \end{bmatrix} = \begin{bmatrix} \Psi + \Phi Q \Phi^T & -G^T - \Phi H^T + \Phi Q \\ -G - H \Phi^T + Q \Phi^T & -H - H^T + Q \end{bmatrix}.$$

Inequality (13.12) thus follows from the fact that the (1,1) matrix block of the latter matrix is the left side of (13.12). ∎

Taking $H = G[I_n \ 0 \ 0 \ 0]^T$, $R_2 = \epsilon_r H$ where $\epsilon_r > 0$ is a scalar tuning parameter and denoting $M_H = H^T M$, we note that in (13.14) the system matrices, excluding D and F, do not multiply Q. It is thus possible to choose vertex-dependent $Q^{(i)}$ while keeping H and G constant. We thus arrive at the following result.

Corollary 13.1 *The exponential stability in the mean square sense of the system (13.1a) where* $B_1 = 0$ *and* $B_2 = 0$ *and where the system matrices lie within the polytope* $\bar{\Omega}$ *of (13.5) is guaranteed if there exist matrices* $0 < Q_j \in \mathcal{R}^{n \times n}$, $\forall j = 1, \ldots N$, $0 < R_1 \in \mathcal{R}^{n \times n}$, $M_H \in \mathcal{R}^{n \times n}$, $G \in \mathcal{R}^{n \times 4n}$, *and a tuning scalar* $\epsilon_r > 0$ *that satisfy the following set of inequalities:*

$$\Omega_j = \begin{bmatrix} \Psi_j + G^T \Phi_j^T + \Phi_j G & -G^T + \Phi_j H \\ -G + H^T \Phi_j^T & -H - H^T + Q_j \end{bmatrix} < 0, \tag{13.15}$$

$\forall j, \ j = 1, 2, \ldots, N$, *where* $H \in \mathcal{R}^{n \times n} = G[I_n \ 0 \ 0 \ 0]^T$,

$$\Psi_j \triangleq \begin{bmatrix} \Psi_{j,1,1} & 0 & 0 & h\epsilon_r(A_0^{j,T}H + M_H^T) - \epsilon_r hH \\ * & \Psi_{j,2,2} & 0 & h\epsilon_r(A_1^{j,T}H - M_H^T) \\ * & * & -\epsilon_r H & -h\epsilon_r M_H^T \\ * & * & * & -\epsilon_r H \end{bmatrix}, \quad \Phi_j = \begin{bmatrix} A_0^{j,T} + M^T \\ A_1^{j,T} - M^T \\ M^T \\ 0 \end{bmatrix}, \tag{13.16}$$

and where

$$M_H = MH$$
$$\Psi_{j,1,1} = -Q_j + D^T(Q_j + h^2 \epsilon_r H)D + R_1,$$
$$\Psi_{j,2,2} = -R_1 + F^T(Q_j + h^2 \epsilon_r H)F.$$

13.4 The Bounded Real Lemma

Based on the stability result of Corollary 13.1, the following result is readily obtained where we consider the system (13.1a) with $z_k = C_1 x_k$ and the following index of performance:

$$J_B \overset{\Delta}{=} ||z_k||^2_{l_2} - \gamma^2 ||w_k||^2_{l_2},$$

(for detailed treatment see [1, 2]).

Theorem 13.2 *[1, 2] Consider the system (13.1a) and (13.2) with $B_2 = 0$ and $D_{12} = 0$. The system is exponentially stable in the mean square sense and, for a prescribed scalar $\gamma > 0$ and a given scalar tuning parameter $\epsilon_b > 0$, the requirement of $J_B < 0$ is achieves for all nonzero $w \in \tilde{l}^2_{\mathcal{F}_k}([0, \infty); \mathcal{R}^q)$, if there exist $n \times n$ matrices $Q > 0$, $R_1 > 0$, and a $n \times n$ matrix Q_m that satisfy the following LMI:*

$$\begin{bmatrix} \tilde{\Gamma}_{11} & \tilde{\Gamma}_{12} & 0 & 0 & \tilde{\Gamma}_{15} & 0 & C_1^T \\ * & -Q & \tilde{\Gamma}_{23} & Q_m & 0 & QB_1 & 0 \\ * & * & \tilde{\Gamma}_{33} & 0 & \tilde{\Gamma}_{35} & 0 & 0 \\ * & * & * & -\epsilon_b Q & -h\epsilon_b Q_m^T & 0 & 0 \\ * & * & * & * & -\epsilon_b Q & \epsilon_b hQB_1 & 0 \\ * & * & * & * & * & -\gamma^2 I_q & 0 \\ * & * & * & * & * & * & -I_r \end{bmatrix} < 0, \qquad (13.17)$$

where

$$\tilde{\Gamma}_{11} = -Q + D^T Q[1 + \epsilon_b h^2]D + R_1,$$
$$\tilde{\Gamma}_{12} = A_0^T Q + Q_m^T,$$
$$\tilde{\Gamma}_{15} = \epsilon_b h[A_0^T Q + Q_m^T] - \epsilon_b hQ,$$
$$\tilde{\Gamma}_{23} = QA_1 - Q_m,$$
$$\tilde{\Gamma}_{33} = -R_1 + (1 + \epsilon_b h^2)F^T QF,$$
$$\tilde{\Gamma}_{35} = \epsilon_b h[A_1^T Q - Q_m^T].$$

Similarly to the stability condition for the uncertain case of Sect. 13.3, we obtain two results for the robust BRL solution. The first one, which is referred to as the quadratic solution, is given in [1, 2]. A new, possibly less conservative, condition is obtained by applying the following vertex-dependent Lyapunov function:

$$\begin{bmatrix} \Psi & \begin{bmatrix} 0 & C_1^T \\ 0 & 0 \\ 0 & 0 \\ hR_2B_1 & 0 \\ -\gamma^2 I_q & 0 \end{bmatrix} \\ \begin{bmatrix} * & * & * & * \\ * & * & * & * \end{bmatrix} & \begin{bmatrix} -\gamma^2 I_q & 0 \\ * & I_r \end{bmatrix} \end{bmatrix} + \begin{bmatrix} \Phi \\ B_1^T \\ 0 \end{bmatrix} Q[\Phi^T \; B_1 0] \overset{\Delta}{=} \hat{\Psi} + \hat{\Phi} Q \hat{\Phi}^T < 0,$$

where Ψ and Φ are given in (13.13a, b).

Following the derivation of the LMI of Corollary 13.1, the following result is readily derived:

$$\begin{bmatrix} \hat{\Psi} + \hat{G}^T\hat{\Phi}^T + \hat{\Phi}\hat{G} & -\hat{G}^T + \hat{\Phi}H \\ -\hat{G} + H^T\hat{\Phi}^T & -H - H^T + Q \end{bmatrix} < 0, \qquad (13.18)$$

where now $\hat{G} \in \mathcal{R}^{n \times 4n+q+r}$ and $H \in \mathcal{R}^{n \times n}$. We thus arrive at the following result for the uncertain case, taking $H = \hat{G}[I_n\ 0\ 0\ 0]^T$, $R_2 = \epsilon_r H$, $M_H = H^T M$.

Corollary 13.2 *Consider the system (13.1a) and (13.2) with $B_2 = 0$ and $D_{12} = 0$ and where the system matrices lie within the polytope $\bar{\Omega}$ of (13.5). The system is exponentially stable in the mean square sense and, for a prescribed $\gamma > 0$ and given tuning parameter ϵ_r, the requirement of $J_B < 0$ is achieved for all nonzero $w \in \tilde{l}^2_{\mathcal{F}_k}([0, \infty); \mathcal{R}^q)$, if there exist $0 < Q \in \mathcal{R}^{n \times n}$, $0 < R_1 \in \mathcal{R}^{n \times n}$, $M_H \in \mathcal{R}^{n \times n}$, and $\hat{G} \in \mathcal{R}^{n \times 4n+q+r}$, that satisfy the following set of LMIs:*

$$\begin{bmatrix} \hat{\Psi}_j + \hat{G}^T\hat{\Phi}^{j,T} + \hat{\Phi}_j\hat{G} & -\hat{G}^T + \hat{\Phi}_j H \\ -\hat{G} + H^T\hat{\Phi}^{j,T} & -H - H^T + Q_j \end{bmatrix} < 0, \qquad (13.19)$$

where $H \in \mathcal{R}^{n \times n} = \hat{G}[I_n\ 0\ 0\ 0]^T$,

$$\hat{\Psi}_j = \begin{bmatrix} \Psi_j & \begin{bmatrix} 0 & C_1^{j,T} \\ 0 & 0 \\ 0 & 0 \\ h\epsilon_r B_{j,1} & 0 \end{bmatrix} \\ \begin{bmatrix} * & * & * & * \\ * & * & * & * \end{bmatrix} & \begin{bmatrix} -\gamma^2 I_q & 0 \\ * & I \end{bmatrix} \end{bmatrix}, \quad and\ \hat{\Phi}_j H = \begin{bmatrix} \Phi_j H \\ B_1^{j,T} H \\ 0 \end{bmatrix}, \forall j,\ j = 1, 2, \ldots, N,$$

where Ψ_j and Φ_j are given in (13.16).

13.5 Static Output-Feedback Control

13.5.1 Static Output-Feedback Control—The Nominal Case

In this section, we address the static output-feedback control problem of the delayed state-multiplicative nominal noisy system. We consider the system of (13.1a, b), (13.2) and the constant controller of (13.3).

Denoting $\xi_k^T \overset{\Delta}{=} [x_k^T\ y_k^T]$ and $\bar{w}_k^T \overset{\Delta}{=} [w_k^T\ n_k^T]$, we obtain the following augmented system:

$$\begin{aligned} \xi_{k+1} &= \hat{A}_0\xi_k + \tilde{B}\bar{w}_k + \hat{A}_1\xi(k-\tau_k) + \tilde{D}\xi_k v_k + \\ &\quad \tilde{F}\xi(k-\tau_k)\mu_k, \quad \xi(\theta) = 0,\ over[-h\ 0], \\ \tilde{z}_k &= \tilde{C}\xi_k, \end{aligned} \qquad (13.20)$$

with the following matrices:

$$\hat{A}_0 = \begin{bmatrix} A_0 & B_2 K \\ C_2 A_0 & C_2 B_2 K \end{bmatrix}, \ \hat{A}_1 = \begin{bmatrix} A_1 & 0 \\ C_2 A_1 & 0 \end{bmatrix}, \ \tilde{F} = \begin{bmatrix} F & 0 \\ C_2 F & 0 \end{bmatrix}, \ \tilde{D} = \begin{bmatrix} D & 0 \\ C_2 D & 0 \end{bmatrix},$$

$$\tilde{C} = [C_1 \ \ D_{12}K], \text{ and } \tilde{B} = \begin{bmatrix} B_1 & 0 \\ C_2 B_1 & D_{21} \end{bmatrix}.$$

$$(13.21)$$

Using the BRL result of Theorem 13.2, we find that (13.4) is satisfied if there exist matrices $\tilde{Q}$, $\tilde{R}_1$, $\tilde{R}_2$ and $\tilde{M}$ of the appropriate dimensions that satisfy the following inequality $\hat{\Gamma} \overset{\Delta}{=}$

$$\begin{bmatrix} \hat{\Gamma}_{11} & \hat{\Gamma}_{12} & 0 & 0 & h(\hat{A}_0^T + \tilde{M}^T)\tilde{R}_2 - \tilde{R}_2 h & 0 & \tilde{C}^T \\ * & -\tilde{Q} & \tilde{Q}(\hat{A}_1 - \tilde{M}) & \tilde{Q}\tilde{M} & 0 & \tilde{Q}\tilde{B} & 0 \\ * & * & -\tilde{R}_1 + \tilde{F}^T(\tilde{Q} + h^2\tilde{R}_2)\tilde{F} & 0 & h(\hat{A}_1^T - \tilde{M}^T)\tilde{R}_2 & 0 & 0 \\ * & * & * & -\tilde{R}_2 & -h\tilde{M}^T\tilde{R}_2 & 0 & 0 \\ * & * & * & * & -\tilde{R}_2 & h\tilde{R}_2\tilde{B} & 0 \\ * & * & * & * & * & -\gamma^2 I_{q+l} & 0 \\ * & * & * & * & * & * & -I_r \end{bmatrix} < 0, \quad (13.22)$$

where $\hat{\Gamma}_{11} = -\tilde{Q} + \tilde{D}^T(\tilde{Q} + h^2\tilde{R}_2)\tilde{D} + \tilde{R}_1$, $\hat{\Gamma}_{12} = (\hat{A}_0 + \tilde{M})^T\tilde{Q}$.
  Taking $\tilde{R}_2 = \epsilon_b \tilde{Q}$, where $\epsilon_b > 0$ is a tuning scalar parameter, the following inequality is obtained:

$$\begin{bmatrix} \hat{\Gamma}_{11} & \hat{A}_0^T\tilde{Q} + \tilde{M}^T\tilde{Q} & 0 & 0 & \epsilon_b h[\hat{A}_0^T\tilde{Q} + \tilde{M}^T\tilde{Q}] - \epsilon_b h\tilde{Q} & 0 & \tilde{C}^T \\ * & -\tilde{Q} & \tilde{Q}\hat{A}_1 - \tilde{Q}\tilde{M} & \tilde{Q}\tilde{M} & 0 & \tilde{Q}\tilde{B} & 0 \\ * & * & \hat{\Gamma}_{33} & 0 & \epsilon_b h[\hat{A}_1^T\tilde{Q} - \tilde{M}^T\tilde{Q}] & 0 & 0 \\ * & * & * & -\epsilon_b\tilde{Q} & -h\epsilon_b\tilde{M}^T\tilde{Q} & 0 & 0 \\ * & * & * & * & -\epsilon_b\tilde{Q} & \epsilon_b h\tilde{Q}\tilde{B} & 0 \\ * & * & * & * & * & -\gamma^2 I_{q+l} & 0 \\ * & * & * & * & * & * & -I_r \end{bmatrix} < 0,$$

$$(13.23)$$

where $\hat{\Gamma}_{11} = -\tilde{Q} + \tilde{D}^T\tilde{Q}[1 + \epsilon_b h^2]\tilde{D} + \tilde{R}_1$, $\hat{\Gamma}_{33} = -\tilde{R}_1 + (1 + \epsilon_b h^2)\tilde{F}^T\tilde{Q}\tilde{F}$.
  Denoting $\tilde{P} = \tilde{Q}^{-1}$ and choosing the following partition:

$$\tilde{P} = \tilde{Q}^{-1} = \begin{bmatrix} P & C_2^T\hat{P} \\ \hat{P}C_2 & \alpha\hat{P} \end{bmatrix}, \quad (13.24)$$

where $Q > 0, P > 0 \in \mathcal{R}^{n \times n}$ and $\hat{Q} > 0, \hat{P} > 0 \in \mathcal{R}^{m \times m}$, where α is a tuning scalar parameter, we multiply (13.23) by $diag\{\tilde{P}, \tilde{P}, \tilde{P}, \tilde{P}, \tilde{P}, I_{q+l}, I_r\}$ from the left and the right and we obtain the following inequality:

$$
\begin{bmatrix}
-\tilde{P}+\tilde{R}_p & \bar{\Upsilon}_{12} & 0 & 0 & \bar{\Upsilon}_{15} & 0 & \bar{\Upsilon}_{17} & 0 & \bar{\Upsilon}_{19} \\
* & -\tilde{P} & \bar{\Upsilon}_{23} & \tilde{M}_P & 0 & \tilde{B} & 0 & 0 & 0 \\
** & * & -\tilde{R}_p & 0 & \bar{\Upsilon}_{35} & 0 & 0 & \bar{\Upsilon}_{38} & 0 \\
** & * & * & -\epsilon_b\tilde{P} & -h\epsilon_b\tilde{M}_P^T & 0 & 0 & 0 & 0 \\
** & * & * & * & -\epsilon_b\tilde{P} & \epsilon_b h\tilde{B} & 0 & 0 & 0 \\
** & * & * & * & * & -\gamma^2 I_{q+l} & 0 & 0 & 0 \\
** & * & * & * & * & * & -I_r & 0 & 0 \\
** & * & * & * & * & * & * & -\tilde{P} & 0 \\
** & * & * & * & * & * & * & * & -\tilde{P}
\end{bmatrix} < 0, \qquad (13.25)
$$

where

$$
\bar{\Upsilon}_{12} = \begin{bmatrix} PA_0^T + C_2^T\hat{P}_K^T B_2^T & PA_0^T C_2^T + C_2^T\hat{P}_K^T B_2^T C_2^T \\ \hat{P}C_2 A_0^T + \alpha\hat{P}_K^T B_2^T & \hat{P}C_2 A_0^T C_2^T + \alpha\hat{P}_K^T B_2^T C_2^T \end{bmatrix}
$$
$$
+ \tilde{M}_P^T, \ \bar{\Upsilon}_{17} = \begin{bmatrix} PC_1^T + C_2^T\hat{P}_K^T D_{12}^T \\ \hat{P}C_2 C_1^T + \alpha\hat{P}_K^T D_{12}^T \end{bmatrix}, \ \bar{\Upsilon}_{15} = -\epsilon_b h\tilde{P} +
$$
$$
\epsilon_b h \begin{bmatrix} PA_0^T + C_2^T\hat{P}_K^T B_2^T & PA_0^T C_2^T + C_2^T\hat{P}_K^T B_2^T C_2^T \\ \hat{P}C_2 A_0^T + \alpha\hat{P}_K^T B_2^T & \hat{P}C_2 A_0^T C_2^T + \alpha\hat{P}_K^T B_2^T C_2^T \end{bmatrix} + \epsilon_b h\tilde{M}_P^T,
$$
$$
\bar{\Upsilon}_{19} = \sqrt{1+\epsilon_b h^2} \begin{bmatrix} PD^T & PD^T C_2^T \\ \hat{P}C_2 D^T & \hat{P}C_2 D^T C_2^T \end{bmatrix},
$$
$$
\bar{\Upsilon}_{23} = \begin{bmatrix} A_1 P & A_1 C_2^T \hat{P} \\ C_2 A_1 P & C_2 A_1 C_2^T \hat{P} \end{bmatrix} - \tilde{M}_P,
$$
$$
\bar{\Upsilon}_{35} = \epsilon_b h \begin{bmatrix} PA_1^T & PA_1^T C_2^T \\ \hat{P}C_2 A_1^T & \hat{P}C_2 A_1^T C_2^T \end{bmatrix} - \epsilon_b h\tilde{M}_P^T,
$$
$$
\bar{\Upsilon}_{38} = \sqrt{1+\epsilon_b h^2} \begin{bmatrix} PF^T & PF^T C_2^T \\ \hat{P}C_2 F^T & \hat{P}C_2 F^T C_2^T \end{bmatrix},
$$

and where we denote

$$
\tilde{R}_p = \tilde{P}R_1\tilde{P}, \quad \tilde{M}_P = \tilde{M}\tilde{P}, \quad \hat{P}_K = K\hat{P}. \qquad (13.26)
$$

We arrive at the following theorem.

Theorem 13.3 *Consider the system (13.1a, b) and (13.2). There exists a controller of the structure of (13.3) that achieves negative J_{OF} for all nonzero $w \in \tilde{L}^2_{\mathcal{F}_t}([0, \infty); \mathcal{R}^q)$, $n \in \tilde{L}^2_{\mathcal{F}_t}([0, \infty); \mathcal{R}^p)$, for a prescribed scalar $\gamma > 0$, a given upper-bound h and positive tuning parameters α and ϵ_b if there exist matrices $P > 0$, $\hat{P} > 0$, $\tilde{M}_P$, $\tilde{R}_P > 0$ and $\hat{P}_K$, that satisfy (13.25). In the latter case, the controller gain matrix is*

$$
K = \hat{P}^{-1}\hat{P}_K. \qquad (13.27)
$$

We note that the result of [15] for the case with no delay is recovered from (13.22) by substituting $R_1 = R_2 = 0$, $M = 0$, and $A_1 = 0$ and multiplying the second row and column in (13.22) by $\tilde{Q}^{-1}$.

Remark 13.1 We note that a simple strategy is applied for the numerical solution of (13.25). The solution of (13.25) involves a search for two scalar variables: α_b and ϵ_b. One may start by taking arbitrary values for the latter two parameters and seek, using line-searching, values for these tuning parameters that leads to a stabilizing controller of minimum γ. Once such a controller is obtained, standard optimization techniques can be used, say MATLAB function "fminsearch", which seek the combination of the two scalar parameters that bring γ to a local minimum.

13.5.2 Static Output-Feedback Control—The Uncertain Case

In the polytopic uncertain case, one may obtain a solution, applying the so-called quadratic solution. By applying a single Lyapunov function and exploiting the convex nature of the LMI of (13.25) (see Theorem 13.3), a set of LMIs can be solved where to each vertex of the uncertainty polytope a single LMI is applied. Obviously, the resulting solution is quite conservative. However, similarly to the robust stability case, a less conservative result maybe obtained by applying vertex-dependent Lyapunov function approach. The result of (13.22) can be written as

$$
\left[\tilde{\Psi} \begin{bmatrix} 0 & \tilde{C}^T \\ 0 & 0 \\ 0 & 0 \\ hR_2\tilde{B} & 0 \\ \begin{bmatrix} * & * & * & * \\ * & * & * & * \end{bmatrix} \begin{bmatrix} -\gamma^2 I_{q+p} & 0 \\ * & I \end{bmatrix} \end{bmatrix} \right] + \begin{bmatrix} \tilde{\Phi} \\ \tilde{B}^T \\ 0 \end{bmatrix} \tilde{Q} \begin{bmatrix} \tilde{\Phi}^T & \tilde{B} & 0 \end{bmatrix}
$$

$$
\stackrel{\Delta}{=} \hat{\Upsilon} + \hat{\Theta}\tilde{Q}\hat{\Theta}^T < 0, \quad \text{where} \quad \tilde{\Phi} \stackrel{\Delta}{=} \begin{bmatrix} \hat{A}_0^T + M^T \\ \hat{A}_1^T - M^T \\ M^T \\ 0 \end{bmatrix},
$$

and where

$$
\tilde{\Psi} \stackrel{\Delta}{=} \begin{bmatrix} \tilde{\Psi}_{1,1} & 0 & 0 & h(\hat{A}_0^T + M^T)R_2 - R_2h \\ * & \tilde{\Psi}_{2,2} & 0 & h(\hat{A}_1^T - M^T)R_2 \\ * & * & -R_2 & -hM^TR_2 \\ * & * & * & -R_2 \end{bmatrix}, \tag{13.28}
$$

with

$$
\tilde{\Psi}_{1,1} = -\tilde{Q} + \tilde{D}^T(\tilde{Q} + h^2R_2)\tilde{D} + R_1,
$$
$$
\tilde{\Psi}_{2,2} = -R_1 + \tilde{F}^T(\tilde{Q} + h^2R_2)\tilde{F}.
$$

Similarly to the derivation of the LMI of Corollary 13.2, the following result is readily derived:

$$\begin{bmatrix} \hat{\Upsilon} + \hat{G}^T \hat{\Theta}^T + \hat{\Theta} \hat{G} & -\hat{G}^T + \hat{\Theta} H \\ -\hat{G} + H^T \hat{\Theta}^T & -H - H^T + \tilde{Q} \end{bmatrix} < 0, \tag{13.29}$$

where now $\hat{G} \in \mathcal{R}^{(n+m) \times (4[n+m]+q+p+r)}$, $H \in \mathcal{R}^{n+m \times n+m}$, and $R_2 \in \mathcal{R}^{n+m \times n+m}$.

Taking $H = \hat{G}[\epsilon_s I_{n+m} \; 0 \; 0 \; 0]^T$, $R_2 = \epsilon_r H$ and denoting $M_H = H^T M$, we arrive at the following result for the uncertain case.

Theorem 13.4 *Consider the system (13.1a, b) and (13.2) and where the system matrices lie within the polytope $\bar{\Omega}$ of (13.5). The system is exponentially stable in the mean square sense and, for a prescribed $\gamma > 0$ and given tuning parameters $\epsilon_r > 0$, $\epsilon_s > 0$, the requirement of $J_{OF} < 0$ is achieved for all nonzero $w \in \tilde{l}^2_{\mathcal{F}_k}([0, \infty); \mathcal{R}^q)$, $n \in \tilde{l}^2_{\mathcal{F}_k}([0, \infty); \mathcal{R}^p)$ if there exist $0 < \tilde{Q}_j \in \mathcal{R}^{n+m \times n+m}$, $0 < R_1 \in \mathcal{R}^{n+m \times n+m}$, $\hat{G} \in \mathcal{R}^{(n+m) \times (4[n+m]+q+p+r)}$, and $K \in \mathcal{R}^{l \times m}$ that satisfy the following set of inequalities:*

$$\begin{bmatrix} \hat{\Upsilon}_j + \hat{G}^T \hat{\Theta}^{j,T} + \hat{\Theta}_j \hat{G} & -\hat{G}^T + \hat{\Theta}_j H \\ -\hat{G} + H^T \hat{\Theta}^{j,T} & -H - H^T + \tilde{Q}_j \end{bmatrix} < 0, \tag{13.30}$$

$$\forall j, j = 1, \dots N,$$

where $H \in \mathcal{R}^{(n+m) \times (n+m)} = \hat{G}[\epsilon_s I_{n+m} \; 0 \; 0 \; 0]^T$,

$$\hat{\Upsilon}_j = \begin{bmatrix} \tilde{\Psi}_j & \begin{bmatrix} 0 & \tilde{C}^{j,T} \\ 0 & 0 \\ 0 & 0 \\ h\epsilon_r H \tilde{B}^j & 0 \end{bmatrix} \\ \begin{bmatrix} * & * & * & * \\ * & * & * & * \end{bmatrix} & \begin{bmatrix} -\gamma^2 I_{q+p} & 0 \\ * & I_r \end{bmatrix} \end{bmatrix}, \quad \hat{\Theta}_j = \begin{bmatrix} \tilde{\Phi}_j \\ \vdots \\ \tilde{B}^{j,T} \\ 0 \end{bmatrix},$$

$\forall j, j = 1, 2, \dots, N$, where $\tilde{\Psi}_j \overset{\Delta}{=}$

$$\tilde{\Psi} \overset{\Delta}{=} \begin{bmatrix} \tilde{\Psi}_{11,j} & 0 & 0 & h\epsilon_r(\hat{A}_0^{j,T} H + M_H^T) - h\epsilon_r H \\ * & \tilde{\Psi}_{22,j} & 0 & h\epsilon_r[\hat{A}_1^{j,T} H - M_H^T] \\ * & * & -\epsilon_r H & -h\epsilon_r M_H^T \\ * & * & * & -\epsilon_r H \end{bmatrix},$$

and where

$$\tilde{\Psi}_{11,j} = -\tilde{Q}_j + \tilde{D}^T(\tilde{Q}_j + h^2 \epsilon_r H)\tilde{D} + R_1, \quad \tilde{B}^j = \begin{bmatrix} B_1^j & 0 \\ C_2 B_1^j & D_{21}^j \end{bmatrix}.$$
$$\tilde{\Psi}_{22,j} = -R_1 + \tilde{F}^T(\tilde{Q}_j + h^2 \epsilon_r H)\tilde{F},$$

Theorem 13.4 provides a solution to the static output-feedback control problem in the uncertain case by means of BMI (Bilinear Matrix Inequality) (see, for example, the product of of $\hat{A}^T R_2$ in (13.28)). In this theorem, the static gain matrix K is a decision variable. In Theorem 13.3 (and in the quadratic solution for the robust case), it was possible to define the product $\hat{P}K$ in (13.26c) as a new decision variable and the inequality that provides the solution to the problem becomes linear. Such a procedure could not provide linear matrix inequality in the vertex-dependent approach of Theorem 13.4. We note, however, that by Theorem 13.4 the L_2 gain of the uncertain system can be readily found using LMIs. It can be used in open-loop analysis (for $K = 0$) and in verifying the L_2 gain of the system for a given $K \neq 0$.

13.6   Example

We consider a modified version of a third-order, two-outputs, three-inputs physical example taken from [15] where we seek static output-feedback controllers for both the nominal system and the uncertain polytopic system. We consider the nominal system of (13.1a, b) and (13.2) where

$$
A = \begin{bmatrix} 0.8500 & 0.3420 & 1.3986 \\ 0.0052 & 0.8500 & -0.1656 \\ 0 & 0 & 0.5488 \end{bmatrix}, \quad B_1 = \begin{bmatrix} 0.0198 & 0.0034 & 0.0156 \\ 0.0001 & 0.0198 & -0.0018 \\ 0 & 0 & 0.0150 \end{bmatrix}, \quad B_2 = \begin{bmatrix} -1.47 \\ -0.0604 \\ 0.4512 \end{bmatrix},
$$

$$
C_2 = \begin{bmatrix} 1 & 0 & 0 \\ 0 & 1 & 0 \end{bmatrix}, \quad C_1 = \begin{bmatrix} 1 & 0 & 0 \\ 0 & 1 & 0 \\ 0 & 0 & 0 \end{bmatrix}, \quad D = \begin{bmatrix} 0 & 0 & 0 \\ 0 & 0 & 0 \\ 0 & 0 & 0.4 \end{bmatrix}, \quad A_1 = \begin{bmatrix} 0 & 0 & 0 \\ 0 & 0.1615 & 0 \\ 0 & 0 & 0.0950 \end{bmatrix} \quad (13.31)
$$

with $D_{21} = 0$, $D_{12} = \begin{bmatrix} 0 \\ 0 \\ 1 \end{bmatrix}$, $F = 0$. We obtain the following results.

- **The stochastic H_∞ controller for the nominal system**: Using Theorem 13.3, a near minimum value of $\gamma = 0.738$ is obtained for $h = 11$, $\alpha = 3.9$, $\epsilon = 1 \times 10^{-6}$. The corresponding static output-feedback controller gain matrix is

$$
K = \begin{bmatrix} 0.0289 & 0.7799 \end{bmatrix},
$$

and the corresponding closed-loop poles are 0.855, $0.652 + 0.186i$, and $0.652 - 0.186i$.

In Fig. 13.1, two step response curves are given for the above nominal system from the third input to the first output with $h = 11$ s and $\alpha = 3.9$. The first curve describes the step response for 10 different random sequences of the multiplicative noise, while the second curve describes the average and the standard deviation of these responses.

- **The robust stochastic H_∞ controller:** We consider the system of (13.1a, b) and (13.2) where the system matrices A, B_1, and B_2 reside in a polytope of four vertices. The system matrices include the matrices of (13.31a-i) as vertex no. 1 and the following three sets of $A^{(j)}$, $B_1^{(j)}$, and $B_2^{(j)}$ for the vertices $j = 2, 3, 4$:

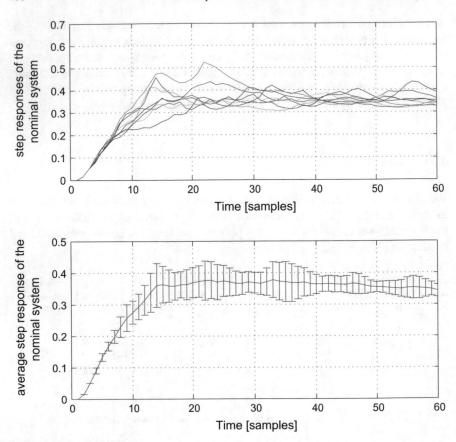

Fig. 13.1 Step responses of the nominal system for $h = 11$ s. Shown are, in the upper curve, the step responses for 10 different random sequences of the multiplicative noise. The lower curve depicts the average and the standard deviation of these 10 responses

$$A^{(2)} = \begin{bmatrix} 0.8500 & 0.3575 & 1.2273 \\ 0.0016 & 0.7872 & -0.1603 \\ 0 & 0 & 0.5488 \end{bmatrix}, \quad B_1^{(2)} = \begin{bmatrix} 0.0199 & 0.0036 & 0.0137 \\ 0.0000 & 0.0199 & -0.0018 \\ 0 & 0 & 0.0150 \end{bmatrix},$$

$$B_2^{(2)} = \begin{bmatrix} -4.9990 \\ -0.0576 \\ 0.4512 \end{bmatrix},$$

$$A^{(3)} = \begin{bmatrix} 0.8500 & 0.9840 & 3.6304 \\ 0.0043 & 0.8142 & -0.4647 \\ 0 & 0 & 0.5488 \end{bmatrix}, \quad B_1^{(3)} = \begin{bmatrix} 0.0197 & 0.0099 & 0.0412 \\ 0.0000 & 0.0197 & -0.0052 \\ 0 & 0 & 0.0150 \end{bmatrix},$$

$$B_2^{(3)} = \begin{bmatrix} -0.4376 \\ -0.1589 \\ 0.4512 \end{bmatrix},$$

$$A^{(4)} = \begin{bmatrix} 0.8500 & 0.5881 & 2.5226 \\ -0.0135 & 0.8517 & -0.4702 \\ 0 & 0 & 0.5488 \end{bmatrix}, \quad B_1^{(4)} = \begin{bmatrix} 0.0199 & 0.0059 & 0.0284 \\ -0.0001 & 0.0197 & -0.0051 \\ 0 & 0 & 0.0150 \end{bmatrix},$$

$$B_2^{(4)} = \begin{bmatrix} -1.4700 \\ -0.0604 \\ 0.4512 \end{bmatrix}.$$

Applying the so-called quadratic solution by solving (13.25) for each of the polytope vertices using the same decision variables with $h = 11$ s, a minimum value of $\gamma = 4.01$ is obtained for $\alpha = 4.0$ and $\epsilon = 1 \times 10^{-6}$. The corresponding static output-feedback controller gain matrix is

$$K = \begin{bmatrix} -0.0052 & 0.2494 \end{bmatrix}. \tag{13.32}$$

We note that by seeking a static controller for each of the four subsystems separately, we obtain a near minimum levels of $\gamma = 0.738, 0.30, 1.58,$ and 1.03 for $i = 1, 2, 3, 4,$ respectively. Obviously, the expected minimum upper-bound for γ in the case where the parameters may reside anywhere in the polytope is higher than the above former values.

Table 13.1 demonstrates the effect of the time delay upper-bound h on the attenuation level of the above uncertain system.

As mentioned above, Theorem 13.4 cannot be readily applied as LMI in the design. However, it should provide a lower bound γ for the controller found by the quadratic solution. Indeed, substituting K from (13.32) in (13.30), a guaranteed upper-bound of $\gamma = 3.56$ is obtained for $h = 11$ s.

Table 13.1 The effect of the time delay upper-bound h on the attenuation level γ for the uncertain system

h (delay upper-bound in s)	γ, (attenuation level)	α	ϵ_b
11	4.01	4.0	1×10^{-6}
200	4.29	3.1	1×10^{-6}
400	5.57	4.1	1×10^{-6}
600	17.44	4.0	1×10^{-6}
650	110	3.0	1×10^{-6}

13.7   Conclusions

In this chapter, the theory of H_∞ static output-feedback control of state-multiplicative noisy systems is developed for discrete-time delayed systems, for both nominal and uncertain polytopic systems. In these systems, the stochastic state-multiplicative uncertainties are encountered in both the delayed and the non-delayed states in the state-space model of the system. The delay is assumed to be unknown and time varying where only the bound on its size is given. Delay-dependent synthesis methods are developed which are based on the input–output approach, in accordance with the approach taken for the solution of the state-feedback and filtering problems previously solved. This approach transforms the delayed system to a non-retarded system with norm-bounded operators. Compared to previous results, the delay-dependent analysis is extended here to the robust polytopic case where a vertex-dependent approach is applied, based on Finsler lemma.

Based on the BRL derivation, the static output-feedback control problem is formulated and solved for the nominal case where a single LMI is obtained. The robust case is also solved where a linear quadratic solution is obtained where a vertex-dependent solution is suggested. Some overdesign is admitted to our solution due to the use of the bounded operators which enable us to transform the retarded system to a norm-bounded one. Some additional overdesign is also admitted in our solution due to the special structure imposed on R_2. The numerical example shows the efficiency of the proposed results. It also demonstrates the applicability of the BMI in Theorem 13.4 in tightening the bound on the guaranteed attenuation level of the closed-loop system.

References

1. Gershon, E., Shaked, U.: Advanced Topics in Control and Estimation of State-Multiplicative Noisy Systems. Lecture Notes in Control and Information Sciences, LNCIS, vol. 439. Springer, Berlin (2013)
2. Gershon, E., Shaked, U.: Stochastic H_∞ control and estimation of state-multiplicative discrete-time systems with delay. In: Proceedings of the European Control Conference (ECC13). Zurich, Swiss (2013)
3. Moheimani, R.: Perspectives in Robust Control. LNCIS, vol. 268. Springer, Berlin (2001)
4. Kar, I.M.: Design of static output feedback controller for uncertain systems. Automatica **35**, 169–175 (1999)
5. Syrmos, V.L., Abdallah, C.T., Dorato, P., Grigoriadis, K.: Static output feedback—a survey. Automatica **33**(2), 125–137 (1997)
6. Leibfritz, F.: An LMI-based algorithm for designing suboptimal static H_2/H_∞ output feedback controllers. SIAM J. Control Optim. **39**, 1711–1735 (2001)
7. Bara, G.I., Boutayeb, M.: Static output feedback stabilization with H_∞ performance for linear discrete-time systems. IEEE Trans. Autom. Control **50**(2), 250–254 (2005)
8. Gershon, E., Shaked, U., Yaesh, I.: H_∞ Control and Estimation of State-Multiplicative Linear Systems. Lecture Notes in Control and Information Sciences, LNCIS, vol. 318. Springer, Berlin (2005)

9. Dragan, V., Morozan, T.: Mixed input-output optimization for time-varying Ito systems with state dependent noise. Dyn. Contin., Discret. Impuls. Syst. **3**, 317–333 (1997)
10. Hinriechsen, D., Pritchard, A.J.: Stochastic H_∞. SIAM J. Control Optim. **36**(5), 1504–1538 (1998)
11. Chen, B.S., Zhang, W.: Stochastic H_2/H_∞ control with state-dependent noise. IEEE Transections Autom. Control **49**(1), 45–57 (2004)
12. Ugrinovsky, V.A.: Robust H_∞ control in the presence of of stochastic uncertainty. Int. J. Control **71**, 219–237 (1998)
13. Gershon, E., Shaked, U., Yaesh, I.: Static output-feedback of state-multiplicative systems with application to altitude control. Int. J. Control **77**, 1160–1170 (2005)
14. Dragan, V., Stoica, A.: A γ Attenuation Problem for Discrete-Time Time-Varying Stochastic Systems with Multiplicative Noise. Reprint Series of the Institute of Mathematics of the Romanian Academy, vol. 10 (1998)
15. Gershon, E., Shaked, U.: Static H_2 and H_∞ output-feedback of discrete-time LTI systems with state-multiplicative noise. Syst. Control Lett. **55**(3), 232–239 (2006)
16. Florchinger, P., Verriest, E.I.: Stabilization of nonlinear stochastic systems with delay feedback. In: Proceeding of the 32nd IEEE Conference Decision and Control, San Antonio, TX, pp. 859–860, Dec 1993
17. Chen, W.H., Guan, Z., Lu, X.: Delay-dependent exponential stability of uncertain stochastic systems with multiple delays: an LMI approach. Syst. Control Lett. **54**, 547–555 (2005)
18. Gershon, E., Shaked, U.: Robust H_∞ output-feedback control of state-multiplicative stochastic systems with delay. Int. J. Robust Nonlinear Control **21**(11), 1283–1296 (2011)
19. Gershon, E.: Robust reduced-order H_∞ output-feedback control of retarded state-multiplicative linear systems. IEEE Trans. Autom. Control **58**(11), 2898–2904 (2013)
20. Xu, S., Lam, J., Chen, T.: Robust H_∞ control for uncertain discrete stochastic time-delay systems. Syst. Control Lett. **51**, 203–215 (2004)
21. Yue, D., Tian, E., Zhang, Y.: Stability analysis of discrete systems with stochastic delay and its applications. Int. J. Innov. Comput., Inf. Control **5**(8), 2391–2403 (2009)
22. de Oliveira, M.C., Skelton, R.E.: Stability test for constrained linear systems. In: Reza Moheimani, S.O. (ed.) Perspectives in Robust Control. Lecture Notes in Control and Information Sciences, vol. 268. Springer, London (2001)

Chapter 14
H_∞ Control of Stochastic Discrete-Time Switched Systems with Dwell

Abstract Linear discrete-time switched stochastic systems are considered, where the issues of mean square stability, stochastic l_2-gain, and state-feedback control design problems are treated and solved. Solutions are obtained for both: nominal and polytopic uncertain systems. In all these problems, the switching obeys a dwell time constraint. In our solution, to each subsystem of the switched system, a Lyapunov function is assigned that is nonincreasing at the switching instants. The latter function is allowed to vary piecewise linearly, starting at the end of the previous switch instant, and it becomes time invariant as the dwell proceeds. In order to guarantee asymptotic stability, we require the Lyapunov function to be negative between consecutive switchings. We thus obtain a linear matrix inequality condition. Based on the solution of the stochastic l_2-gain problem, we derive a solution to the state-feedback control design, where we treat a variety of special cases. Being affine in the system matrices, all the above solutions are extended to the uncertain polytopic case. The proposed theory is demonstrated by a flight control example.

14.1 Introduction

In this chapter, we consider linear, switched, stochastic, state-multiplicative systems characterized by dwell time constraints. For these systems, we derive solutions to the mean square stability, stochastic l_2-gain, and state-feedback control problems.

Stochastic state-multiplicative systems have been treated in the literature, in the control and signal processing context, in the last five decades (see [1] and the references therein [2, 3]). These systems belong to a special class of nonlinear systems, traditionally referred to as stochastic bilinear systems. Following the vast research in the stochastic H_2 framework, the research interest has been shifted toward H_∞ and other optimization indices already in the early 80s. A large body of knowledge has accumulated spanning also to delayed systems for both continuous- and discrete-time systems (see [4] and the references therein [5–9]).

The problems of stability, control, and estimation of deterministic switched systems have been extensively investigated in the past [10, 11]—focusing mainly on switching without constraints. Unfortunately, the realistic case is usually

© Springer Nature Switzerland AG 2019
E. Gershon and U. Shaked, *Advances in H_∞ Control Theory*,
Lecture Notes in Control and Information Sciences 481,
https://doi.org/10.1007/978-3-030-16008-1_14

characterized by dwell time during which the system is not allowed to switch or cannot be switched. Imposing dwell time on the switched subsystem may guarantee stability in cases where the system is unstable without dwell [12].

Switching of stochastic systems has been considered in the literature mainly in the context of stochastic Markov jumps [13–17], whereas the issue of stochastic-switched systems with dwell time has hardly been treated. Markov jump systems with state-multiplicative noise have also been treated in the literature [18]. Only recently, a solution was obtained [19] for the problem of stabilizing a stochastic continuous-time switched system with dwell time for both nominal- and polytopic-type uncertain systems. To the best of our knowledge, the issues of stability and control of discrete-time switched systems with dwell time has not been considered before in the literature.

In [20], the solution of the deterministic counterpart of our problems has been found for the continuous-time case. In [21], the deterministic discrete-time version of [20] has been solved. The deterministic result of [20] has been extended to the stochastic case in [19]. A piecewise linear in time Lyapunov Function (LF) was applied there, which enabled the solution of the robust polytopic case where the Ito calculus approach was adopted.

In the present chapter, we solve the stochastic discrete-time counterpart problems of [19] where special and new emphasis on the robust case is put which leads to less conservative solutions. We apply a LF which is decreasing at the switching instances, and it varies piecewise linearly in time during the dwell. After the dwell, this LF becomes time invariant. A special structure is assigned to the LF which depends on the vertices. This LF enables the derivation of sufficient conditions for the uncertain case which, similarly to the solution for the nominal case, are expressed in form of Linear Matrix Inequalities (LMIs).

In the present chapter, we assume that the switching signal is unknown in real-time. The case where the switching signal is known with some delay is referred to as "asynchronous switching." Switching-dependent Lyapunov function method, similar to the ones presented in this chapter, can be applied to asynchronous switching, see, for example, [22, 23].

The chapter is organized as follows: In Sect. 14.2, we bring the problem formulation followed, in Sect. 14.3, by the investigation of the stochastic stability, where sufficient conditions are obtained for the stability of the stochastic-switched systems. We first derive bounds on the stochastic l_2-gain for the nominal case, in Sect. 14.4, and then we extend these results to the case of polytopic-type uncertainty. We also present, in Sect. 14.4, an alternative, less conservative, condition for the bound on the stochastic l_2-gain for uncertain systems. The latter result is then applied to the solution of the robust stochastic state-feedback control problem in Sect. 14.5. A practical control engineering example, taken from the aviation field, is brought where a switched state-feedback controller is derived for robust polytopic H_∞ control of the F4E fighter aircraft.

Notation: Throughout the chapter, the superscript "'" stands for matrix transposition, $\mathcal{R}^n$ denotes the n-dimensional Euclidean space, $\mathcal{R}^{n\times m}$ is the set of all $n \times m$

real matrices, $\mathcal{N}$ is the set of natural numbers, and the notation $P > 0$, (respectively, $P \geq 0$) for $P \in \mathcal{R}^{n \times n}$ means that P is symmetric and positive definite (respectively, semi-definite). We denote by $l^2(\Omega, \mathcal{R}^n)$ the space of square-summable $\mathcal{R}^n-$ valued functions on the probability space $(\Omega, \mathcal{F}, \mathcal{P})$, where Ω is the sample space, $\mathcal{F}$ is a σ algebra of a subset of Ω called events, and $\mathcal{P}$ is the probability measure on $\mathcal{F}$. By $(\mathcal{F}_k)_{k \in \mathcal{N}}$, we denote an increasing family of σ-algebras $\mathcal{F}_k \subset \mathcal{F}$. We also denote by $\tilde{l}^2(\mathcal{N}; \mathcal{R}^n)$ the n-dimensional space of nonanticipative stochastic processes $\{f_k\}_{k \in \mathcal{N}}$ with respect to $(\mathcal{F}_k)_{k \in \mathcal{N}}$ where $f_k \in l^2(\Omega, \mathcal{R}^n)$. On the latter space, the following l^2-norm is defined:

$$\|\{f_k\}\|_{\tilde{l}_2}^2 = E\{\sum_0^\infty \|f_k\|^2\} = \sum_0^\infty E\{\|f_k\|^2\} < \infty,$$
$$\{f_k\} \in \tilde{l}_2(\mathcal{N}; \mathcal{R}^n), \tag{14.1}$$

where $\| \cdot \|$ is the standard Euclidean norm and where $E\{*\}$ is the expectation operator. Throughout the manuscript, we refer to the notation of exponential l^2 stability, or internal stability, in the sense of [24] (see Definition 2.1, page 927, there). Switching law of a switched system is denoted by $\sigma(t)$. A symmetric matrix $\begin{bmatrix} Q & R^T \\ R & P \end{bmatrix}$ is denoted by $\begin{bmatrix} Q & * \\ R & P \end{bmatrix}$. We denote by δ_{ij} the Kronecker delta function.

14.2 Problem Formulation

We consider the following linear stochastic state-multiplicative switched system:

$$x(t+1) = A_{\sigma(t)}x(t) + B_{\sigma(t)}w(t) + F_{\sigma(t)}x(t)\beta(t) + \bar{G}_{\sigma(t)}x(t)\nu(t), \quad x(0) = x_0,$$
$$z(t) = C_{\sigma(t)}x(t) + D_{\sigma(t)}w(t),$$
$$\tag{14.2}$$

which is defined for all $t \in \mathcal{N}$, where $x(t) \in \mathcal{R}^n$ is the system state vector, $w(t) \in \mathcal{R}^q$ is the exogenous disturbance in $\tilde{l}_{\mathcal{F}_t}^2([0, \infty); \mathcal{R}^q)$, and $z(t) \in \mathcal{R}^l$ is the objective vector. The variables $\beta(t)$ and $\nu(t)$ are zero-mean real scalar white-noise sequences that satisfy

$$E\{\beta_k\beta_j\} = \delta_{kj}, \quad E\{\nu_k\nu_j\} = \delta_{kj} \quad E\{\beta_k\nu_j\} = 0, \quad \forall k, j \in \mathcal{N}. \tag{14.3}$$

We note that we deliberately inserted two stochastic terms in (14.2a) (instead of using one term only) in order to set the right platform for the state-feedback design in Sect. 14.5. The switching rule $\sigma(t) \in \{1 \ldots M\}$, for each $t \in \mathcal{N}$, is such that $A_{\sigma(t)} \in \{A_1, \ldots, A_M\}$, $F_{\sigma(t)} \in \{F_1, \ldots, F_M\}$, $B_{\sigma(t)} \in \{B_1, \ldots, B_M\}$, $\bar{G}_{\sigma(t)} \in \{\bar{G}_1, \ldots, \bar{G}_M\}$, and $C_{\sigma(t)} \in \{C_1, \ldots, C_M\}$, $D_{\sigma(t)} \in \{D_1, \ldots, D_M\}$, where $A_i \in \mathcal{R}^{n \times n}, i = 1, \ldots M$ is a stable matrix. The matrices $A_i, B_i, C_i, D_i,$ and $F_i, \bar{G}_i$ are assumed to reside within the following polytope:

$$\Omega_i = \sum_{j=1}^{N_i} \eta^j \Omega_i^{(j)}, \quad \sum_{j=1}^{N_i} \eta^j = 1, \quad \eta^j \geq 0, \tag{14.4}$$

where

$$\Omega_i = \begin{bmatrix} A_i & B_i & F_i & \bar{G}_i \\ C_i & D_i & 0 & 0 \end{bmatrix} \quad \text{and} \quad \Omega_i^{(j)} = \begin{bmatrix} A_i^{(j)} & B_i^{(j)} & F_i^{(j)} & \bar{G}_i^{(j)} \\ C_i^{(j)} & D_i^{(j)} & 0 & 0 \end{bmatrix}, i = 1, \dots M, \ j = 1, \dots N. \tag{14.5}$$

The above model naturally imposes discontinuity in $A_{\sigma(t)}$ since this matrix jumps instantaneously from A_{i_1} to A_{i_2} for some $i_1 \neq i_2$ at the switching instant. The latter property also applies to all the system matrices.

For the stochastic state-feedback control problem, we consider the following system:

$$\begin{aligned} x(t+1) &= A_{\sigma(t)}x(t) + B_{\sigma(t)}w(t) + B_{2,\sigma(t)}u(t) + F_{\sigma(t)}x(t)\beta(t) + H_{\sigma(t)}u(t)\nu(t), \\ x(0) &= 0, \\ z(t) &= C_{1,\sigma(t)}x(t) + D_{11,\sigma(t)}w(t) + D_{12,\sigma(t)}u(t), \end{aligned} \tag{14.6}$$

where $x(t)$, $w(t)$, $\beta(t)$, $\nu(t)$, and $\sigma(t)$ are defined above, $u(t) \in \mathcal{R}^l$ is the control signal, and A_i, B_i, $C_{1,i}$, $B_{2,i}$, $D_{11,i}$, and $D_{12,i}$ are assumed to reside in the following polytope:

$$\bar{\Omega}_i = \sum_{j=1}^{N_i} \bar{\eta}^j(t)\bar{\Omega}_i^{(j)}, \quad \sum_{j=1}^{N_i} \bar{\eta}^j(t) = 1, \quad \bar{\eta}^j(t) \geq 0, \tag{14.7}$$

where

$$\bar{\Omega}_i = \begin{bmatrix} A_i & B_i & B_{2,i} & F_i & H_i \\ C_{1,i} & D_{11,i} & D_{12,i} & 0 & 0 \end{bmatrix} \quad \text{and} \quad \bar{\Omega}_i^{(j)} = \begin{bmatrix} A_i^{(j)} & B_i^{(j)} & B_{2,i}^{(j)} & F_i^{(j)} & H_i^{(j)} \\ C_{1,i}^{(j)} & D_{11,i}^{(j)} & D_{12,i}^{(j)} & 0 & 0 \end{bmatrix}, \tag{14.8}$$

where $i = 1, \dots M, \ j = 1, \dots N.$

Note that in the system (14.6a), A_i is no longer required to be a stable matrix and that in the polytope description of (14.7)–(14.8), we require that in each subsystem $i = 1, \dots M$ the matrices F and H are constant matrices.

14.3   Stochastic Stability

We introduce a LF which provides an efficient way to deal with uncertainties and with controller synthesis. In order to derive the LF, the following result, which is a stochastic version of the time-convexity method presented in [25], is first applied.

Lemma 14.1 *Assume that for some time interval $t \in [t_0, t_f]$, of length $\delta = t_f - t_0 > 0$ there exist two symmetric matrices P_1 and P_2 of compatible dimensions that satisfy the following:*

$$P_1, P_2 > 0, \quad \begin{bmatrix} -P_1 & * & * \\ P_1 A + \frac{P_2-P_1}{\delta} A - P_1 - \frac{P_2-P_1}{\delta} & * & * \\ P_1 F + \frac{P_2-P_1}{\delta} F & 0 & -P_1 - \frac{P_2-P_1}{\delta} F \end{bmatrix} < 0,$$

$$\begin{bmatrix} -P_2 + \frac{P_2-P_1}{\delta} & * & * \\ P_2 A & -P_2 & * \\ P_2 F & 0 & -P_2 \end{bmatrix} < 0. \tag{14.9}$$

Then, for the system $x(t+1) = Ax(t) + Fx(t)\beta(t)$ where $\beta(t)$ is a white-noise sequence as described above, the LF

$$V(t) = E\{x'(t)P(t)x(t)\}, \quad \text{with } P(t) = P_1 + (P_2 - P_1)\frac{t - t_0}{\delta} \tag{14.10}$$

is strictly decreasing over the time interval $t \in [t_0, t_f]$.

Proof The condition for $V(t)$ to decrease for every value of $x(t)$ is $A'P(t+1)A + F'P(t+1)F - P(t) < 0$. Using Schurs complements [26], this condition is equivalent to

$$\begin{bmatrix} -P(t) & * & * \\ P(t+1)A & -P(t+1) & * \\ P(t+1)F & 0 & -P(t+1) \end{bmatrix} < 0. \tag{14.11}$$

We note that on the time interval $t \in [t_0, t_f - 1]$ we have that $P(t) = P_1 + (P_2 - P_1)\frac{t-t_0}{\delta}$ and therefore that $P(t+1) = P(t) + \frac{P_2-P_1}{\delta}$. Substituting for $t = t_0$, $P(t) = P_1$, and $P(t+1) = P_1 + \frac{P_2-P_1}{\delta}$ in (14.11), condition (14.9b) is obtained. Similarly, substituting for $t = t_f - 1$, $P(t) = P_2 - \frac{P_2-P_1}{\delta}$ and $P(t+1) = P_2$ in (14.11), condition (14.9c) is obtained. The conditions of (14.9) are affine in P_1 and P_2, and therefore they guarantee that the LF will decrease for any convex combination of P_1 and P_2. Noting that since $P(t)$ changes linearly in time, $P(t)$ is a convex combination of P_1, P_2, for any $t \in [t_0, t_f]$. Since the conditions of (14.9) are affine in P_1, P_2, this implies that the LF will decrease at any point between the extremes of the considered time interval.

Note that the choice of the LF of (14.10) introduces an extra degree of freedom in comparison to the LF with a constant P matrix. The time affinity of $P(t)$ allows us to solve the stability problem here over the time interval $[t_0, t_f - 1]$ using the time-convexity method of [25].

Choosing below K to be an integer in $[1, T]$, according to the allowed computational complexity and taking for simplicity $\delta = T/K$, the following is obtained.

Theorem 14.1 *The nominal switched stochastic system (14.2a) with $B_{\sigma(t)} = 0$, $\bar{G}_{\sigma(t)} = 0$ is globally asymptotically stable in the mean square sense for any switching law with dwell time greater than or equal to $T > 0$ if there exists a collection of symmetric matrices $P_{i,k}, i = 1, \ldots M; \ k = 0, \ldots K$ of compatible dimensions such that, for all $i = 1, \ldots M$, the following holds:*

$$
\begin{bmatrix}
-P_{i,k} & * & * \\
P_{i,k}A_i + \frac{P_{i,k+1}-P_{i,k}}{\delta}A_i & -P_{i,k} - \frac{P_{i,k+1}-P_{i,k}}{\delta} & * \\
P_{i,k}F_i + \frac{P_{i,k+1}-P_{i,k}}{\delta}F_i & 0 & -P_{i,k} - \frac{P_{i,k+1}-P_{i,k}}{\delta}
\end{bmatrix} < 0,
$$

$$
\begin{bmatrix}
-P_{i,k+1} + \frac{P_{i,k+1}-P_{i,k}}{\delta} & * & * \\
P_{i,k+1}A_i & -P_{i,k+1} & * \\
P_{i,k+1}F_i & 0 & -P_{i,k+1}
\end{bmatrix} < 0 \quad
\begin{bmatrix}
-P_{i,K} & * & * \\
P_{i,K}A_i & -P_{i,K} & * \\
P_{i,K}F_i & 0 & -P_{i,K}
\end{bmatrix} < 0,
$$

$$
\begin{bmatrix}
-P_{q,K} & * & * \\
P_{i,0}A_q & -P_{i,0} & * \\
P_{i,0}F_q & 0 & -P_{i,0}
\end{bmatrix} < 0, \quad k = 0, \ldots K-1, \ i = 1 \ldots M, \ q = 1, \ldots i-1, i+1, \ldots M.
$$

$$(14.12)$$

Proof Let $i_0 = \sigma(0)$, and let $\tau_1, \tau_2, \ldots$ be the switching instants, where $\tau_{h+1} - \tau_h > T$, $\forall h = 1, 2, \ldots$. Define $\tau_{h,k} = \tau_h + \sum_{j=1}^{k} \delta_j$ for $k \geq 1$, and $\tau_{h,0} = \tau_h$. Note that the dwell time constraint implies $\tau_{h,K} + 1 \leq \tau_{h+1,0} = \tau_{h+1}$. Choose the LF, $V(t) = E\{x'(t)P(t)x(t)\}$, where

$$
P(t) = \begin{cases}
P_{i,k} + (P_{i,k+1} - P_{i,k})\frac{t - \tau_{h,k}}{\delta} & t \in [\tau_{h,k}, \tau_{h,k+1}) \\
P_{i,K} & t \in [\tau_{h,K}, \tau_{h+1,0}) , h = 1, 2, \ldots \\
P_{i_0,K} & t \in [0, \tau_1)
\end{cases} \quad (14.13)
$$

and where i is the index of the subsystem that is active at time t. Assume that at some switching instant τ_h the system switches from the qth subsystem to the ith subsystem, meaning that τ_h is the last time step in which the system follows the qth dynamics. At the switching instant τ_h, we have

$$
V(\tau_h) = E\{x(\tau_h)'P_{q,K}x(\tau_h)\} \text{ and } V(\tau_h + 1)
$$

$$
= E\{x(\tau_h)'(A_q + \beta(t)F_q)'P_{i,0}(A_q + \beta(t)F_q)x(\tau_h),
$$

therefore, for the LF to decrease at the switching instants for any $x(\tau_h)$ we require that $A_q'P_{i,0}A_q + F_q'P_{i,0}F_q - P_{q,K} < 0$ which is condition (14.12e).

After the dwell time and before the next switching occurs, we have that $V(t) = E\{x(t)'P_{i,K}x(t)\}$, where $x(t+1) = A_i x(t) + F_i \beta(t)x(t)$. Therefore, $V(t+1) - V(t) = E\{x(t)'[A_i'P_{i,K}A_i + F_i'P_{i,K}F_i - P_{i,K}]x(t)\}$ and (14.12d) guarantees that this expression is negative for any $x(t) \neq 0$. During the dwell time, we consider the time intervals $t \in [\tau_{h,k}, \tau_{h,k+1} - 1]$ where $\delta_{k+1} = \tau_{h,k+1} - \tau_{h,k}$. The matrix $P(t)$ changes then linearly from $P_{i,k}$ to $P_{i,k+1}$, and applying the conditions of the above lemma, conditions (14.12b, c) guarantee that in these intervals $V(t)$ is strictly decreasing which implies the asymptotic stability of the system.

Remark 14.1 We note that the computation burden increases with K. In most cases, one may apply $K \ll T$. In the case where $K = T$ we have $\delta = 1$, $\forall k = 1 \ldots K$. In this case, $P_{i,k}$ are decision variables at each step during the dwell time. Unless the computational cost is too high, this choice of K should provide the least conservative result.

Noting that the LMIs in Theorem 14.1 are affine in the system matrices, one can readily extend the latter result to systems with polytopic-type uncertainty. The following result is thus obtained for the robust case.

Corollary 14.1 *The system (14.2a) with $B_{\sigma(t)} = 0$ and $\bar{G}_{\sigma(t)} = 0$ with the uncertainty of (14.5) is globally asymptotically mean square stable for any switching law with dwell time greater than or equal to $T > 0$ if there exists a collection of symmetric matrices $P_{i,k}$, $i = 1, \ldots M$; $k = 0, \ldots K$ of compatible dimensions, where K is a prechosen integer in $[1, T]$ such that, for all $i = 1, \ldots M$ the following holds:*

$$
\begin{bmatrix}
-P_{i,k} & * & * \\
P_{i,k}A_i^{(j)} + \frac{P_{i,k+1}-P_{i,k}}{\delta} A_i^{(j)} & -P_{i,k} - \frac{P_{i,k+1}-P_{i,k}}{\delta} & * \\
P_{i,k}F_i^{(j)} + \frac{P_{i,k+1}-P_{i,k}}{\delta} F_i^{(j)} & 0 & -P_{i,k} - \frac{P_{i,k+1}-P_{i,k}}{\delta}
\end{bmatrix} < 0,
$$

$$
\begin{bmatrix}
-P_{i,k+1} + \frac{P_{i,k+1}-P_{i,k}}{\delta} & * & * \\
P_{i,k+1}A_i^{(j)} & -P_{i,k+1} & * \\
P_{i,k+1}F_i^{(j)} & 0 & -P_{i,k+1}
\end{bmatrix} < 0
$$

$$
\begin{bmatrix}
-P_{i,K} & * & * \\
P_{i,K}A_i^{(j)} & -P_{i,K} & * \\
P_{i,K}F_i & 0 & -P_{i,K}
\end{bmatrix} < 0, \qquad
\begin{bmatrix}
-P_{q,K} & * & * \\
P_{i,0}A_q^{(j)} & -P_{i,0} & * \\
P_{i,0}F_q^{(j)} & 0 & -P_{i,0}
\end{bmatrix} < 0,
$$

$$
k = 0, \ldots K - 1, \quad i = 1 \ldots M, \quad q = 1, \ldots i - 1, i + 1, \ldots M, \quad j = 1 \ldots N.
$$

$$(14.14)$$

14.4 Stochastic l_2-gain Analysis

In order to analyze the l_2-gain of the stochastic-switched state-multiplicative system, we first consider the stochastic l_2-gain of a general LTV system without switching:

$$
\begin{aligned}
x(t + 1) &= A(t)x(t) + B(t)w(t) + F(t)x(t)\beta(t), \quad x(0) = 0 \\
z(t) &= C(t)x(t) + D(t)w(t)
\end{aligned}
$$

$$(14.15)$$

with the cost function

$$J(\bar{t}) = E\{\sum_{t=0}^{\bar{t}}[z'(t)z(t) - \gamma^2 w'(t)w(t)]\},\qquad (14.16)$$

where $\beta(t)$ is a white-noise sequence described above and where it is required that $J(\bar{t}) < 0$ for $\bar{t} \to \infty$, $\forall w \in \tilde{l}_2$. It is well known that the condition for this requirement to hold is the existence of $P(t) > 0$, $t \in \mathcal{N}$ that satisfies the following [1]:

$$\begin{bmatrix} -P(t) & * & * & * & * \\ 0 & -\gamma^2 I & * & * & * \\ P(t+1)A(t) & P(t+1)B(t) & -P(t+1) & * & * \\ C(t) & D(t) & 0 & -I & * \\ P(t+1)F(t) & 0 & 0 & 0 & -P(t+1) \end{bmatrix} < 0$$

or equivalently that there exists $Q(t) > 0$, $t \in \mathcal{N}$ that satisfies the following [1]:

$$\begin{bmatrix} -Q(t) & * & * & * & * \\ 0 & -\gamma^2 I & * & * & * \\ A(t)Q(t) & B(t) & -Q(t+1) & * & * \\ C(t)Q(t) & D(t) & 0 & -I & * \\ F(t)Q(t) & 0 & 0 & 0 & -Q(t+1) \end{bmatrix} < 0,\qquad (14.17)$$

where $Q(t) = P^{-1}(t)$.

Taking $P(t)$ to be as in (14.13), the latter result is readily applied for the switched systems of (14.2a).

Theorem 14.2 *The stochastic l_2-gain of the system (14.2a) with $\bar{G}_{\sigma(t)} = 0$ and with the uncertainty (14.5) is less than a prescribed scalar γ, for any switching law with dwell time greater than or equal to $T > 0$, if there exists a collection of symmetric matrices $P_{i,k}, i = 1, \ldots M; \ k = 0, \ldots K$ of compatible dimensions, where K is a prechosen integer in $[1, T]$ such that, for all $i = 1, \ldots M$ the following holds:*

$$\begin{bmatrix} -P_{i,k} & * & * & * & * \\ 0 & -\gamma^2 I & * & * & * \\ P_{i,k}A_i^{(j)} + \frac{P_{i,k+1}-P_{i,k}}{\delta}A_i^{(j)} & P_{i,k}B_i^{(j)} + \frac{P_{i,k+1}-P_{i,k}}{\delta}B_i^{(j)} & \Upsilon_{3,3} & * & * \\ C_i^{(j)} & D_i^{(j)} & 0 & -I & * \\ P_{i,k}F_i^{(j)} + \frac{P_{i,k+1}-P_{i,k}}{\delta}F_i^{(j)} & 0 & 0 & 0 & \Upsilon_{5,5} \end{bmatrix} < 0,$$

where $P_{i,k} > 0$, $\Upsilon_{3,3} = -P_{i,k} - \frac{P_{i,k+1}-P_{i,k}}{\delta}$, $\Upsilon_{5,5} = -P_{i,k} - \frac{P_{i,k+1}-P_{i,k}}{\delta}$

$$\begin{bmatrix} -P_{i,k+1} + \frac{P_{i,k+1}-P_{i,k}}{\delta} & * & * & * & * \\ 0 & -\gamma^2 I & * & * & * \\ P_{i,k+1}A_i^{(j)} & P_{i,k+1}B_i^{(j)} & -P_{i,k+1} & * & * \\ C_i^{(j)} & D_i^{(j)} & 0 & -I & * \\ P_{i,k+1}F_i^{(j)} & 0 & 0 & 0 & -P_{i,k+1} \end{bmatrix} < 0$$

$$\begin{bmatrix} -P_{i,K} & * & * & * & * \\ 0 & -\gamma^2 I & * & * & * \\ P_{i,K}A_i^{(j)} & P_{i,K}B_i^{(j)} & -P_{i,K} & * & * \\ C_i^{(j)} & D_i^{(j)} & 0 & -I & * \\ P_{i,K}F_i^{(j)} & 0 & 0 & 0 & -P_{i,K} \end{bmatrix} < 0,$$

$$\begin{bmatrix} -P_{q,K} & * & * & * & * \\ 0 & -\gamma^2 I & * & * & * \\ P_{i,0}A_q^{(j)} & P_{i,0}B_q^{(j)} & -P_{i,0} & * & * \\ C_q^{(j)} & D_q^{(j)} & 0 & -I & * \\ P_{i,0}F_q^{(j)} & 0 & 0 & 0 & -P_{i,0} \end{bmatrix} < 0$$

$$k = 0, \ldots K-1, \quad i = 1 \ldots M, \quad q = 1, \ldots i-1, i+1, \ldots M, \quad j = 1 \ldots N.$$

$$(14.18)$$

Proof Let $P(t)$ be as in (14.13), and assume that at some switching instant τ_h the system switches from the qth subsystem to the ith subsystem. At the switching instant τ_h, we have

$$P(t) = P_{q,K}, \quad P(t+1) = P_{i,0},$$

and the dynamics matrices are those of the qth subsystem. Substation into (14.17a) yields (14.18d). After the dwell time, and before the next switching instant, we have that $P(t) = P_{i,K}$, and the dynamic matrices are those of the ith subsystem. Substitution in (14.17a) leads then to (14.18c). During the dwell time, we consider the time intervals $t \in [\tau_{h,k}, \tau_{h,k+1} - 1]$ where $\delta = \tau_{h,k+1} - \tau_{h,k}$. The matrix $P(t)$ changes then linearly from $P_{i,k}$ to $P_{i,k+1}$, and the dynamics matrices are those of the ith subsystem. Substitution in (14.17a) yields (14.18b, c).

We note that it is possible to write the conditions of Theorem 14.2 in terms of $Q = P^{-1}$ as follows.

Corollary 14.2 *The stochastic l_2-gain of the system (14.2a) with $\bar{G}_{\sigma(t)} = 0$ and with the uncertainty (14.5) is less than a prescribed positive γ, for any switching law with dwell time greater than or equal to $T > 0$, if there exists a collection of symmetric matrices $Q_{i,k}, i = 1, \ldots M, k = 0, \ldots K$ of compatible dimensions, where $K \in [1, T]$, such that for all $i = 1, \ldots M$ the following holds:*

$$Q_{i,k} > 0,$$

$$\begin{bmatrix} -Q_{i,k} & * & * & * & * \\ 0 & -\gamma^2 I & * & * & * \\ A_i^{(j)} Q_{i,k} & B_i^{(j)} & -Q_{i,k} - \frac{Q_{i,k+1}-Q_{i,k}}{\delta} & * & * \\ C_i^{(j)} Q_{i,k} & D_i^{(j)} & 0 & -I & * \\ F_i^{(j)} Q_{i,k} & 0 & 0 & 0 & -Q_{i,k} - \frac{Q_{i,k+1}-Q_{i,k}}{\delta} \end{bmatrix} < 0$$

$$\begin{bmatrix} -Q_{i,k+1} + \frac{Q_{i,k+1}-Q_{i,k}}{\delta} & * & * & * & * \\ 0 & -\gamma^2 I & * & * & * \\ A_i^{(j)} Q_{i,k+1} - A_i^{(j)} \frac{Q_{i,k+1}-Q_{i,k}}{\delta} & B_i^{(j)} & -Q_{i,k+1} & * & * \\ C_i^{(j)} Q_{i,k+1} - C_i^{(j)} \frac{Q_{i,k+1}-Q_{i,k}}{\delta} & D_i^{(j)} & 0 & -I & * \\ F_i^{(j)} Q_{i,k+1} - F_i^{(j)} \frac{Q_{i,k+1}-Q_{i,k}}{\delta} & 0 & 0 & 0 & -Q_{i,k+1} \end{bmatrix} < 0$$

$$\begin{bmatrix} -Q_{i,K} & * & * & * & * \\ 0 & -\gamma^2 I & * & * & * \\ A_i^{(j)} Q_{i,K} & B_i^{(j)} & -Q_{i,K} & * & * \\ C_i^{(j)} Q_{i,K} & D_i^{(j)} & 0 & -I & * \\ F_i^{(j)} Q_{i,K} & 0 & 0 & 0 & -Q_{i,K} \end{bmatrix} < 0,$$

$$\begin{bmatrix} -Q_{q,K} & * & * & * & * \\ 0 & -\gamma^2 I & * & * & * \\ A_q^{(j)} Q_{q,K} & B_q^{(j)} & -Q_{i,0} & * & * \\ C_q^{(j)} Q_{q,K} & D_q^{(j)} & 0 & -I & * \\ F_q^{(j)} Q_{q,K} & 0 & 0 & 0 & -Q_{i,0} \end{bmatrix} < 0$$

$$k = 0, \ldots K - 1, \quad i = 1 \ldots M, \quad q = 1, \ldots i - 1, i + 1, \ldots M, \quad j = 1 \ldots N.$$
(14.19)

14.5   State-Feedback Control

We consider the following controller form:

$$u(t) = G(t)x(t). \tag{14.20}$$

The system (14.2) with the controller (14.20) is equivalent to the one of (14.6), where $A_{\sigma(t)}$ and $C_{1,\sigma(t)}$ are replaced by $A_{\sigma(t)} + B_{2,\sigma(t)} G(t)$ and $C_{1,\sigma(t)} + D_{12,\sigma(t)} G(t)$, respectively. Therefore, if $G(t)$ is taken to be constant, denoting $A_{cl,i}^{(j)} = A_i^{(j)} + B_{2,i}^{(j)} G$, $C_{cl,1,i}^{(j)} = C_{1,i}^{(j)} + D_{12,i}^{(j)} G$, the system (14.2) with the controller (14.20) is equivalent to the one of (14.6), where $A_{\sigma(t)}$ and $C_{1,\sigma(t)}$ are replaced by $A_{cl,i}^{(j)}$ and $C_{cl,1,i}^{(j)}$ respectively and $\bar{G} = HG$. Assuming still that $\bar{G} = 0$ and applying the latter substitutions to the conditions of Corollary 14.2, a set of Bilinear Matrix Inequality (BMI) conditions is obtained a solution to which, if exists, guarantees a stochastic l_2-gain bound of γ for the closed-loop system (14.6). These conditions are obtained via BMIs because the same state-feedback gain should be applied to all the subsystems.

Remark 14.2 BMIs are known to be non-convex. They can be solved either by using a solver that is able to solve non-convex problems, or by applying local iterations, where at each step LMIs are solved, which allows the use of standard solvers.

The case where the switching signal is measured online can be solved by replacing the constant state-feedback gain matrix G by switching-dependent gains G_i, $i = 1, \ldots, M$. In this case, a computationally efficient result can be obtained that applies time-varying gains via solutions of LMIs. The latter is achieved by allowing the gains to be time varying during the dwell time and by seeking a matrix $Q(t)$ as in Corollary 14.2. In the theorem below, we use a time-varying gain $G_{\sigma(t)}(t)$, apply Corollary 14.2, such that $Y_{i,k} = G_{i,k} Q_{i,k}$, and $Y_{i,k}$ is piecewise linear in time in the same manner as $Q(t)$.

Theorem 14.3 *The stochastic l_2-gain of the system (14.6), where $H_{\sigma(t)} = 0$, with the uncertainty (14.7), and the controller (14.20) is less than a prescribed positive scalar γ, for any switching law with dwell time greater than or equal to $T > 0$, if there exists: a collection of matrices $Q_{i,k} = Q'_{i,k}$, $Y_{i,k}$, $i = 1, \ldots M$, $k = 0, \ldots K$ of compatible dimensions, where $K \in [1, \ T]$ such that, for all $i = 1, \ldots M$ the following holds:*

$$
Q_{i,k} > 0, \quad
\begin{bmatrix}
-Q_{i,k} & & * & * & * & * \\
0 & & -\gamma^2 I & * & & * & * \\
A_i^{(j)} Q_{i,k} + B_{2,i}^{(j)} Y_{i,k} & B_{1,i}^{(j)} & -\bar{Q}_{i,k} & * & * \\
C_{1,i}^{(j)} Q_{i,k} + D_{12,i}^{(j)} Y_{i,k} & D_{11,i}^{(j)} & 0 & -I & * \\
F_i^{(j)} Q_{i,k} & & 0 & 0 & 0 & -Q_{i,k} - \frac{Q_{i,k+1} - Q_{i,k}}{\delta}
\end{bmatrix} < 0
$$

$$
\begin{bmatrix}
-\tilde{Q}_{i,k} & & * & * & * & * \\
0 & & -\gamma^2 I & * & & * & * \\
A_i^{(j)} \tilde{Q}_{i,k} + B_{2,i}^{(j)} \tilde{Y}_{i,k} & B_{1,i}^{(j)} & -Q_{i,k+1} & * & * \\
C_{1,i}^{(j)} \tilde{Q}_{i,k} + D_{12,i}^{(j)} \tilde{Y}_{i,k} & D_{11,i}^{(j)} & 0 & -I & * \\
F_i^{(j)} \tilde{Q}_{i,k} & & 0 & 0 & 0 & -Q_{i,k+1}
\end{bmatrix} < 0
$$

$$
\begin{bmatrix}
-Q_{i,K} & & * & * & * & * \\
0 & & -\gamma^2 I & * & & * & * \\
A_i^{(j)} Q_{i,K} + B_{2,i}^{(j)} Y_{i,K} & B_{1,i}^{(j)} & -Q_{i,K} & * & * \\
C_{1,i}^{(j)} Q_{i,K} + D_{12,i}^{(j)} Y_{i,K} & D_{11,i}^{(j)} & 0 & -I & * \\
F_i^{(j)} Q_{i,K} & & 0 & 0 & 0 & -Q_{i,K}
\end{bmatrix} < 0,
$$

$$\begin{bmatrix} -Q_{q,K} & & * & * & * & * \\ 0 & -\gamma^2 I & * & & * & * \\ A_q^{(j)} Q_{q,K} + B_{2,q}^{(j)} Y_{q,K} & B_{1,q}^{(j)} & -Q_{i,0} & * & * \\ C_{1,q}^{(j)} Q_{q,K} + D_{12,q}^{(j)} Y_{q,K} & D_{11,q}^{(j)} & 0 & -I & * \\ F_q^{(j)} Q_{q,K} & & 0 & 0 & 0 & -Q_{i,0} \end{bmatrix} < 0 \tag{14.21}$$

$$k = 0, \ldots K - 1, \quad i = 1 \ldots M, \quad q = 1, \ldots i-1, i+1, \ldots M, \quad j = 1 \ldots N,$$

where $\bar{Q}_{i,k} = Q_{i,k} + \frac{Q_{i,k+1} - Q_{i,k}}{\delta}$, $\tilde{Q}_{i,k} = Q_{i,k+1} - \frac{Q_{i,k+1} - Q_{i,k}}{\delta}$, and $\tilde{Y}_{i,k} = Y_{i,k+1} - \frac{Y_{i,k+1} - Y_{i,k}}{\delta}$.

We note that in order to preserve the convexity for time values in the interval $[\tau_{h,k}, \tau_{h,k+1})$, $Y_\sigma(t)$ must change linearly over this time interval, which implies $Y_\sigma(t) = Y_{i,k} + \frac{t - \tau_{h,k}}{\delta}(Y_{i,k+1} - Y_{i,k})$. Therefore, the controller gain $G_\sigma(t)$ will be

$$G_{\sigma(t)} =$$

$$\begin{cases} [Y_{i,k} + \frac{t - \tau_{h,k}}{\delta}(Y_{i,k+1} - Y_{i,k})][Q_{i,k} + (Q_{i,k+1} - Q_{i,k})\frac{t - \tau_{h,k}}{\delta}]^{-1} & t \in [\tau_{h,k}, \tau_{h,k+1}) \\ Y_{i,K} Q_{i,K}^{-1} & t \in [\tau_{h,K}, \tau_{h+1,0}) \\ Y_{i_0,K} Q_{i_0,K}^{-1} & t \in [0, \tau_1) \end{cases} \tag{14.22}$$

A simpler controller that is constant between the switching instances can be obtained by applying the method of [27]. This controller, which is derived in the following corollary, may lead, however, to more conservative results.

Corollary 14.3 *The stochastic l_2-gain of the system (14.6) with the uncertainty (14.7), where $H_{\sigma(t)} = 0$, and the controller (14.20) is less than a prescribed positive scalar γ, for any switching law with dwell time greater than or equal to $T > 0$, if there exist: a collection of matrices $Q_{i,k} = Q_{i,k}'$, Y_i, $M_{i,k}$, $i = 1, \ldots M, k = 0, \ldots K$ of compatible dimensions, where $K \in [1, T]$, and an integer α such that, for all $i = 1, \ldots M$ the following holds:*

$$Q_{i,k} > 0,$$

$$\begin{bmatrix} -0.5 M_{i,k} - 0.5 M_{i,k}' & * & * & * & * & * \\ 0 & -\gamma^2 I & * & * & * & * \\ A_i^{(j)} M_{i,k} + B_{2,i}^{(j)} Y_{i,k} & B_{1,i}^{(j)} & -\bar{Q}_{i,k} & * & * & * \\ C_{1,i}^{(j)} M_{i,k} + D_{12,i}^{(j)} Y_{i,k} & D_{11,i}^{(j)} & 0 & -I & * & * \\ F_i^{(j)} M_{i,k} & 0 & 0 & 0 & -\bar{Q}_{i,k} & * \\ M_{i,k} - Q_{i,k} + 0.5\alpha M_{i,k}' & 0 & -\alpha[M_{i,k}' A_i^{(j)'} + Y_{i,k}' B_{2,i}^{(j)'}] & \Gamma_{6,4a} & \alpha M_{i,k}' F_i^{(j)'} & \Gamma_{6,6a} \end{bmatrix} < 0,$$

where $\Gamma_{6,4a} = \alpha[M_{i,k}' C_{1,i}^{(j)'} + Y_{i,k}' D_{12,i}^{(j)'}],$

$$\begin{bmatrix} -0.5 \tilde{M}_{i,k} - 0.5 \tilde{M}_{i,k}' & * & * & * & * & * \\ 0 & -\gamma^2 I & * & * & * & * \\ A_i^{(j)} \tilde{M}_{i,k} + B_{2,i}^{(j)} \tilde{Y}_{i,k} & B_{1,i}^{(j)} & -Q_{i,k+1} & * & * & * \\ C_{1,i}^{(j)} \tilde{M}_{i,k} + D_{12,i}^{(j)} \tilde{Y}_{i,k} & D_{11,i}^{(j)} & 0 & -I & * & * \\ F_i^{(j)} \tilde{M}_{i,k} & 0 & 0 & 0 & -Q_{i,k+1} & * \\ \tilde{M}_{i,k} - \tilde{Q}_{i,k} + 0.5\alpha \tilde{M}_{i,k}' & 0 & -\alpha[\tilde{M}_{i,k}' A_i^{(j)'} + \tilde{Y}_{i,k}' B_{2,i}^{(j)'}] & \Gamma_{6,4b} & \alpha \tilde{M}_{i,k}' F_i^{(j)'} & \Gamma_{6,6b} \end{bmatrix} < 0,$$

where $\Gamma_{6,4b} = \alpha[\tilde{M}'_{i,k} C^{(j)'}_{1,i} + \tilde{Y}'_{i,k} D^{(j)'}_{12,i}]$,

$$\begin{bmatrix} -0.5M_{i,K} - 0.5M'_{i,K} & * & * & * & * & * \\ 0 & -\gamma^2 I & * & * & * & * \\ A^{(j)}_i M_{i,K} + B^{(j)}_{2,i} Y_{i,K} & B^{(j)}_{1,i} & -Q_{i,K} & * & * & * \\ C^{(j)}_{1,i} M_{i,K} + D^{(j)}_{12,i} Y_{i,K} & D^{(j)}_{11,i} & 0 & -I & * & * \\ F^{(j)}_i M_{i,K} & 0 & 0 & 0 & -Q_{i,K} & * \\ M_{i,K} - Q_{i,K} + 0.5\alpha M'_{i,K} & 0 & -\alpha[M'_{i,K} A^{(j)'}_i + Y'_{i,K} B^{(j)'}_{2,i}] & \Gamma_{6,4c} & \alpha M'_{i,K} F^{(j)'}_i & \Gamma_{6,6c} \end{bmatrix} < 0,$$

where $\Gamma_{6,4c} = \alpha[M'_{i,K} C^{(j)'}_{1,i} + Y'_{i,K} D^{(j)'}_{12,i}]$,

$$\begin{bmatrix} -0.5M_{q,K} - 0.5M'_{q,K} & * & * & * & * & * \\ 0 & -\gamma^2 I & * & * & * & * \\ A^{(j)}_q M_{q,K} + B^{(j)}_{2,i} Y_{q,K} & B^{(j)}_{1,q} & -Q_{i,0} & * & * & * \\ C^{(j)}_{1,q} M_{q,K} + D^{(j)}_{12,q} Y_{q,K} & D^{(j)}_{11,q} & 0 & -I & * & * \\ F^{(j)}_q M_{q,K} & 0 & 0 & 0 & -Q_{i,0} & * \\ M_{q,K} - Q_{q,K} + 0.5\alpha M'_{q,K} & 0 & -\alpha[M'_{q,K} A^{(j)'}_q + Y'_{q,K} B^{(j)'}_{2,q}] & \Gamma_{6,4d} & \alpha M'_{q,K} F^{(j)'}_q & \Gamma_{6,6d} \end{bmatrix} < 0$$

$$k = 0, \ldots K-1, \quad i = 1 \ldots M, \quad q = 1, \ldots i-1, i+1, \ldots M, \quad j = 1, \ldots N,$$

$$(14.23)$$

where $\Gamma_{6,4d} = \alpha[M'_{q,K} C^{(j)'}_{1,q} + Y'_{q,K} D^{(j)'}_{12,q}]$ *and where*

$$\Gamma_{6,6a} = -\alpha(M'_{i,k} + M_{i,k}), \quad \Gamma_{6,6b} = -\alpha(\tilde{M}'_{i,k} + \tilde{M}_{i,k}),$$
$$\Gamma_{6,6c} = -\alpha(M'_{i,K} + M_{i,K}), \quad \Gamma_{6,6d} = -\alpha(M'_{q,K} + M_{q,K}),$$
$$\tilde{M}_{i,k} = M_{i,k+1} - \frac{M_{i,k+1} - M_{i,k}}{\delta},$$

and where $\tilde{Q}_{i,k}$, $\tilde{Y}_{i,k}$, *and* $\bar{Q}_{i,k}$ *are given following* (14.21).

In this case, the controller gain $G_\sigma(t)$ will be

$$G_{\sigma(t)} = \begin{cases} \hat{G}_{i,k} & t \in [\tau_{h,k}, \tau_{h,k+1}) \\ Y_{i,K} M^{-1}_{i,K} & t \in [\tau_{h,K}, \tau_{h+1,0}) \\ Y_{i_0,K} M^{-1}_{i_0,K} & t \in [0, \tau_1) \end{cases}, \qquad (14.24)$$

where $\hat{G}_{i,k} = [Y_{i,k} + \frac{t - \tau_{h,k}}{\delta}(Y_{i,k+1} - Y_{i,k})][M_{i,k} + (M_{i,k+1} - M_{i,k})\frac{t - \tau_{h,k}}{\delta}]^{-1}$. If the controller gain is required to be constant between the switching instances, $Y_{i,k}$ and $M_{i,k}$ are chosen to be independent of k, which leads to the controller gain: $G_\sigma(t) = Y_i M^{-1}_i, \forall t \in [\tau_h, \tau_{h+1})$.

Remark 14.3 In Theorem 14.3 and Corollary 14.3, $H_{\sigma(t)}$ was taken to be zero in order to simplify the resulting LMIs. Including the second state-multiplicative white-noise sequence in the dynamics of (14.6) results in a mere addition to the LMIs of (14.21b–e). Considering the LMI of (14.21b), for example, one should then replace the latter LMI with the following inequality:

$$\begin{bmatrix} \Gamma_{11,i}^{j} & * \\ \Gamma_{22,i}^{j} & -Q_{i,k} \end{bmatrix} < 0,$$

where $\Gamma_{22,i}^{j} = [H_i^{(j)} Y_{i,k} \ 0 \ 0 \ 0]$ and where $\Gamma_{11,i}^{j}$ is the LMI of (14.21b). The latter modification should be applied similarly to the LMIs of (14.21b–e) and to the LMIs of (14.23b–e).

Remark 14.4 We note that in Corollary 14.3 (and also in Corollary 14.2) the decision variables $Q_{i,k}$ are taken to be vertex independent. The latter restriction imposes some conservatism on the possible solution of the state-feedback control problem. Taking, in the case where the system parameters are time invariant, a vertex-dependent solution in (14.23), $Q_{i,k}$, may be replaced by $Q_{i,k}^j$, $j = 1, \ldots N$ in (14.23a–e) (noting that it also appears in the definitions of $\bar{Q}_{i,k}$ and $\tilde{Q}_{i,k}$ following (14.21)).

14.6   Example—State-Feedback Control

We consider the problem of stabilizing and attenuating disturbances acting on the longitudinal short period mode of the F4E fighter aircraft with additional canards, taken from [28].

The continuous-time state-space model for this aircraft [written in Ito form] is given by

$$d \begin{bmatrix} N_z \\ q \\ \delta_e \end{bmatrix} = [\begin{bmatrix} a_{11} & a_{12} & a_{13} \\ a_{21} & a_{22} & a_{23} \\ 0 & 0 & -30 \end{bmatrix} \begin{bmatrix} N_z \\ q \\ \delta_e \end{bmatrix} + \begin{bmatrix} b_1 \\ 0 \\ 30 \end{bmatrix} u + I_3 w]dt + \begin{bmatrix} 0 \\ 0 \\ 4.5\delta_e \end{bmatrix} d\beta,$$

where $H = 0$ and where u is the control input and where the objective and measurement vectors are given by $z = \begin{bmatrix} N_z & q & u \end{bmatrix}'$ and $y = \begin{bmatrix} N_z & q \end{bmatrix}'$, respectively. In this model, N_z is the *normal acceleration*, q is the *pitch rate*, and δ_e is the *elevator angle*. We note that the stochastic term in the system is caused by a flutter of the elevon which is a phenomenon related to the aeroelasticity of the elevon. The parameters of the model for four Operating Points (O.Ps) are described in Table 14.1 and it is assumed that between the O.Ps the system parameters are a convex combination of the four parameter sets of the table.

Using Euler method of discretization, with sampling time of 0.01 s, a descritized version of the continuous-time dynamics is obtained (The resulting F is thus $\begin{bmatrix} 0 & 0 & 0 \\ 0 & 0 & 0 \\ 0 & 0 & 4.5 \end{bmatrix} \sqrt{0.01}$). A state-feedback controller is then sought that achieves a minimum bound on the stochastic ℓ_2-gain of the resulting closed-loop system.

The standard way of solving the latter stochastic state-feedback problem is to apply the method of [1] where a single LF is applied and to find a solution over the uncertainty polytope. The latter polytope is quite large, and the resulting ℓ_2-gain

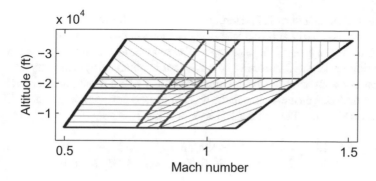

Fig. 14.1 The overlapping subpolytopes

bound may be too large. An alternative way of solving the problem is to split the parameters polytope into subregions as depicted in Fig. 14.1.

We note that to prevent control chattering, the subregions overlap. The size of the overlap is dictated by the bandwidth of the airplane which is taken here to be equivalent to 2 s. It is assumed that there is an online indication of what subregion the system parameters reside. We solve the problem by considering it to be a switched system with dwell time of 2 s.

We use the theory of Corollary 14.3 and design robust state-feedback controllers, assuming that the switching signal is measured online. In Table 14.2, we compare our results with those achieved by the Gain Scheduling (GS) control design method [29].

The first result is obtained by applying a vertex-dependent LF where a single controller is sought via the method of [27]. This approach yields a minimum attenuation level $\gamma = 7.85$. The second result is obtained by applying the GS control design [29]

Table 14.1 The four operating points (O.Ps) of the F4E aircraft

O.P.	Mach	Altitude (ft)	a_{11}	a_{12}	a_{13}	a_{21}	a_{22}	a_{23}	b_1
1	0.5	5000	0.9896	17.41	96.15	0.2648	−0.8512	−11.39	−97.78
2	0.9	35000	−0.6607	18.11	84.34	0.08201	−0.6587	−10.81	−272.2
3	0.85	5000	−1.702	50.72	263.5	0.2201	−1.418	−31.99	−85.09
4	1.5	35000	−0.5162	29.96	178.9	−0.6896	−1.225	−30.38	−175.6

Table 14.2 Min. values of the attenuation level γ

Result no.	Controller type	γ
1	A single controller	7.85
2	GS	22.87
3	Corollary 14.3 ($\alpha = 10^{-5}$)	5.3

with the improvement of [27]. Here, due to the uncertainty in B_2, an augmentation of the dynamic matrix A is required in order to avoid bilinearity [30]. This augmentation introduces a delay of one step in the control u, and it thus deteriorates the attenuation level. The last result is the one that is obtained by Corollary 14.3 where different constant state-feedback gains and single LFs are obtained for each of the subsystems. The constant gains make this controller easily implementable. The four state-feedback gains that are derived by Corollary 14.3 are

$$K_{OP1} = [0.2803 \quad 2.6735 \quad -0.4336], \quad K_{OP2} = [0.3158 \quad 2.9249 \quad -0.7011],$$
$$K_{OP3} = [0.2519 \quad 2.4834 \quad -0.707] \quad K_{OP4} = [0.4131 \quad 3.9091 \quad -0.9298].$$

Table 14.2 clearly demonstrates the advantage of applying switching in order to reduce the conservatism of the existing methods of robust stochastic discrete-time state-feedback control.

14.7   Conclusions

A method for analyzing the mean square stability of stochastic, state-multiplicative, linear discrete-time switched systems is introduced which applies a switching-dependent LF. The method is applied to both nominal and uncertain polytopic-type systems.

The corresponding stochastic l_2-gain criterion is obtained for the first time via a set of LMIs. Based on the latter criterion, the H_∞ state-feedback control problem is solved where a time-varying switched state-feedback gain is derived. Constant and linear-in-time feedback gains are obtained then as special cases.

In the case where the discrete-time system encounters polytopic-type parameter uncertainties, a vertex-dependent switched LF is applied which significantly reduces the conservatism of the solution.

The theory presented can be easily extended to include norm-bounded uncertainties, on top of the polytopic-type uncertainties. We note that taking the stochastic parameters of the system to be zeros, the deterministic results of [21] are readily recovered. These latter results have been shown in [21] to be superior to all the other results that have been obtained for uncertain switching systems with dwell.

The example in Sect. 14.6 demonstrates the tractability and the applicability of the present chapter to real engineering systems. In this example, a significant reduction of the bound on the achievable attenuation level was obtained by applying the conditions of Corollary 14.3 which results in a simpler controller, compared to the gain-scheduling design. The result of Corollary 14.3 requires only a partial information on the systems' parameters, and the resulting controller gains are constant for each of the resulting subsystems.

References

1. Gershon, E., Shaked, U., Yaesh, I.: H_∞ Control and Estimation of State-Multiplicative Linear Systems. Lecture Notes in Control and Information Sciences, LNCIS, vol. 318. Springer (2005)
2. Ugrinovsky, V.A.: Robust H_∞ control in the presence of of stochastic uncertainty. Int. J. Control **71**, 219–237 (1998)
3. Dragan, V., Stoica, A.: A γ Attenuation Problem for Discrete-time Time-Varying Stochastic Systems with Multiplicative Noise. Reprint Series of the Institute of Mathematics of the Romanian Academy, vol. 10 (1998)
4. Gershon, E., Shaked, U.: Advanced Topics in Control and Estimation of State-Multiplicative Noisy Systems. Lecture Notes in Control and Information Sciences, vol. 439. Springer (2013)
5. Mao, X.: Robustness of exponential stability of stochastic differential delay equations. IEEE Trans. Autom. Control **41**, 442–447 (1996)
6. Yong, J., Zhou, X.Y.: Stochastic controls: Hamiltonian systems and HJB equations. In: Applications of Mathematics: Stochastic Modelling and Applied Probability, vol. 43. Springer (1999)
7. Yue, D., Won, S.: Delay-dependent robust stability of stochastic systems with time delay and nonlinear uncertainties. IEE Electron. Lett. **37**, 992–993 (2001)
8. Xu, S., Chen, T.: Robust H_∞ control for uncertain stochastic systems with state-delay. IEEE Trans. Autom. Control **47**, 2089–2094 (2002)
9. Xu, S., Lam, J., Mao, X., Zou, Y.: A new LMI condition for delay-dependent robust stability of stochastic time-delay systems. Asian J. Control **7**, 419–423 (2005)
10. Zhendong, S., Sam Shuzhi, G.: Stability theory of switched dynamical systems. Springer's Communications and Control Engineering Series, Berlin (2011)
11. Vu, L., Liberzon, D.: Invertibility of switched linear systems. Automatica **44**, 949–958 (2008)
12. Sun, Z., Ge, S.S.: Analysis and synthesis of switched linear control systems. Automatica **41**, 181–195 (2005)
13. Costa, O.L.V., Guilherme, R.A.M.: Linear minimum mean square filter for discrete-time linear systems with Markov jumps and multiplicative noises. Automatica **47**(3), 466–476 (2011)
14. Li, L., Ugrinovskii, V.A.: On necessary and sufficient conditions for H-infinity output feedback control of Markov jump linear systems. IEEE Trans. Autom. Control **52**(7), 1287–1292 (2007)
15. Huang, Y., Zhang, W., Feng, G.: Infinite horizon $H_2 - H_\infty$ control for stochastic systems with Markovian jumps. Automatica **44**, 857–863 (2008)
16. Mao, X., Yuan, C.: Stochastic Differential Equations with Markovian Switching. Imperial College Press (2006)
17. Zhang, L., Lam, J.: Necessary and sufficient conditions for analysis and synthesis of Markov Jump linear systems with incomplete transition descriptions. IEEE Trans. Autom. Control **55**, 1695–1701 (2002)
18. Costa, O., de Paulo, W.: Indefinite quadratic with linear costs optimal control of Markov jump with multiplicative noise systems. Automatica **43**, 587–597 (2007)
19. Shaked, U., Gershon, E.: Robust H_∞ control of stochastic linear switched systems with dwell time. Int. J. Robust Nonlinear Control **24**(11), 1664–1676 (2013)
20. Allerhand, L.I., Shaked, U.: Robust stability and stabilization of linear switched systems with dwell time. IEEE Trans. Autom. Control **56**, 381–386 (2001)
21. Allerhand, L.I., Shaked, U.: Robust stability and controller synthesis of discrete linear switched systems with dwell time. Int. J. Control **85**(6), 735–745 (2012)
22. Zhng, L., Gao, H.: Asynchronously switched control of switched linear systems with average dwell time. Automatica **46**(5), 953–958 (2010)
23. Allerhand, L.I., Shaked, U.: Robust switching-based fault tolerant control. IEEE Trans. Autom. Control **60**(8), 2272–2276 (2015)
24. Bouhtouri, E.A., Hinrichsen, D., Pritchard, A.J.: H_∞ type control for discrete-time stochastic systems. Int. J. Robust Nonlinear Control **9**, 923–948 (1999)
25. Boyarski, S., Shaked, U.: Time-convexity and time-gain-scheduling in finite-horizon robust H_∞-control. In: Proceedings of the 48th CDC09, Shanghai, China (2009)

26. Boyd, S., El Ghaoui, L., Feron, E., Balakrishnan, V.: Linear Matrix Inequalities in System and Control Theory. SIAM, Philadelphia, PA (1994)
27. de Oliveira, M.C., Skelton, R.E.: Stability test for constrained linear systems. In: Reza Moheimani, S.O. (ed.) Perspectives in Robust Control. Lecture Notes in Control and Information Sciences, vol. 268. Springer, London (2001)
28. Petersen, I.R.: Quadratic stabilizability of uncertain linear systems: existence of a nonlinear stabilizing control does not imply existence of a stabilizing control. IEEE Trans. Autom. Control **30**(3), 291–293 (1985)
29. Apkarian, P., Gahinet, P.: A convex characterization of gain-scheduled H_∞ conterollers. IEEE Trans. Autom. Control **40**(5), 853–864 (1995)
30. Gahinet, P., Nemirovski, A., Laub, A.J., Chilali, M.: LMI Control Toolbox For Use with MATLAB. The MathWorks, Inc., MA, USA (1995)

Part III
Optimal Control of Biochemical Systems

Chapter 15
Introduction and Literature Survey

Abstract This chapter is the introduction to part III of the book. This chapter contains a short survey of the current research approaches to the field of **control of biochemical pathways** which is dominated by the BST [Biochemical System Theory] and the MCA [Metabolic Control Analysis] theories, which constitute the mainstream of system biology in the latter aspect. Alongside these theoretical frameworks, there is a host of other research frameworks that are aided by a great number of numerical software packages and which are based on different and various aspects of optimization techniques. In Chap. 16, a humble attempt is conducted in order to explore the possible optimality of biochemical pathways via the worst-case approach of the H_∞ control theory. Starting from simple enzymatic reactions and applying linearization, a simple pathway is considered which contains a negative feedback loop, upon which various aspects of linear modern control theory are applied. The theory developed is demonstrated via the threonine synthesis pathway which contains three negative feedback loops. In Chap. 17, a further study of other optimal measures is conducted, including the peak-to-peak and the energy-to-peak optimal measures. The latter optimal measures are applied and demonstrated on the threonine synthesis pathway and the glycolytic pathway in yeast. The third part of the book culminates in a conjecture according to which biochemical feedback control systems may apply either a worst-case strategy or a peak-to-peak strategy accounting for their inherent robustness in the face of parameter uncertainties.

Abbreviations

BRL	Bounded Real Lemma
BST	Biochemical System Theory
LF	Lyapunov Function
LMI	Linear Matrix Inequality
LTI	Linear Time Invariant
LTV	Linear Time Variant
MCA	Metabolic Control Analysis
M	Molar, likewise mM (milli-Molar)
MM	Michaelis–Menten [enzyme]
OPs	Operating Points

© Springer Nature Switzerland AG 2019
E. Gershon and U. Shaked, *Advances in H_∞ Control Theory*,
Lecture Notes in Control and Information Sciences 481,
https://doi.org/10.1007/978-3-030-16008-1_15

P2P   Peak to Peak
SOF   Static Output Feedback
[ ]   Chemical Concentration
ADP   Adenosine Diphosphate
ASP   Aspartate
ASPP β-aspartyl phosphate
ATP   Adenosine Triphosphate
G6P   Glucose 6 Phosphate
HDH   Homoserine Dehydrogenase
HK   Homoserine Kinase
HKI   Aspartate Kinase I
THR   Threonine
SISO   Single-Input Single-Output (system)

15.1   Introduction

All life forms on Planet Earth, with no exception, are organic by nature and are being
transformed, developed, and evolved through a series of remarkably complicated
biochemical pathways (see [1] for basic introductory of the material). These pathways, whether they occur in microorganisms such as bacteria, or in higher organism,
such as the animal kingdom or plants, are composed of pathways of enzymatic reactions where typically, each step in the reaction pathway is carried out via a single
enzyme. A typical biochemical pathway may include several enzymes and is highly
organized in both space and time (i.e., in certain compartments of the cell interior and
in synchronization with other pathways). Most biochemical pathways are branched
and some are circular (see for example the citric acid cycle [1], which is a principle
pathway in both the anabolic and catabolic phases of the cell metabolism).

In principle, the function of a single enzyme is exerted by increasing the rate by
which a certain (and highly specific) reaction system is driven from a nonequilibrium
initial state to a final chemical (dynamical) equilibrium state. Once an equilibrium is
reached, the presence of the enzyme (say in a test tube) cannot change the ratios of
the reaction components (i.e., reactants to products) [1–3]. Remarkably, during the
entire course of the chemical enzymatic reaction, the enzyme molecule possesses
a series of transitory states; however, at the final stage of the overall reaction, the
enzyme molecule is freely released, ready to begin a second cycle.

The study of the chemical kinetics of enzymatic reactions within the biochemistry
sciences was greatly advanced since the pioneering works of the first half of the last
century [1–3]. The reaction scheme for the simplest case is described by the following
reaction:

$$S + E \underset{k_2}{\overset{k_1}{\rightleftarrows}} ES \xrightarrow{k_p} E + P,$$

where E is the enzyme, S is the substrate, ES is the enzyme–substrate complex, and k_1 and k_2, k_p are the second-order and first-order rate constants, respectively. Note that the left-hand step is reversible, whereas the second step is irreversible (the latter can be easily relaxed). The above reaction scheme accounts for a small fraction, although important, of the large variety of enzyme reactions [1, 2]. Typically, key enzymes in a given biochemical pathway are allosteric; most of them are composed of several subunits with multiple sites that regulate the enzyme kinetic and thermodynamic features. The latter more involved cases include enzymes that require specific coenzymes and ions and that act simultaneously on two or more substrate molecules to produce one or more products.

The biochemical pathways are obviously higher in the biochemical hierarchy than single enzymes and are highly regulated by feedback loops, some of which are exerted by a product of the same pathway—a mechanism typically denoted in the biochemical literature as a "feedback inhibition" phenomenon (see [4–6]). The feedback can be either positive or negative and in some cases both types are involved in the same pathway. The research of biochemical pathways stimulates, naturally, questions that aim at the understanding the kinetic and dynamic properties of the complete pathways (either branched, unbranched, or circular ones). One such major question is how robust is a given pathway in the presence of the system parameter uncertainties.

In the last four decades, classical and traditional engineering theory and practice were adopted and applied to many fields of biology, including many different types of biochemical networks among them metabolic, signaling, and genetic pathways (see [4–7] and the references therein). Specifically, we note that the application of feedback control theory to biological systems has been focused on both the molecular and cellular levels. For example, in [8], it has been shown that gradient sensing and chemotaxis in a simple biochemical network is essentially robust as a result of integral control. A study of similar phenomena in the social amoebae *Dictyostelium Discoideum* has shown that the aggregation of these species during starvation is performed by utilizing traditional principles of engineering feedback control theory [9].

The issue of robustness of (metabolic) biochemical pathways to changes in the system parameters such as varied concentrations of the enzymes involved has been studied by the seminal work of Savageau [10–13] since the late 60s (see also [14] and the references therein). By applying a power law technique for the analysis of enzyme-driven pathways, which are composed of essentially nonlinear units, Savageau and coworkers have extensively studied the robustness of biochemical enzyme pathways resulting in the well known and powerful **BST framework (Biochemical System Theory)** (see [14–17] and especially [18] for a thorough study of BST).

A different approach to BST, aiming at the study of biochemical pathways (including metabolic signaling and genetic pathways) is the **Metabolic Control Analysis (MCA)** framework which was formulated in the mid-70s (see [5, 6] for a comprehensive account of this topic). MCA is a powerful framework that has been shown to be equivalent to classical control theory in most of its properties [19]. Evidently, the issue of robustness of biochemical pathways (or networks) is an already a vast and established field of research where one of its goals is the issue of optimality.

Indeed, the study of the optimality of biochemical pathways has become a vast and extensive field of research (see [6, 17, 18, 20–23]) which is aided by a host of numerical optimization techniques and computer software programs (see [22] and the references therein). To the best of our knowledge, no previous attempt was carried out in order to evaluate the optimality of an enzymatic pathway, either (physically realizable) theoretical one or real one, in the H_∞ sense or some related control-oriented optimal measures which are key elements in the optimal control field. The motivation for such a study stems from the powerful physical meanings of the optimal control measures such as the H_∞, or the P2P (Peak to Peak) norm measures (to name a few), within the modern control engineering field. It is conceivable that biochemical feedback systems may fall into one of these categories. The question whether a real biochemical pathway "applies" a worst-case strategy in the face of the disturbances and various parameter uncertainties or whether it "applies" a peak-to-peak strategy is a key question that can be addressed by the powerful and established fields of modern control techniques.

In this part of the book, a humble attempt is made to look into the above concerns by exploiting some of the tools of traditional and modern engineering feedback control theory—culminating in a different approach to the study of the optimal robustness of biochemical pathways. This part includes two chapters: In Chap. 16, we present the basic method that we apply in order to explore the possible optimality of a given biochemical pathway. We start with simple enzymatic reactions where we linearize the kinetic dynamics of each reaction in order to apply classical and modern linear feedback control tools. We then turn to a simple theoretical four-enzyme system that is subject to a negative feedback loop [going from the last product to the first enzyme] and look at the open- and closed-loop behavior of the system. We show in the latter study that the closed-loop configuration can be modeled as a SOF (Static Output Feedback) control system. We then show that following linearization of the system components enables us to account for the nonlinearity of the reaction system as a polytopic-type uncertain linear system. At this point, we formulate the basic H_∞ test that we use in order to asses the robustness of the closed-loop system to the systems parameter uncertainties [i.e., in the kinetic constants and concentrations of the enzyme involved] followed by application of the latter tool to the threonine synthesis pathway. The latter pathway is relatively short, and it contains three negative feedback loops that are exerted by the end product of the pathway, threonine. We explore, following linearization, the optimality of the threonine synthesis pathway. This pathway in the bacteria E. coli has been studied extensively by [24–26] and has been shown to contain three negative feedback loops. We explore the optimality of this pathway in the H_∞ sense and obtain results that are in line with the experimental behavior of its components [27–30].

In Chap. 17, we extend the approach that we have developed in Chap. 16, to explore the possible optimality of biochemical pathways via other system norms including the H_2, P2P (Peak to Peak), and energy-to-peak norms. We probe the effect of the various uncertainties in the system parameters via the latter norms in an attempt to identify the proper norm(s) that may account for the system robustness in the face of various uncertainties and disturbances. We demonstrate the results of our study via the threonine synthesis and the glycolytic pathway [in yeasts]. We show, in Chap. 17, that similar results are achieved for the latter two completely different pathways [31–33].

References

1. Lehninger, A.L.: Principles of Biochemistry. Worth Publishers (1982)
2. Segel, I.R.: Enzyme Kinetics—Behavior and Analysis of Rapid Equilibrium and Steady-State Enzyme Systems. Wiley (1975)
3. Wilkinson, F.: Chemical Kinetics and Reaction Mechanisms. Van Nostrand Reinhold Company (1980)
4. Wolkenhauer, O., Ghosh, B.K., Cho, K.H.: Control and coordination in biochemical networks. IEEE Control Syst. **24**(4), 30–34 (2004)
5. Fell, D.D.: Metabolic control analysis: a survey of it's theoretical and experimental development. Biochem. J. **286**, 313–330 (1992)
6. Fell, D.: Frontiers in Metabolism: Understanding the Control of Metabolism. Portland Press (1997)
7. Saez-Rodriguez, J., Kremling, A., Conzelmman, H., Bettenbrock, K., Gilles, E.D.: Modular analysis of signal transduction networks. IEEE Control Syst. **24**(4), 35–52 (2004)
8. Yi, T.M., Huang, Y., Simon, M.I., Doyle, J.: Robust perfect adaptation in bacterial chemotaxis through integral feedback control. Proc. Natl. Acad. Sci. U. S. A. **97**, 4649–4653 (2000)
9. Iglesias, P.A.: Feedback control in intracelluar signaling pathways: regulating chemotaxis in Disctyostelium Discoideum. Eur. J. Control **9**(2–3), 227–236 (2003)
10. Savageau, M.A.: Biochemical systems analysis. I. Some mathematical properties of the rate law for the component enzymatic reactions. J. Theor. Biol. **25**, 365–369 (1969)
11. Savageau, M.A.: Biochemical systems analysis. II. The steady-state solutions for an n-pool system using a power-law approximation. J. Theor. Biol. **25**, 370–379 (1969)
12. Savageau, M.A.: Biochemical systems analysis. 3. Dynamic solutions using a power-law approximation. J. Theor. Biol. **26**, 215–226 (1970)
13. Savageau, M.A.: Biochemical Systems Analysis: A Study of Function and Design in Molecular Biology. Addison Wesley, Reading, MA (1976)
14. Alves, R., Savageau, M.A.: Effect of overall feedback inhibition in unbranched biosynthetic pathways. Biophys. J. **79**, 2290–2304 (2000)
15. Voit, E.O.: Computational Analysis of Biochemical Systems: A Practical Guide for Biochemists and Molecular Biologists. Cambridge University Press, Cambridge, UK (2000)
16. Goel, G.: Reconstructing Biochemical Systems. Systems Modeling and Analysis Tools for Decoding Biological Designs. VDM Verlag Dr. Mller, Saarbrcken, Germany (2008)
17. Torres, N.V., Voit, E.O.: Pathway Analysis and Optimization in Metabolic Engineering. Cambridge University Press, Cambridge, UK (2005)
18. Voit, E.O.: Biochemical systems theory: a review. ISRN Biomath. **2013**(Article ID 897658), 53 (2013). https://doi.org/10.1155/2013/897658
19. Ingalls, B.P.: A frequency domain approach to sensitivity analysis of biochemical systems. J. Phys. Chem. B **108**, 1143–1152 (2004)
20. Schuster, S., Heinrich, R.: Minimization of intermediate concentrations as a suggested optimality principle for biochemical networks, I. Theoretical analysis. J. Math. Biol. **29**, 425–442 (1991)
21. Savinell, J.M., Palsson, B.O.: Network analysis of intermediary metabolizm using linear optimization, I. Developmenmt of mathematical formalism. J. Theor. Biol. **154**, 421–454 (1992)

22. Mends, P., Kell, D.B.: Non-linear optimaization of biochemical pathways: applications to metabolic engineering and parameter estimation. Bioinformatics **14**(10), 869–883 (1998)

23. Morohashi, M., Winn, A.E., Borisuk, M.T., Bolouri, H., Doyle, L., Kitano, H.: Robustness as a measure of plausibility in models of biochemical networks. J. Theor. Biol. **216**, 19–30 (2002)

24. Chassagnole, C., Rais, B., Quentin, E., Fell, D., Mazat, J.P.: An integrated study of Threonine-pathway enzyme kinetics in Echerichia coli. Biochem. J. **356**, 415–423 (2001)

25. Chassagnole, C., Rais, B., Quentin, E., Fell, D., Mazat, J.P.: Threonine synthesis from aspartate in Echerichia coli cell-free extract: pathway dynamics. Biochem. J. **356**, 425–432 (2001)

26. Chassagnole, C., Rais, B., Quentin, E., Fell, D., Mazat, J.P.: Control of Threonine-synthesis pathway in Echerichia coli: theoretical and experimental approach. Biochem. J. **356**, 433–444 (2001)

27. Gershon, E., Hiler, R., Shaked, U.: Classical control theory approach to enzymatic reactions. In: Proceedings of the European Control Conference (ECC03), Cambridge, England (2003)

28. Gershon, E., Shaked, U.: H_∞ feedback control of biochemical pathways via system uncertainty. In: Proceedings of the 5rd IFAC Symposium on Robust Control Design (ROCOND), Toulouse, France, July 2006

29. Gershon, E., Shaked, U.: H_∞ feedback-control theory in biochemical systems. Int. J. Robust Nonlinear Control **18**, 18–50 (2008)

30. Gershon, E., Yokev, O., Shaked, U.: H_∞ feedback-control of the Threonine Synthesis pathway via system uncertainty. In: Proceedings of the European Control Conference (ECC07), Kos, Greece, June 2007

31. Gershon, E., Shaked, U.: Robust polytopic analysis of the feedback-control of Glycolysis in Yeasts via some system norms. In: Proceedings of the 20th Mediterranean Conference on Control and Automation (MED12), Barcelona, Spain, July 2012

32. Gershon, E., Navon, M.: Robust feedback-control analysis of the Threonine Synthesis Pathway via various system norms. In: Proceedings of the 22nd Mediterranean Conference on Control and Automation (MED14), Palermo, Sicily, June 2014

33. Gershon, E., Navon, M., Shaked, U.: Robust peak to peak and H_∞ static output-feedback control of the Threonine Synthesis Pathway. In: Proceedings of the European Control Conference (ECC15), Linz, Austria, July 2015

Chapter 16
H_∞ Feedback Control Theory in Biochemical Systems

Abstract A new approach is presented for the study of H_∞ optimal control of biochemical pathways. Starting with various linear models of single enzymatic reaction systems, a simple unbranched four-block enzyme system that contains a negative feedback loop is analyzed in the open- and the closed-loop configurations, where it is shown that the closed-loop configuration is one of the static output-feedback control systems. The original nonlinear four-block system is modeled as a polytopic-type uncertain linear system, where the extent of the nonlinearity can be "tuned" by a fictitious uncertainty interval, thus better capturing the nonlinear nature of the systems under study. The sensitivity of the latter enzymatic system to variations in certain variables is explored via the optimal H_∞ control approach. Based on this approach, the threonine synthesis pathway that contains three negative feedback loops is analyzed and is shown to be optimal in the H_∞ sense.

16.1 Introduction

In the present chapter, we bring the results of our study concerning the possible optimality of biochemical pathways in the H_∞ sense. We extend our approach to other performance measures in Chap. 17, based on the full scope of the derivations that are given here. Our approach, in the present chapter, is to explore the optimality of biochemical pathways by starting with the modeling of simple enzymatic reactions, applying a simple Taylor expansion [1–7]. Following this step, we apply our basic approach to account for the nonlinearity of the systems explored (since enzymatic reactions are nonlinear by nature). We model the original nonlinear system as a polytopic-type uncertain linear system, where the extent of the nonlinearity can be "tuned" by a fictitious uncertainty interval, thus better capturing the nonlinear nature of the systems under study. We then use the simple linear models to construct a simple four-block enzyme pathway. The pathway that we explore could be easily extended to include any reasonable number of enzymes and multiple points of regulation within the pathway; we have chosen, however, a four-block pathway in order to demonstrate the basic idea that we have developed. We proceed by simulating the system open- and closed-loop transient and the steady-state responses to a stimulus

© Springer Nature Switzerland AG 2019
E. Gershon and U. Shaked, *Advances in H_∞ Control Theory*,
Lecture Notes in Control and Information Sciences 481,
https://doi.org/10.1007/978-3-030-16008-1_16

step function at the input of the pathway. Specifically, going from an open-loop system to a closed-loop one, we replace the first enzyme model by a different model, applicable to a simple Michaelis–Menten (MM) competitively inhibited enzyme [8–10]. We then show that, in fact, in the closed-loop configuration the system operates in a static output-feedback control mode. The numerical results of the simulations that we perform for the closed-loop system are verified to be significantly close to the results obtained by numerical integration of the nonlinear differential equations of the enzymes involved. At this point, we present an H_∞ optimal analysis for the assessment of the optimality of the feedback gain found for the closed-loop system of the four-block pathway, including the linear analysis, via system uncertainty, in the presence of real physical uncertainties.

We apply a similar analysis to the study of a real, well-studied, biochemical pathway: the threonine synthesis pathway. This pathway in the bacteria E. Coli has been studied extensively by [11–13]. The kinetics and regulation of the individual enzyme involved in the pathway were studied by several researches (see [11] and the references therein). In this chapter, we bring the major components of the threonine synthesis pathway. The interesting fact, which is abundant in metabolism, is that this pathway is internally regulated by feedback inhibition in three out of its five enzymes. The inhibition is exerted by the end product of the pathway—threonine, which is also consumed or removed by other routes. In our study, we show, numerically, that the threonine synthesis pathway is optimal in the H_∞ sense, in face of various parameter uncertainties of the system.

This chapter is organized as follows: In Sect. 16.2, we bring the modeling of several simple enzymatic reactions and the basic linearization procedure that we use in our analysis. In Sect. 16.3, we present our approach where the nonlinear nature of the systems under study is approximated by treating the nonlinear systems as polytopic uncertain linear systems. In Sect. 16.4, we analyze, for simplicity, the dynamics of a four-block enzymatic pathway, in both the open- and closed-loop configurations. In Sect. 16.5, we treat the case where a simple transport delay is incorporated in the system and in Sects. 16.3 and 16.7 we present the basic uncertainty analysis which is latter applied to the study of real biochemical pathway. In Sect. 16.8, we present the optimal H_∞ analysis that we use and we apply it there to the closed-loop four-block pathway whose dynamics has been analyzed in Sect. 16.4. In Sect. 16.9, we apply the various tools that we developed to the optimal analysis of the threonine synthesis pathway. We conclude our study in Sect. 16.10.

Notation: Throughout the chapter, the sign "[ ]" stands for chemical concentration (usually in molar units, but note that enzyme concentration is typically around micro-Molar units and that of the substrate around mM (milli-Molar)). Derivatives of reactants or products are chemical velocities. A positive velocity represents rate of formation and a negative one represents the rate of degradation (note that no distinction is made here between rate and velocity). Small "k's" represent chemical rate constants, where large "K's" are chemical equilibrium constants.

16.2 Simple Models of Enzymatic Reactions

In this section, we bring the model construction of two basic enzymatic reactions needed to construct, at a later stage, a simple hypothetical pathway that is composed of these models (see [8] and also [9, 10] for further involved models).

16.2.1 Simple Michaelis–Menten Enzymatic Mechanism

In this section, we introduce the basic tool of **linearization** in the study of various enzyme reaction schemes. We consider the following Michaelis–Menten enzymatic reaction system [9]:

$$S + E \underset{k_{-1}}{\overset{k_1}{\rightleftarrows}} ES \underset{k_{-2}}{\overset{k_2}{\rightleftarrows}} E + P, \tag{16.1}$$

where ES is the enzyme–substrate complex and where we note that in the simplest case the reaction of $E + P \overset{k_{-2}}{\rightarrow} ES$ is abolished. The rate of the product P formation is described by the following kinetic equation:

$$v = \frac{d[P]}{dt} = \frac{[S]V_{max}}{[S] + K_M}, \tag{16.2}$$

where $[\cdot]$ denotes chemical concentration, $K_M = \frac{k_2 + k_{-1}}{k_1}$ and $V_{max} = [E_{total}] \times k_2$ (note that k_2 is usually assigned as $k_{catalytic}$). Equation (16.2) expresses the velocity of the product formation as a function of the substrate concentration where the total concentration of the enzyme is conserved (i.e., $[E_{total}] = [E_{free}] + [ES] = constant$). The derivation of (16.2) can be found in most introductory books of General Biochemistry (see [8], page 214). Traditionally, three cases of substrate concentrations are considered by the experimental biochemist.

16.2.1.1 Case-1: Low Substrate Concentration

In the case where $[S] \ll K_M$ we obtain $v = \frac{d[P]}{dt} = \frac{[S]V_{max}}{K_M} = K[S]$, $K = \frac{V_{max}}{K_M}$. Applying Laplace transform [14] to the latter equation, we obtain $P(s) = \frac{KS(s)}{s}$, where $P(s)$ and $S(s)$ are the Laplace transforms of $[P]$ and $[S]$, respectively. In the case where the amount of S (i.e., number of moles) at a given concentration $[S]$ is practically unlimited (say a constant supply of S) one obtains a single integrator behavior of the reaction system, where $[S]$ is taken as a step function input signal. In the (usual) case where the amount of S is limited (i.e., $[S]$ decreases as the reaction proceeds) the relation between the initial concentration of $[S]$ and $[P]$ is described by the block diagram of Fig. 16.1. Note that $\delta(t)$ denotes unit impulse function.

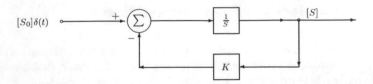

Fig. 16.1 Block diagram of the system of Sect. 16.2.1.1 (Case 1) with $S_0\delta(t)$ representing the initial concentration of the substrate

16.2.1.2   Case-2: Mid-substrate Concentration

In the case where $[S] \cong K_M$, the initial formation rate of the product concentration can be derived around a given set point, say $[S_0]$, for example $[S_0] = K_M$ for which $v = \frac{V_{max}}{2}$. The following linearization procedure is applied:

Denoting

$$F([S]) \overset{\Delta}{=} v = \frac{d[P]}{dt} \quad \text{and} \quad \bar{S} \overset{\Delta}{=} [S] - [S_0],$$

we obtain, using a Taylor series expansion around $[S_0]$, the following linear approximation:

$$v = \frac{d[P]}{dt} = \tilde{V} + \alpha\bar{S}, \tag{16.3}$$

where

$$\tilde{V} = F([S_0]) = \frac{V_{max}[S_0]}{K_M + [S_0]}, \quad \alpha \overset{\Delta}{=} \frac{\partial F}{\partial[S]}\Big|_{[S_0]} = \frac{V_{max}[[S_0] + K_M] - V_{max}[S_0]}{([S_0] + K_M)^2}. \tag{16.4}$$

We note that the set point of the system can be defined as the fraction f of the maximal velocity:

$$\tilde{V} = f \times V_{max}. \tag{16.5}$$

Considering, for example, $[S_0] = K_M$ and performing the partial derivative, we have $f = 0.5$, $\tilde{V} = \frac{V_{max}}{2}$, and $\alpha = \frac{V_{max}}{4K_M}$, which leads to the following kinetic equation around the chosen set point: $F(S) = 0.5V_{max} + \frac{V_{max}}{4K_M}\bar{S}$. Applying then the Laplace transform to (16.3) the following is obtained:

$$\bar{P}(s) \overset{\Delta}{=} P(s) - \frac{P_0}{s} = \frac{\tilde{V}}{s^2} + \frac{\bar{S}(s)\alpha}{s}. \tag{16.6}$$

The block diagram that describes the linear approximation of this case is given in Fig. 16.2.

Remark 16.1 In Fig. 16.2, and in all the figures that follow, the initial concentration values are considered to be a unit-step function of time multiplied by the corresponding initial concentration.

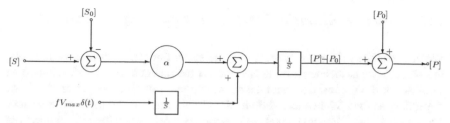

Fig. 16.2 Block diagram of the system of Sect. 16.2.1.2 (Case 2) with $f V_{max} \delta(t)$ representing the initial velocity of the enzyme. $[S_0]$ and $[P_0]$ are the initial concentrations of the substrate and the product, respectively

Remark 16.2 The case where $[S] \cong K_M$ was shown to be the basal intracellular concentration, at the level of the cell interior (the so-called "in vivo") [8, 9]. In this case, the enzyme works around half of its catalytic power (recall that $v = 0.5 V_{max}$)— an experimental fact that shows the flexibility and adaptability of the system to respond to changing conditions and demands of the cell metabolic needs [8, 9]. Once the product P is highly "needed" by the cell, the formation rate "jumps" from the basal state to higher levels in relatively short time. Likewise, in the case where P is hardly needed, the rate relaxes to lower rate values (say $0.1 V_{max}$), again in a relatively short time. We note this fact to emphasize the point that the modeling of Case 2, though restricted to small perturbation fits, in many cases, the normal biochemistry of the cell. Evidently, as the substrate concentration deviates strongly from K_M (say $\bar{S}$ increases), the nonlinear behavior of the system is heavily accentuated and, therefore, the simple linearization procedure of this section is no longer valid. In the latter case, one may resort to describing function analysis [14], a possibility which is left for further study.

Remark 16.3 A question may arise whether the latter linear approximation is bio-chemically relevant. Considering both (16.2) and (16.3) we note that at the "natural" set point of $[S_0] = K_M$ the substrate concentration may fluctuate by as much as 50%, causing only $\approx 5\%$ error when using the linearized approximation of (16.3). The latter fluctuation has been shown to cover the significant range of variation, **at a basal operating mode of the cell** [8, 9], in the intracellular concentration of $[S]$.

16.2.1.3   Case-3: High Substrate Concentration

In this case $[S] \gg K_M$. Therefore, the enzyme reaction works at the maximum velocity of V_{max} (the so-called "full catalytic power"). In living cells, this case may occur quite frequently, depending on the metabolic needs of the cell.

16.2.2   Simple Michaelis–Menten Enzymatic Mechanism—II

We consider the Michaelis–Menten enzyme system of (16.1), where we note that the product can associate with the enzyme via the rate constant k_{-2}. The later fact implies that the net reaction can take the direction of either $S \longrightarrow P$ or $P \longrightarrow S$, depending on the initial concentrations of both substances and the kinetic constants. Strictly speaking, all enzyme-catalyzed reactions are reversible; therefore, the system of (16.1) is more realistic then the one analyzed by neglecting the direction $P \longrightarrow S$. In fact, the system of (16.1) involves only one central complex (i.e., ES), whereas, in reality, the system involves a second central complex of EP. Restricting our inquiry to (16.1), the kinetic equation for the later reaction is (see [9], page 29)

$$v_{net} = \dot{y} \triangleq \frac{d[P]}{dt} = \frac{\frac{V_{max_f}}{K_s}u - \frac{V_{max_r}}{K_p}y}{\Delta}, \quad \Delta \triangleq 1 + \frac{u}{K_s} + \frac{y}{K_p},$$

where v_{net} is the velocity in the forward direction (i.e., $S \longrightarrow P$), $u \triangleq [S]$, $y \triangleq [P]$, $K_s = \frac{k_{-1}+k_2}{k_1}$, $K_p = \frac{k_2+k_{-1}}{k_{-2}}$, $V_{max_f} = k_2[E]_t$ and $V_{max_r} = k_{-1}[E]_t$, and where $[E]_t$ is the total concentration of the enzyme. Applying similar arguments to those of Sect. 16.2.1 we obtain the following linear model:

$$\tilde{F}(u, y) \triangleq \dot{y}(u, y) = \tilde{v}_{net_0} + \frac{\partial \tilde{F}}{\partial u}|_{u_0, y_0}(u - u_0) + \frac{\partial \tilde{F}}{\partial y}|_{u_0, y_0}(y - y_0),$$

where

$$\tilde{v}_{net_0} = \frac{\frac{V_{max_f}}{K_s}[S_0] - \frac{V_{max_r}}{K_p}[P_0]}{1 + \frac{[S_0]}{K_s} + \frac{[P_0]}{K_p}}.$$

Denoting

$$\bar{u} \triangleq u - u_0, \quad \bar{y} \triangleq y - y_0, \quad \bar{\beta} \triangleq \frac{\partial \tilde{F}}{\partial u}|_{u_0, y_0}, \quad \text{and} \quad \bar{\alpha} \triangleq -\frac{\partial \tilde{F}}{\partial y}|_{u_0, y_0},$$

we obtain

$$\dot{\bar{y}} = \tilde{v}_{net_0} - \bar{\alpha}\bar{y} + \bar{\beta}\bar{u}, \tag{16.7}$$

and where

$$\bar{\alpha} \triangleq \frac{-[\frac{V_{max_r}}{K_p} + \frac{V_{max_f}+V_{max_r}}{K_s K_p}u_0]}{\Delta_0^2}, \quad \text{and} \quad \bar{\beta} \triangleq \frac{\frac{V_{max_f}}{K_s} + \frac{V_{max_f}+V_{max_r}}{K_s K_p}y_0}{\Delta_0^2},$$

where $\Delta_0 = \Delta|_{u_0, y_0}$. Choosing, as an example, the special case where $[S_0] = [P_0]$ and where the equilibrium constant of $S \overset{K_1}{\underset{}{\rightleftarrows}} P$ is such that the net reaction proceeds in the

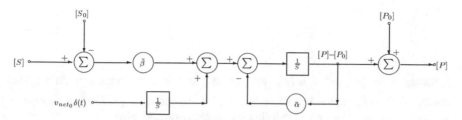

Fig. 16.3 Block diagram of the system of Sect. 16.2.2 with $v_{net_0}\delta(t)$ representing the initial velocity of the enzyme. $[S_0]$ and $[P_0]$ are the initial concentrations of the substrate and the product, respectively

forward direction of $S \longrightarrow P$, one obtains that $\tilde{v}_{net_0} = \dfrac{\frac{V_{max_f}}{K_s}[S_0] - \frac{V_{max_r}}{K_p}[S_0]}{1 + \frac{[S_0]}{K_s} + \frac{[S_0]}{K_p}}$. Applying Laplace transform to (16.7), the following is found:

$$\bar{Y}(s) \triangleq Y(s) - \frac{Y_0}{s} = \frac{\tilde{v}_{net_0}}{(s + \bar{\alpha})s} + \frac{\bar{\beta}\bar{U}(s)}{s + \bar{\alpha}},$$

where $Y(s)$ and $\bar{U}(s)$ are the Laplace transforms of $[P]$ and $[\bar{S}]$, respectively. This case is described by the block diagram of Fig. 16.3.

16.2.3 Simple Inhibitory Systems—Competitive Inhibition

An inhibitor is any substance that reduces the velocity of an enzyme-catalyzed reaction. Among the simple inhibitory mechanisms the most abundant one is that of reversible nature where the inhibitor molecule competes with the substrate on the same active site of the enzyme (the so-called "competitive inhibitor"). Once the inhibitor binds to the free enzyme, the substrate can no longer bind to the enzyme (that is, both are *mutually exclusive*). However, although the presence of the competitive inhibitor reduces the apparent affinity of the enzyme towards the substrate (i.e., K_M increases), this type of inhibition is negligible when the substrate concentration is high enough (that is V_{max} is not changed). We consider the following simple enzyme system:

$$E + S \underset{}{\overset{K_s}{\leftarrow}} ES \xrightarrow{k_p} E + P$$
$$E + I \underset{}{\overset{K_i}{\leftarrow}} EI, \tag{16.8}$$

where we note that the inhibitor–enzyme complex EI dissociates to its components with the equilibrium constant of K_i. The kinetic equation for the above reaction is (see [9], page 103, [10])

$$\dot{y} = \frac{d[P]}{dt} = v = \frac{V_{max} \frac{[S]}{K_s}}{1 + \frac{[S]}{K_s} + \frac{[I]}{K_i}}. \qquad (16.9)$$

Defining $u \triangleq [S]$ and $I = [I]$, we obtain the following nonlinear differential equation: $\dot{y} + K_s^{-1}u\dot{y} + K_i^{-1}I\dot{y} = V_{max}K_s^{-1}u$. Denoting $\tilde{F}(u, I) \triangleq \dot{y}$, $\bar{u} = u - u_0$, and $\bar{I} = I - I_0$, we have the following linear approximation:

$$\dot{y} = \tilde{V}_0 + \alpha\bar{u} - \beta\bar{I}, \quad \tilde{V}_0 = \frac{V_{max}\frac{[S_0]}{K_s}}{1 + \frac{[S_0]}{K_s} + \frac{[I_0]}{K_i}}, \qquad (16.10)$$

where

$$\alpha \triangleq \frac{\partial \tilde{F}}{\partial u}|_{u_0, I_0} = V_{max} \frac{[K_s^{-1}\Upsilon - K_s^{-2}u_0]}{\Upsilon^2},$$

$$\beta \triangleq \frac{\partial F}{\partial I}|_{u_0, I_0} = V_{max} \frac{K_s^{-1}K_i^{-1}u_0}{\Upsilon^2},$$

$$\Upsilon \triangleq 1 + \frac{u_0}{K_s} + \frac{I_0}{K_i}.$$

Applying Laplace transform to (16.10), the following is thus obtained:

$$\bar{P}(s) \triangleq P(s) - \frac{P_0}{s} = \frac{\alpha}{s}\bar{U}(s) - \frac{\beta}{s}\bar{I}(s) + \frac{\tilde{V}_0}{s^2},$$

where $\bar{U}(s)$ and $\bar{I}(s)$ are the Laplace transforms of $\bar{u}$ and $\bar{I}$, respectively. The set point in the above analysis can be easily determined once a pair of $[S_0]$, $[I_0]$ is chosen. Choosing, for example, $\tilde{V}_0 = 0.5V_{max}$ and a given $[S_0]$, the inhibitor concentration is easily recovered from (16.9). We have $[I_0] = K_i[\frac{[S_0]}{K_s} - 1]$. The block diagram of Fig. 16.4 is thus obtained.

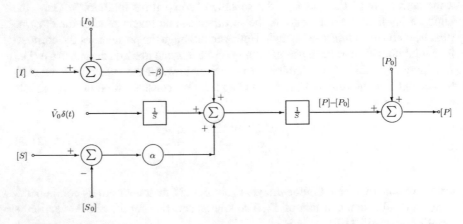

Fig. 16.4 Block diagram of the system of Sect. 16.2.3 with $\tilde{V}_0\delta(t)$ representing the initial velocity of the enzyme. $[S_0]$, $[P_0]$ and $[I_0]$ are the initial concentrations of the substrate, the product and the inhibitor of the enzyme, respectively

Note that in the case where $[I] \ll K_i$ the results of Sect. 16.2.1 are recovered with $K_s = K_M$.

16.2.4 Regulation via Energy Charge—Product Inhibition

An interesting case in enzyme biochemistry pathways is one where the total amount of the substrate and the product is constant (i.e., $[S] + [P] = C$, where C is constant) during the basal state of the cell normal life [8, 9]. We consider the system of (16.1) and obtain (see [9], page 120)

$$v = \frac{V_{max}[S]}{K_s[1 + \frac{[P]}{K_p}] + [S]}.$$

Defining

$$y \overset{\Delta}{=} [P], \ u \overset{\Delta}{=} [S], \ F \overset{\Delta}{=} \dot{y}(t) \ \text{ and } \ \gamma \overset{\Delta}{=} \frac{K_s}{K_p},$$

the following linearized kinetics is obtained:

$$F(y) = \tilde{K} + \frac{\partial \Gamma}{\partial y}|_{y_0} \bar{y},$$

where $\bar{y} = y - y_0$,

$$\frac{\partial F}{\partial y}|_{y_0} \overset{\Delta}{=} \bar{K} = \frac{-V_{max}[C + K_s + (\gamma - 1)y_0] - (\gamma - 1)[C - y_0]V_{max}}{[C + K_s + (\gamma - 1)y_0]^2},$$

and

$$\tilde{K} \overset{\Delta}{=} -\bar{K}y_0 + \frac{(C - y_0)V_{max}}{C + K_s + (\gamma - 1)y_0}. \tag{16.11}$$

Rearranging the latter equation, we have

$$\dot{\bar{y}} = \tilde{K} + \bar{K}\bar{y}. \tag{16.12}$$

Applying Laplace transform to the above equation, the following equation is obtained:

$$\bar{Y}(s) \overset{\Delta}{=} Y(s) - \frac{Y_0}{s} = \frac{\tilde{K}}{s(s - \bar{K})}.$$

The resulting linearized model is described by the block diagram of Fig. 16.5.

Note that dissimilar to Sect. 16.2.2, where the product is also allowed to produce the substrate via the dynamics of (16.1), the feedback loop in Fig. 16.5 is different

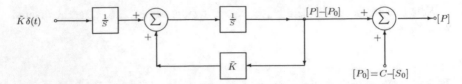

Fig. 16.5 Block diagram of the system of Sect. 16.2.4 with $\tilde{K}\delta(t)$ representing the initial velocity of the enzyme. $[S_0]$ and $[P_0]$ are the initial concentrations of the substrate and the product of the enzyme, respectively

from the one in Fig. 16.3. The latter is **inherent to all reversible chemical reactions**. Here, the accumulation of the product exerts a biochemically significant inhibitory effect, as indicated by the title of this section.

16.2.5 Product Inhibition—Numerical Example

We consider the case of product inhibition of Sect. 16.2.4 and we refer to the example given in ([9], see page 123). This example deals with a typical enzyme reaction which is regulated by the "energy charge" of the cell [15]. The given rate constants and concentrations are of typical value and they were shown to demonstrate this important type of inhibition. We have $K_p = 0.1$, $K_s = 1.0$, $C = [S] + [P] = 10$, where we note that the net reaction can proceed either in the direction of $S \longrightarrow P$, or $P \longrightarrow S$, depending on the ration of $\frac{[S]}{[P]}$. Choosing as a set point $v = 0.5V_{max}$ we have $[S] = 9$, $[P] = 1$ (see [9], page 123) and we obtain, applying (16.11) and (16.12), $\tilde{K} = -0.25V_{max}$ and $\bar{K} = 0.70V_{max}$. Note that $V_{max} > 0$ and it partially depends on the concentration of the enzyme $[E]_t$. The result clearly demonstrates the stability of the feedback loop inherent to this type of inhibition. Applying elementary control result [14], it follows that in this specific type of inhibition any small changes in $\tilde{K}$ (caused by changes in C or by the kinetic constants) are attenuated with a time constant of $\frac{4}{V_{max}}$. Also, the steady-state response level of this process tends to $\frac{4}{V_{max}}$.

16.2.6 Random Bireactant Systems

We consider the kinetics of a multireactant enzyme system where we treat the special case in which the enzyme can bind randomly to either substrates A or B to produce the product P via a complex EAB, as in the following multi-step reaction (see [9], page 274) :

$$E + A \overset{K_A}{\leftrightarrows} EA, \; EA + B \overset{\alpha K_B}{\leftrightarrows} EAB \overset{k_p}{\longrightarrow} E + P$$
$$E + B \overset{K_B}{\leftrightarrows} EB, \; EB + A \overset{\alpha K_A}{\leftrightarrows} EAB \overset{k_p}{\longrightarrow} E + P. \tag{16.13}$$

The kinetic analysis of the above reaction is based on the assumption of a rapid equilibrium of all the reversible steps in (16.13), in comparison with the last catalytic step (the so-called "rate-limiting step"). Note also that the reaction is not sensitive to the order by which both substrates A and B bind to the enzyme. However, once the enzyme binds to one of the substrates say A, the equilibrium constant of, say,

$E + B \overset{K_B}{\leftarrow} EB$ is changed by a factor of α.

For the reaction of (16.13), we have [9]

$$v = \frac{d[P]}{dt} = \frac{V_{max}\frac{[A][B]}{\alpha K_A K_B}}{1 + \frac{[A]}{K_A} + \frac{[B]}{K_B} + \frac{[A][B]}{\alpha K_A K_B}}. \tag{16.14}$$

Rearranging the above, and defining $\bar{\alpha} = \alpha K_A K_B$, $\bar{\beta} = \alpha K_B$, and $\bar{\gamma} = \alpha K_A$, the kinetic equation of the above reaction system is

$$v(t) = \frac{[A][B]V_{max}}{\bar{\Delta}}, \quad \bar{\Delta} = \bar{\alpha} + \bar{\beta}[A] + \bar{\gamma}[B] + [A][B].$$

Denoting

$$\tilde{F}(u_1, u_2) \overset{\Delta}{=} v, \ u_1 \overset{\Delta}{=} [A] \ \text{and} \ u_2 \overset{\Delta}{=} [B],$$

we have the following linearized model around the set point (u_{1_0}, u_{2_0}):

$$\tilde{F}(u_1, u_2) = \tilde{F}(u_{1_0}, u_{2_0}) + \hat{\alpha}\bar{u}_1 + \hat{\beta}\bar{U}_2, \tag{16.15}$$

where $\bar{u}_1 \overset{\Delta}{=} u_1 - u_{1_0}$ and $\bar{u}_2 \overset{\Delta}{=} u_2 - u_{2_0}$ and where

$$\hat{\alpha} \overset{\Delta}{=} \frac{\partial \tilde{F}}{\partial u_1}\Big|_{u_{1_0}, u_{2_0}} = \frac{u_{2_0} V_{max}[\bar{\Delta} - u_{1_0}[\bar{\beta} + u_{2_0}]]}{\bar{\Delta}^2},$$

$$\hat{\beta} \overset{\Delta}{=} \frac{\partial \tilde{F}}{\partial u_2}\Big|_{u_{1_0}, u_{2_0}} = \frac{u_{1_0} V_{max}[\bar{\Delta} - u_{2_0}[\bar{\gamma} + u_{1_0}]]}{\bar{\Delta}^2}.$$

Similarly to Sect. 16.2.1, the set point of the above analysis can be chosen as $\tilde{F}(u_{1_0}, u_{2_0}) = 0.5V_{max}$. Choosing, for example $[A_0] = [B_0]$, the initial concentration of A can be easily recovered from (16.14). We have

$$\tilde{F}(u_{1_0}, u_{2_0}) = 0.5V_{max} = \tilde{F}([A_0], [B_0]) = \frac{V_{max}[A_0]^2}{\bar{\alpha} + [A_0](\bar{\beta} + \bar{\gamma} + [A_0])}.$$

Applying the Laplace transform to (16.15), we obtain

$$\bar{P}(s) \overset{\Delta}{=} P(s) - \frac{P_0}{s} = \frac{\tilde{F}(u_{1_0}, u_{2_0})}{s^2} + \frac{\hat{\alpha}\bar{U}_1(s)}{s} + \frac{\hat{\beta}\bar{U}_2(s)}{s},$$

which leads to the block diagram of Fig. 16.6.

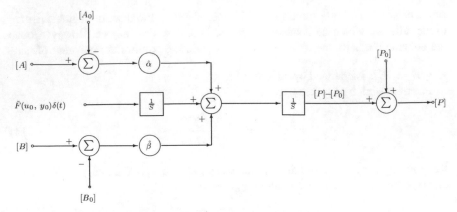

Fig. 16.6 Block diagram of the system of Sect. 16.2.6 with $\tilde{F}(u_0, y_0)\delta(t)$ representing the initial velocity of the enzyme. $[A_0]$, $[B_0]$, and $[P_0]$ are the initial concentrations of the first substrate, the second substrate, and the product of the enzyme, respectively

16.2.7 Hexokinase I Regulation by Feedback Inhibition

In this section, we utilize the advent of our approach by considering one of the most abundant regulatory mechanisms of biochemical pathways, that is, a feedback mechanism around a single enzyme where the inhibitor is the product of the same enzyme (product inhibition, feedback inhibition). Most key enzymes in cell metabolism posses such an "elementary" hardware of feedback control and we maintain that these mechanisms exhibit the known principles of control theory, as generally perceived in control engineering. Hence, the rational for this study as a whole.

Among the most complex of all enzyme-catalyzed mechanisms, we find a large number of enzymes that are generally composed of two or more subunits. These enzymes, generally known as **allosteric enzymes** [8, 9], exhibit a complex behavior. In general, these enzymes are composed of one or more catalytic sites (where the substrates are converted into products) and, in addition, one or more regulatory sites (or modulatory sites) which exert an **internal control** on the active catalytic sites. The effect of the regulatory sites is mediated through special and highly specific molecules which may include metabolites, ATP and related compounds, metal ions, vitamin molecules (sometimes generally known as coenzymes). A very interesting case is the one of (self) feedback inhibition, as mentioned above, where the product of an enzyme acts through the regulatory site of the enzyme and inhibits its catalysis. One of the most studied enzymes of the later kind is an enzyme called "Hexokinase I" (HKI) that catalyzes the transfer of a phosphate group to glucose (sugar) molecule [16–22]. The enzyme binds to both ATP (which is the phosphate donor) and glucose to yield ADP and a product called G6P. The latter is also an inhibitor that binds either to the catalytic site of HKI or to a regulatory site (that resides at a remote terminal of the enzyme molecule). G6P has been shown to compete with the ATP molecule as a competitive inhibitor (see Sect. 16.2.3) and as noncompetitive inhibitor with respect to glucose. Two different mechanisms for the inhibition of Hexokinase I by G6p have been suggested by [20]. In the sequel, we bring the modeling of these two mechanisms.

16.2.7.1 Hexokinase I Feedback Inhibition—Mechanism I

Most investigators believe now that G6p competes with ATP at the active site of the enzyme which resides at the C-terminal half of the polypeptide chain [20]. Considering Scheme I (see also [20], page 31157), we obtain the following enzymatic reaction:

$$EI_2 \overset{K_{ii}}{\to} EI \overset{K_i}{\to} E \overset{K_S}{\to} ES \overset{k_{cat}}{\to} E + P, \tag{16.16}$$

where I and S represent G6p and ATP, respectively, and where we note that Glucose is saturating and therefore does not appear in Scheme I or in the following kinetic rate equation, which was used in the analysis of the kinetic results obtained in [20]:

$$v = \frac{d[P]}{dt} = \frac{V_{max}}{1 + \frac{K_S}{S}[1 + \frac{I}{K_i} + \frac{I^2}{K_i K_{ii}}]}. \tag{16.17}$$

In the above equation, V_{max} is the maximal velocity, K_M is the Michaelis constant for ATP, and K_i and K_{ii} are inhibition constants for the binding of the first and second molecules of G6P, respectively. Scheme I is equally valid for the interpretation of kinetic data for inhibitor binding at independent sites with different affinities or for inhibitor binding to sites with identical affinities coupled by a mechanism of negative cooperativity [20]. Denoting $\alpha \overset{\Delta}{=} \frac{K_S}{K_i}$, $\beta \overset{\Delta}{=} \frac{K_S}{K_i K_{ii}}$ and $\Delta \overset{\Delta}{=} 1 + \frac{K_S}{S} + \alpha \frac{I}{S} + \beta \frac{I^2}{S}$ we obtain

$$\frac{d[P]}{dt} = \frac{V_{max}}{\Delta}. \tag{16.18}$$

Denoting further $F \overset{\Delta}{=} \frac{d[P]}{dt}$, we obtain

$$\frac{\partial F}{\partial S} = \frac{V_{max}[\frac{K_S}{S^2} + \frac{\alpha I}{S^2} + \frac{\beta I^2}{S^2}]}{\Delta^2}, \quad \frac{\partial F}{\partial I} = \frac{-V_{max}[\frac{\alpha}{S} + \frac{2\beta I}{S}]}{\Delta^2}. \tag{16.19}$$

By arguments similar to those of the previous sections, we obtain the following linear model:

$$\frac{d[\bar{P}]}{dt} = F(S_0, I_0) + \bar{\alpha}\bar{S} - \bar{\beta}\bar{I}, \tag{16.20}$$

where $\bar{P} = P - P_0$ and where $\bar{S} \overset{\Delta}{=} S - S_0$, $\bar{I} \overset{\Delta}{=} I - I_0$, and where

$$\bar{\alpha} \overset{\Delta}{=} \frac{\partial F}{\partial S}|_{S_0, I_0}, \quad \text{and} \quad \bar{\beta} \overset{\Delta}{=} -\frac{\partial F}{\partial I}|_{S_0, I_0}.$$

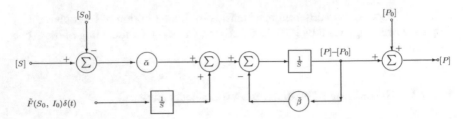

Fig. 16.7 Block diagram of the Hexokinase system of Sect. 16.2.7.1 with $\tilde{F}(S_0, I_0)\delta(t)$ representing the initial velocity of the enzyme. $[S_0]$, $[P_0]$ and $[I_0]$ are the initial concentrations of the substrate, the product, and the inhibitor of the enzyme, respectively

Applying Laplace transform to (16.20), the following is found:

$$\bar{P}(s) \triangleq P(s) - \frac{P_0}{s} = \frac{\bar{\alpha}\bar{S}(s)}{s+\bar{\beta}} + \frac{F(S_0, I_0)}{(s+\bar{\beta})s}, \quad \text{and} \quad F(S_0, I_0) = \frac{V_{max}}{\Delta}\Big|_{S_0, I_0},$$

where we note that, due to the fact that G6P is both a substrate and an inhibitor, the time derivatives of $\frac{d[G6P]}{dt} = \frac{dI}{dt}$ and $\frac{dP}{dt}$ are linearly proportional, depending only on the stoichiometry of the chemical reaction equation. We note that P in Scheme I expresses the concentration of ADP, which is measured experimentally, and which is produced simultaneously with G6P in the reaction assay (although the initial concentration of G6P is a reaction parameter). The linearized model of the kinetic results of Scheme I is depicted in the block diagram of Fig. 16.7.

16.2.7.2   Hexokinase I Feedback Inhibition—Mechanism II

The second mechanism proposed by [20] relates to the suggestion raised by other investigators (see [20, 21] and the references therein) that G6P exerts its effect by binding to an allosteric site that is topologically distinct from the active site. Considering Scheme II on [20] (see page 31157), we obtain the following enzymatic reaction:

$$EI_nI_c \overset{K_{nc}}{\rightleftarrows} EI_n \overset{K_n}{\rightleftarrows} E \overset{K_S}{\rightleftarrows} ES \overset{k_{cat}}{\rightarrow} E + P,$$

$$EI_nI_c \overset{K_{cn}}{\rightleftarrows} EI_c \overset{K_c}{\rightleftarrows} E \overset{K_S}{\rightleftarrows} ES \overset{k_{cat}}{\rightarrow} E + P. \tag{16.21}$$

Scheme II differs from Scheme I of the previous section by the fact that it explicitly defines binding sites for G6P at the N-terminal and the C-terminal halves of the enzyme. Site-specific constants for the dissociation of G6P from the N-terminal and the C-terminal halves are represented by K_n and K_c, respectively. These constants measure the dissociation of G6p from either the N-terminal half or the C-terminal half, whichever applies, when the alternative site is not occupied by G6P [20]. K_{nc} represents the dissociation of G6P from the C-terminal site when the N-terminal site is occupied by another molecule of G6P and K_{cn} represents the dissociation of G6P from the N-terminal site when the C-terminal site is occupied by another molecule of G6P. We note that inhibitor binding to the two sites has a random rapid equilibrium. We also note that if one site is impaired by mutation, inhibition occurs by way of the other site [20]. Considering Scheme II, the following kinetic equation was obtained [20]:

$$v = \frac{d[P]}{dt} = \frac{V_{max}}{1 + \frac{K_S}{S}[1 + \frac{I}{K_n} + \frac{I}{K_c} + \frac{I^2}{K_n K_{nc}}]}. \tag{16.22}$$

Denoting $\tilde{\alpha} \triangleq \frac{K_S[K_n + K_c]}{K_n K_c}$, $\tilde{\beta} \triangleq \frac{K_S}{K_n K_{nc}}$, and $\hat{\Delta} \triangleq 1 + \frac{K_S}{S} + \tilde{\alpha}\frac{I}{S} + \tilde{\beta}\frac{I^2}{S}$ we obtain

$$v = \frac{d[P]}{dt} = \frac{V_{max}}{\Delta}. \tag{16.23}$$

Denoting further $F \triangleq v$ we obtain

$$\frac{\partial F}{\partial S} = \frac{V_{max}[\frac{K_S + \tilde{\alpha}I + \tilde{\beta}I^2}{S^2}]}{\hat{\Delta}^2}, \quad \frac{\partial F}{\partial I} = \frac{-V_{max}[\frac{\tilde{\alpha}}{S} + 2\frac{\tilde{\beta}I}{S}]}{\hat{\Delta}^2}. \tag{16.24}$$

Similarly to the previous section, we obtain

$$\dot{\bar{P}} = F(S_0, I_0) + \hat{\alpha}\bar{S} - \hat{\beta}\bar{I}, \tag{16.25}$$

where

$$\hat{\alpha} \triangleq \frac{\partial F}{\partial S}|_{S_0, I_0}, \quad \text{and} \quad \hat{\beta} \triangleq -\frac{\partial F}{\partial I}|_{S_0, I_0}.$$

Applying Laplace transform to (16.25), the following is found:

$$\bar{P}(s) \triangleq P(s) - \frac{P_0}{s} = \frac{\hat{\alpha}\bar{S}(s)}{s + \hat{\beta}} + \frac{F(S_0, I_0)}{(s + \hat{\beta})s}, \quad \text{and} \quad F(S_0, I_0) = \frac{V_{max}}{\Delta}|_{S_0, I_0},$$

where the arguments of the previous section are applied also here. The block diagram of Fig. 16.8 is obtained.

Remark 16.4 We note that the results reported in [20] were obtained from in vitro assays of the enzyme. HKI in vivo, however, is bound to the outer mitochondrial membrane. It is conjectured in [20] that for HKI, a fail safe mechanism exists with respect to G6P inhibition: if one mode of G6P inhibition is lost (say inhibition at the C-terminal half), then the another mode remains, which assures virtually no diminution in G6P inhibition. It would be interesting to apply our approach to the results of in vivo experiments [22], of HKI, with the aim of probing the feedback strategy at the physiological level.

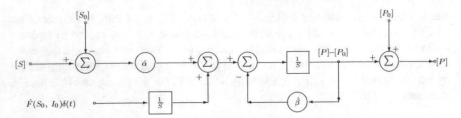

Fig. 16.8 Block diagram of the Hexokinase system of Sect. 16.2.7.2 with $\tilde{F}(S_0, I_0)\delta(t)$ representing the initial velocity of the enzyme. $[S_0]$, $[P_0]$, and $[I_0]$ are the initial concentrations of the substrate, the product, and the inhibitor of the enzyme, respectively

16.3 Linear Analysis via System Uncertainty

The simple linearization procedure that has been applied to Sect. 16.2.1 (Case 2) and Sect. 16.2.3, which results in the linear approximations of (16.3) and (16.10a, b), respectively, will be shown in Sect. 16.4 to model a four-block enzyme system quite accurately (comparing to a pure nonlinear simulation). An alternative, considerably more accurate, approach to the linearization of the nonlinear equation of (16.2) (or (16.9) for the inhibited enzyme) is proposed; it is based on transforming the system to an uncertain one where the "extent of nonlinearity" is manifested via the resulting system uncertainty. Since the local linearization via a Taylor expansion is accurate as long as the system velocities do not deviate considerably from their set point values, a possible approach, which is practiced in nonlinear systems, is to assemble the system around a group of local different set points. This approach extends the range of the resulting system behavior where it performs similarly to the original nonlinear system kinetics.

Considering the simple irreversible MM (Michaelis–Menten) kinetics of (16.3), we note that due to the fact that (16.3) is affine in $\tilde{V}$ and α, the result of Sect. 16.2.1 can be extended to the case where the latter scalars lie in a convex bounded domain. In this case, we require that (16.3) holds for all the operating points (set points) of the uncertainty interval, for a single set of scalars ($\tilde{V}$, α), where the variables $\tilde{V}$ and α are linearly dependent on the set point f. Thus, we obtain, for the two extreme points of the interval, the pairs $\{\tilde{V}_i, \ \alpha_i\}$, $i = 1, 2$ where

$$v_i = \frac{d[P]}{dt} = \tilde{V}_i + \alpha_i \bar{S}. \tag{16.26}$$

We note that the linearity of (16.26) in $\tilde{V}_i$ and α_i, implies that any convex analysis and synthesis that is applied to the latter two points will also hold for all the set points residing in the uncertainty interval. Note that each end point of the uncertainty interval can also be described by the pair of $\{f_i, \ \alpha_i\}$ where f is defined in (16.5).

Similarly to the above, the same technique is readily applied to the inhibitory mechanism of Sect. 16.2.3. Considering (16.10), we note that, since also in this case

$\tilde{V}$ α and β are linearly dependent on the set point f, the uncertainty polytope will be an interval with the vertices $\{\tilde{V}_{0,i}, \ \alpha_i, \ \beta_i\}, \ i = 1, 2$ where for both end points of this interval we have the following:

$$\dot{y} = \tilde{V}_{0,i} + \alpha_i \bar{u} - \beta_i \bar{I}, \quad \tilde{V}_{0,i} = \frac{V_{max} \frac{[S_{0,i}]}{K_s}}{1 + \frac{[S_{0,i}]}{K_s} + \frac{[I_{0,i}]}{K_i}}, \quad i = 1, 2 \qquad (16.27)$$

where $S_{0,i}$ and $I_{0,i}$ are the initial substrate and inhibitor concentrations, respectively.

16.4 Simple Unbranched Enzyme Pathways—Theoretical Considerations

The simple enzyme reactions systems modeled in Sect. 16.2 can be viewed as some of the basic building blocks of the great verity of biochemical pathways of the cell. In the proceeding sections, use will be made of the latter property by analyzing, for simplicity, a pathway of 4 MM (Michaelis–Menten) enzymes, where we begin by analyzing the open-loop pathway and proceed through various aspects of the closed-loop system.

16.4.1 Unbranched Enzyme Pathway—Open-Loop System

We consider a four-block pathway where each block is a simple MM enzyme of the irreversible type and where we assume that each component of the pathway operates around its set point (that is, each enzyme operates around a fraction f of it's V_{max}). We note that in this case the initial state of the total system is given by the initial concentrations of the first substrate $[S_{0,1}]$, the three intermediate substances $[P_{0,1}]$, $[P_{0,2}]$, $[P_{0,3}]$ (which are the initial substrate concentrations of the second, third, and fourth enzymes, respectively), and the end product of the fourth enzyme $P_{0,4}$. This setup is depicted in Fig. 16.9.

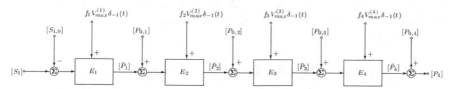

Fig. 16.9 The open-loop pathway connection of 4 MM enzymes modeled as in Sect. 16.4.1. For each enzyme E_j, the set point is indicated by its initial velocity $f_j V_{max}^j$ and $[\bar{P}_j]$ is the concentration deviation of that enzyme from its initial end-product concentration $[P_{0,j}]$

We note that in Fig. 16.9, each enzyme is identified by its kinetic constants and set point and is denoted, for simplicity, by E_j. Note also that for each enzyme, say the jth, $j = 1, \ldots, 4$, the set point condition is determined (once its K_M and $k_{catalytic}$ are given) by either the initial concentration of its substrate $S_{0,j}$ or by a given fraction of its V_{max}^j (i.e., once $f_j V_{max}^j$ is given then $S_{0,j}$ is determined by (16.4a)). For the setup of Fig. 16.9, we obtain the following ordinary differential equations, based on the linearly approximated differential equations of Sect. 16.2.1, where for each block we apply the result of the latter section:

$$\frac{d\bar{P}_1}{dt} = \tilde{V}_1 + \alpha_1 \bar{S}_1 - \tilde{V}_2 - \alpha_2 \bar{P}_1, \qquad \frac{d\bar{P}_2}{dt} = \tilde{V}_2 + \alpha_2 \bar{P}_1 - \tilde{V}_3 - \alpha_3 \bar{P}_2,$$

$$\frac{d\bar{P}_3}{dt} = \tilde{V}_3 + \alpha_3 \bar{P}_2 - \tilde{V}_4 - \alpha_4 \bar{P}_3, \qquad \frac{d\bar{P}_4}{dt} = \tilde{V}_4 + \alpha_4 \bar{P}_3. \tag{16.28}$$

We note that while all four enzymes operate as integrators, the fact that each product of the first three enzymes is removed by its successive enzyme transforms all three subsystems to simple first-order stable systems. The product of the fourth enzyme, however, is not removed but rather accumulates being the product of a pure integrator. Taking the Laplace transform of the latter equations, we obtain

$$\bar{P}_1(s) = \frac{\tilde{V}_1}{s(s+\alpha_2)} - \frac{\tilde{V}_2}{s(s+\alpha_2)} + \frac{\alpha_1 \bar{S}_1(s)}{(s+\alpha_2)}, \qquad \bar{P}_2(s) = \frac{\tilde{V}_2}{s(s+\alpha_3)} - \frac{\tilde{V}_3}{s(s+\alpha_3)} + \frac{\alpha_2 \bar{P}_1(s)}{(s+\alpha_3)},$$

$$\bar{P}_3(s) = \frac{\tilde{V}_3}{s(s+\alpha_4)} - \frac{\tilde{V}_4}{s(s+\alpha_4)} + \frac{\alpha_3 \bar{P}_2(s)}{(s+\alpha_4)}, \qquad \bar{P}_4(s) = \frac{\tilde{V}_4}{s^2} + \frac{\alpha_4 \bar{P}_3(s)}{s}. \tag{16.29}$$

The above system is readily casted into the following state-space representation:

$$\frac{d}{dt}\hat{P} = A\hat{P} + B\delta_{-1}(t) + B_1\delta_{-1}(t), \quad \text{where } \hat{P} = col\{\bar{P}_1, \ldots, \bar{P}_4\}$$

with

$$A = \begin{bmatrix} -\alpha_2 & 0 & 0 & 0 \\ \alpha_2 & -\alpha_3 & 0 & 0 \\ 0 & \alpha_3 & -\alpha_4 & 0 \\ 0 & 0 & \alpha_4 & 0 \end{bmatrix}, \quad B = \begin{bmatrix} \tilde{V}_{12} \\ \tilde{V}_{23} \\ \tilde{V}_{34} \\ \tilde{V}_4 \end{bmatrix}, \quad \text{and} \quad B_1 = \begin{bmatrix} \alpha_1 \bar{S}_1 \\ 0 \\ 0 \\ 0 \end{bmatrix}, \tag{16.30}$$

where $\delta_{-1}(t)$ is the unit-step function, $\bar{S}_1$ is the amplitude of the scaled stimulus step function $[S_1]$ subtracted by the initial value $[S_{0,1}]$ (i.e., $[\bar{S}_1(t)] = ([S_1(t)] - [S_{0,1}(t)])\delta_{-1}(t)$), and where

$$\tilde{V}_{12} \triangleq \tilde{V}_1 - \tilde{V}_2, \quad \tilde{V}_{23} \triangleq \tilde{V}_2 - \tilde{V}_3, \quad \text{and} \quad \tilde{V}_{34} \triangleq \tilde{V}_3 - \tilde{V}_4.$$

The transfer function of the system from $\bar{S}_1$ to $\bar{P}_4$ as depicted in figure is $T(s)_{ol} = \frac{\alpha_1\alpha_2\alpha_3\alpha_4}{s(s+\alpha_2)(s+\alpha_3)(s+\alpha_4)}$. To simulate the kinetic behavior of the above pathway, we apply a constant supply of the first substrate S_1 (a step function) where we probe the various intermediate velocities and concentrations involved in the pathway. At this point, it is

Table 16.1 Kinetic parameters and simulation results of a simple open-loop four-MM enzyme pathway of the irreversible type of Sect. 16.2.1. The set points are designated by f and α of each enzyme is given by (16.4b). The following terms $[S_0]$ (μM), $[P_0]$ (μM), $[\bar{P}_{s.s}]$ (μM) and $[P_{s.s}]$ (μM), $v_{s.s}$ $\left(\frac{M}{s}\right)$ are the, respectively, scaled initial substrate and initial end product concentrations, the steady-state values of the end product deviation $\bar{P}$ from the former term, the total concentration of that end product (i.e., $\bar{P} + P_0$), and the steady-state velocity of the enzyme

Enzyme no.	f	f*Vmax $\left(\frac{M}{s}\right)$	α (s^{-1})	$[S_0]$ (μM)	$[P_0]$ (μM)	$[\bar{P}_{s.s}]$ (μM)	$[P_{s.s}]$ (μM)	$v_{s.s}$ $\left(\frac{M}{s}\right)$
1	0.35	0.035	682.5683	53.8462	42.8571	19.4898	62.3469	0.0395
2	0.3	0.03	490.0000	42.8571	33.3333	25.8667	59.2000	0.0395
3	0.25	0.025	562.5000	33.3333	42.8571	19.4898	62.3469	0.0395
4	0.3	0.03	490.0000	42.8571	100.0000	*	*	0.0395

important to note that the initial condition of each enzyme is considered to be a step function of a given magnitude and **not** an initial value responsible to the so-called ZIR (zero input response) as in electrical or mechanical setups. Thus, the pathway considered here is subject to four step functions of magnitude $[S_{0,1}]$, $[S_{0,2}]$, $[S_{0,3}]$, $[S_{0,4}]$ which are applied at the input of each enzyme, respectively. An additional step function of S_1 stands for the stimulus given at the input of the first enzyme. Note that the outcome of this special setup is that the system will still respond to a zero input stimulus. Depending on the chosen enzyme kinetic parameters and set points, the system trajectory derivatives will tend to converge to a unique enzyme velocity corresponding to the steady-state response of the system.

For simplicity and clearness of the results, we choose four enzymes with the same kinetic constants but with different set points. We consider the case where for all four enzymes we have $[E_{tot}] = 1\,\mu$M, $K_M = 10\,\mu$M, and $k_{catalytic} = 10^5$ /s. The resulting $Vmax = [E_{tot}] \times k_{catalytic}$ is 0.1 $\left(\frac{M}{s}\right)$ for each enzyme. Choosing the following set point fractions $f_1 = 0.35$, $f_2 = 0.3$, $f_3 = 0.25$, $f_4 = 0.3$, and a stimulus step of $[S_1] = 1.2 \times [S_{0,1}]$, we obtain $[\bar{S}_1] = 0.2 \times [S_{0,1}]$. Thus, we obtain the system of (16.30) with the following matrices:

$$A = \begin{bmatrix} -490 & 0 & 0 & 0 \\ 490 & -562.500 & 0 & 0 \\ 0 & 562.5000 & -490.0000 & 0 \\ 0 & 0 & 490.0000 & 0 \end{bmatrix}, \quad \text{and} \quad B = \begin{bmatrix} 0.0095 \\ 0.0050 \\ -0.0050 \\ 0.0300 \end{bmatrix}.$$

In Table 16.1, we bring the kinetic parameters and the steady-state simulation results of the above linearly four-enzyme system. The set points indicated in the table determine the initial substrate concentration of the four enzymes according to (16.4).

To clarify, we note, once again, that in Table 16.1, the initial end-product concentrations of the first three enzymes are the initial substrate concentrations of their successive enzymes (i.e., $[S_{2,0}] = [P_{1,0}]$, $[S_{3,0}] = [P_{2,0}]$, $[S_{4,0}] = [P_{3,0}]$). Note also that since the fourth end product $\bar{P}_4$ is not removed by a successive enzyme, it is accumulated due to the integrator nature of the enzyme. Obviously, the later product will accumulate to biochemically impossible levels. In the biochemical realm, the end product of a pathway usually feeds one or more different pathways within the cell or rather is released into a different compartment, examples being the bloodstream or the intracellular medium, where the latter end product is exposed to various processes. As will be shown latter, we will demonstrate the effect of feedback on the kinetics of the fourth enzyme.

In Fig. 16.10, we depict the transient and the steady-state simulation results of the system descried in Table 16.1. Note that, as shown in Fig. 16.10a, all four enzymes reach the steady-state velocity level of 0.04 $\left(\frac{M}{s}\right)$. In Fig. 16.10b, it is shown that the first three intermediates reach certain steady-state levels (given in percents, relative to the initial value), whereas in Fig. 16.10c it is shown that $\bar{P}_4$ monotonically accumulates, as explained above. Considering (16.28d), we note that the accumulation of the latter product is an outcome of two factors. The first is determined by $\tilde{V}_4$ (which in turn is determined by the set point fraction f_4), whereas the second factor depends on the input $\bar{P}_3$. In Fig. 16.10c, we bring the response to the second term (denoted in the figure as $\bar{P}_{4_N}$) in order to demonstrate later the effect of feedback (in the closed-loop setup) on this part of the total response.

16.4.2   Unbranched Enzyme Pathway—Closed-Loop System

In a typical biochemical pathway, the first enzyme is sometimes a key regulatory enzyme, subject to feedback inhibition either by its product, or by an intermediate or end product of the same pathway or other pathways (see [23–29] for various aspects of the subject). In order to investigate the step response of the closed-loop system, we replace the first enzyme of the pathway with an MM enzyme of the irreversible type which is subject to reversible competitive inhibition as in Sect. 16.2.3. We analyze the case where the product of the third enzyme $\bar{P}_3$ acts as a competitive inhibitor to the first enzyme. This setup is depicted in Fig. 16.11.

For the setup of Fig. 16.11, we obtain the differential equations of (16.28b–d) where the equation for the end product of the first enzyme $\frac{d\bar{P}_1}{dt}$ is replaced by

$$\frac{d\bar{P}_1}{dt} = \tilde{V}_1 + \alpha_1 \bar{S}_1 - \beta \bar{I}_1 - \tilde{V}_2 - \alpha_2 \bar{P}_1,$$

where $\bar{I}_1$ is the inhibitor deviation from its initial value, i.e., $\bar{I}_1 = [I_1] - [I_{0,1}]$, where $[I_{0,1}]$ is the initial concentration of the inhibitor and where $\tilde{V}_1$ is given in Sect. 16.2.1. We note that for the first enzyme, the set point is determined by both $[I_{0,1}]$, $[S_{0,1}]$ as in (16.10b). In our case, $\bar{I}_1 = [\bar{P}_3]$, $[I_{0,1}] = [P_{0,3}]$. Applying the Laplace transform to the above equation, we obtain

$$\bar{P}_1(s) = \frac{\tilde{V}_1}{s(s+\alpha_2)} - \frac{\beta \bar{P}_3(s)}{(s+\alpha_2)} - \frac{\tilde{V}_2}{s(s+\alpha_2)} + \frac{\alpha_1 \bar{S}_1(s)}{(s+\alpha_2)}. \tag{16.31}$$

Considering the latter, we obtain the following state-space representation for the inhibited system:

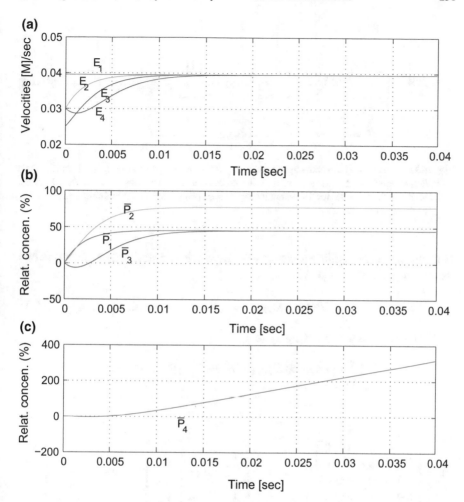

Fig. 16.10 The results of open-loop simulation of the system given in Table 16.1. Shown are from the above: **a** enzyme velocities, **b** the relative deviation (in percents) of $\bar{P}_1$, $\bar{P}_2$, $\bar{P}_3$ from their initial values (i.e., $100 \times (\bar{P}_j / P_{0,j})$) and **c** the relative deviation of $\bar{P}_{4_N}$ from from $P_{0,4}$

$$\frac{d}{dt}\hat{P} = \tilde{A}\hat{P} + \tilde{B}\delta_{-1}(t) + B_1 w(t), \quad \text{where } \hat{P} = col\{\bar{P}_1, \dots, \bar{P}_4\}$$

with

$$\tilde{A} = \begin{bmatrix} -\alpha_2 & 0 & -\beta & 0 \\ \alpha_2 & -\alpha_3 & 0 & 0 \\ 0 & \alpha_3 & -\alpha_4 & 0 \\ 0 & 0 & \alpha_4 & 0 \end{bmatrix}, \quad \tilde{B} = \begin{bmatrix} \tilde{V}_{12} \\ \tilde{V}_{23} \\ \tilde{V}_{34} \\ \tilde{V}_4 \end{bmatrix}, \quad \text{and} \quad B_1 = \begin{bmatrix} \alpha_1 \bar{S}_1 \\ 0 \\ 0 \\ 0 \end{bmatrix}. \quad (16.32)$$

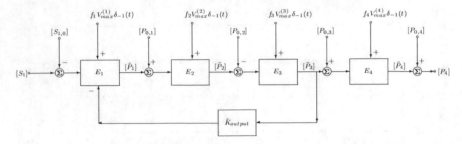

Fig. 16.11 The closed-loop pathway connection of the system depicted in Fig. 16.9. The feedback inhibition of the first enzyme is exerted by the product of the third enzyme. Note that the initial concentration of the end product of the latter enzyme (i.e., $[P_{0,3}]$) is the initial inhibitor concentration $[I_{0,1}]$ of the first enzyme

Note that the above-inhibited system actually employs a static output-feedback control where the control signal is

$$u(t) = K_{output} C \hat{P}(t), \quad K_{output} = -\beta \quad \text{and} \quad C = [0\ 0\ 1\ 0]. \tag{16.33}$$

We thus obtain the following system:

$$\frac{d}{dt}\hat{P} = (\hat{A} + \tilde{B}_2 K_{output} C)\hat{P} + \tilde{B}\delta_{-1}(t) + B_1 w(t),$$
$$Y(t) = C\hat{P}(t)$$
,

where

$$\hat{A} = \begin{bmatrix} -\alpha_2 & 0 & 0 & 0 \\ \alpha_2 & -\alpha_3 & 0 & 0 \\ 0 & \alpha_3 & -\alpha_4 & 0 \\ 0 & 0 & \alpha_4 & 0 \end{bmatrix}, \quad \tilde{B}_2 = \begin{bmatrix} 1 \\ 0 \\ 0 \\ 0 \end{bmatrix}, \quad \tilde{B} = \begin{bmatrix} \tilde{V}_{12} \\ \tilde{V}_{23} \\ \tilde{V}_{34} \\ \tilde{V}_4 \end{bmatrix}, \quad \text{and} \quad B_1 = \begin{bmatrix} \alpha_1 \bar{S}_1 \\ 0 \\ 0 \\ 0 \end{bmatrix},$$

(16.34)

and where K_{output} is the static output gain and $Y(t)$ is the measurement signal. It follows from (16.31) that a constant $\alpha_1 \bar{S}$ implies $w(t) = \delta_{-1}(t)$. The model of (16.34) allows, however, other kinds of "disturbances" such as an exponentially decreasing functions. The resulting system is readily cast into the following block diagram of Fig. 16.12.

To probe the effect of the feedback loop, we consider the system that was analyzed in the open-loop setup. As mentioned above, we replace the first enzyme with the one of Sect. 16.2.3 (noting that for zero inhibition this enzyme retains its original MM irreversible nature of Sect. 16.2.1). We apply the same initial values to all the open-loop system intermediates. We note that since $[I_{0,1}] = [P_{0,3}] \neq 0$, the set point of Enzyme 1 changes as found in Sect. 16.2.3. This change clearly stems from the presence of the inhibitor that lowers the initial velocity of the enzyme for the same given $[S_{0,1}]$. Choosing a typical value for the inhibition equilibrium constant $K_{I,1} = 3 \times 10^{-5}$ M, we obtain the system of (16.32) with the following matrices:

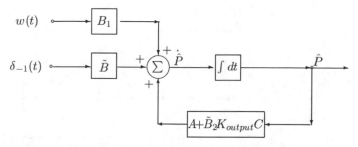

Fig. 16.12 Block diagram of the system (16.34)

Table 16.2 Kinetic parameters and simulation results of the closed-loop system of (16.32) for $K_{I,1} = 3 \times 10^{-5}$ M. The feedback loop is exerted by the third intermediate $\bar{P}_3$ which acts as a reversible competitive inhibitor of the first enzyme

Enzyme no.	f	f*Vmax $\left(\frac{M}{s}\right)$	$\alpha\,(s^{-1})$	$[S_0]\,(\mu M)$	$[P_0]\,(\mu M)$	$[\bar{P}_{s,s}]\,(\mu M)$	$[P_{s,s}]\,(\mu M)$	$v_{s,s}\left(\frac{M}{s}\right)$
1	0.1815	0.0181	445.7178	53.8462	42.8571	−12.7988	30.0583	0.0237
2	0.3	0.03	490.0000	42.8571	33.3333	−2.2603	31.0731	0.0237
3	0.25	0.025	562.5000	33.3333	42.8571	−12.7988	30.0583	0.0237
4	0.3	0.03	490.0000	42.8571	100.0000	*	*	0.0237

$$\tilde{A} = \begin{bmatrix} -490 & 0 & -203.88 & 0 \\ 490 & -562.500 & 0 & 0 \\ 0 & 562.5000 & -490.0000 & 0 \\ 0 & 0 & 490.0000 & 0 \end{bmatrix}, \quad \text{and} \quad \tilde{B} = \begin{bmatrix} -0.0089 \\ 0.0050 \\ -0.0050 \\ 0.0300 \end{bmatrix}.$$

For the above system $V_{max} = 0.1$ M/s as was the case in the open-loop case, since the inhibition is reversibly competitive [9]. The parameters α_1, β of Enzyme 1 are given in Sect. 16.2.1. In Table 16.2, we bring the closed-loop system parameters and the steady-state results of the system subject to a unit-step stimulus as in Table 16.2.

Note that the set point of the first enzyme is now 0.185 (compared to 0.35 of the open loop) with $\alpha_1 = 275.871$. Also note that the set point of the first enzyme is now 0.185 (compared to 0.35 of the open loop) with $\alpha_1 = 275.871$.

In Fig. 16.13, we bring, similarly to Fig. 16.13, the transient and steady-state results of the closed-loop system response to the same stimulus input as in the open-loop case. The effect of the feedback, given the current value of $K_{I,1}$, is shown (Fig. 16.13a) to reduce the steady-state velocities of all four enzymes to 0.0237 M/s. Note that the product deviations of $\bar{P}_1$, $\bar{P}_2$, $\bar{P}_3$ (Fig. 16.13b) are now negative leading to lower values for the total concentration of these products (i.e., $P_{ss,1}$, $P_{ss,2}$ and $P_{ss,3}$, see Table 16.2). Choosing a higher value of $K_{I,1}$ (i.e., weaker inhibition) would result in a less negative product deviation so that for a negligible inhibition (very high values of $K_{I,1}$) one obtains the results of the open-loop setup. Similarly to the latter result, in Fig. 16.13c, it is shown (contrary to Fig. 16.10c) that the feedback impact causes a negative deviation of the end product $\bar{P}_4$. Evidently, one concludes

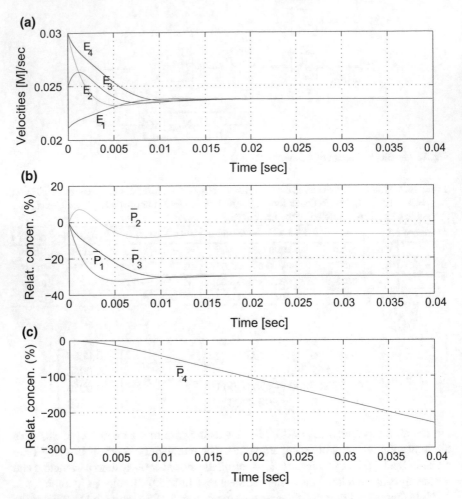

Fig. 16.13 The results of closed-loop simulation of the system given in Table 16.2. Shown are from the above: **a** enzyme velocities, **b** the relative deviation (in percents) of $\bar{P}_1$, $\bar{P}_2$, $\bar{P}_3$ from their initial value (i.e., $100 \times (\bar{P}_j / P_{0,j})$) and **c** the relative deviation of $\bar{P}_{4_N}$ from from $P_{0,4}$

that the role of the feedback loop is to regulate the transient and the steady-state levels of the pathway intermediates.

In order to evaluate the results obtained by the linearized model, we compare the results with those obtained by solving the following nonlinear differential equations:

$$
\begin{aligned}
\frac{d\bar{P}_1}{dt} &= \frac{Vmax_1 \times (S_{0,1} + \bar{S}_1(t))}{K_{M_1} + (S_{0,1} + \bar{S}_1(t))} - \frac{Vmax_2 \times (S_{0,2} + \bar{P}_1(t))}{K_{M_2} + (S_{0,2} + \bar{P}_1(t))}, \quad [P_{0,1}] = [S_{0,2}], \\
\frac{d\bar{P}_2}{dt} &= \frac{Vmax_2 \times (S_{0,2} + \bar{P}_1(t))}{K_{M_2} + (S_{0,2} + \bar{P}_1(t))} - \frac{Vmax_3 \times (S_{0,3} + \bar{P}_2(t))}{K_{M_3} + (S_{0,3} + \bar{P}_2(t))}, \quad [P_{0,2}] = [S_{0,3}], \\
\frac{d\bar{P}_3}{dt} &= \frac{Vmax_3 \times (S_{0,3} + \bar{P}_2(t))}{K_{M_3} + (S_{0,3} + \bar{P}_2(t))} - \frac{Vmax_4 \times (S_{0,4} + \bar{P}_3(t))}{K_{M_4} + (S_{0,4} + \bar{P}_3(t))}, \quad [P_{0,3}] = [S_{0,4}], \\
\frac{d\bar{P}_4}{dt} &= \frac{Vmax_4 \times (S_{0,4} + \bar{P}_3(t))}{K_{M_4} + (S_{0,4} + \bar{P}_3(t))}, \quad [S_{0,4}] = [P_{0,3}].
\end{aligned}
\qquad (16.35)
$$

Similarly to the linear case, each component of the pathway, in the nonlinear setup, operates around a unique set point that is determined by its initial substrate concentration, which is identical to the one chosen for the linearized case. In the closed-loop setup, we obtain the system of (16.35b–d) where the equation for $\frac{d\bar{P}_1}{dt}$ is now

$$
\frac{d\bar{P}_1}{dt} = \frac{Vmax_1 \frac{(S_{0,1}+\bar{S}_1(t))}{K_{M_1}}}{1+\frac{(S_{0,1}+\bar{S}_1(t))}{K_{M_1}}+\frac{P_{0,3}+\bar{P}_3(t)}{KI_1}} - \frac{Vmax_2 \times (S_{0,2}+\bar{P}_1(t))}{K_{M_2}+(S_{0,2}+\bar{P}_1(t))} , \tag{16.36}
$$
$$
[S_{0,2}] = [P_{0,1}], \; [P_{0,3}] = [I_{0,1}], \; \bar{P}_3(t) = \bar{I}_1(t),
$$

where we note that the set point of the first enzyme is determined, as in the linear case, by $[I_{0,1}]$, $[S_{0,1}]$. In Fig. 16.14, we bring the result of the latter analysis where

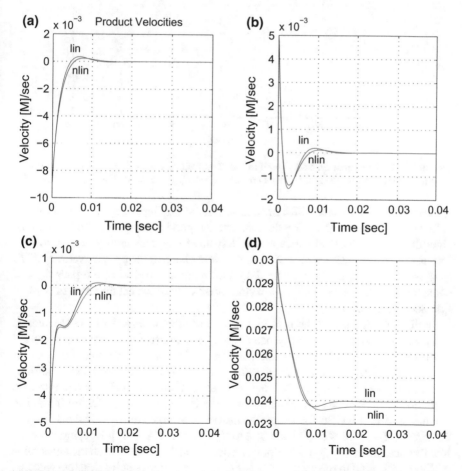

Fig. 16.14 Comparison of the closed-loop product velocities obtained by our linear treatment (lin) with the nonlinear (nlin) solution obtained by solving for (16.35) and (16.36). **a** Velocity of the first product $\frac{d\bar{P}_1}{dt}$, **b** velocity of the second product $\frac{d\bar{P}_2}{dt}$, **c** velocity of the third product $\frac{d\bar{P}_3}{dt}$, **d** velocity of the forth product $\frac{d\bar{P}_4}{dt}$

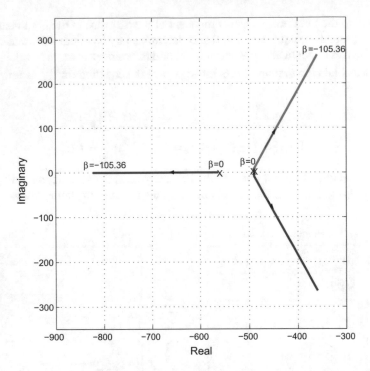

Fig. 16.15 Root loci plot of the system (16.32). The feedback gain β varies from $\beta = 0$ (i.e., open-loop system) to $\beta = -105.36$ which corresponds to $K_{I,1} = 5 \times 10^{-6}$ M, $I_{0,1} = 42.85\,\mu$M

we depict the product velocities of both solutions: the linearized and the nonlinear solutions. All subfigures clearly demonstrate the good agreement of both solutions. Note that the product velocities of the first three enzymes converge to zero thus forming the steady-state values of $\bar{P}_1$ to $\bar{P}_3$. Note also that the product velocity of $\bar{P}_4$ must be nonzero since the enzyme delivers the output product of the pathway and could not be negative because the enzyme works only in the net direction of $S \rightarrow P$ (see Sect. 16.2.1).

In Fig. 16.15, we plot the root loci of a subsystem of Fig. 16.13 [14] which is composed of the first three enzymes (namely the closed-loop part). Note that the feedback gain $\bar{K}_{output}$ is β in (16.33). In Fig. 16.15, we plot the pole locations for $\beta = [-105.36 \quad 0]$ where the lower value is obtained for the inhibition conditions obtained for $K_{I,1} = 5 \times 10^{-6}$ M, $I_{0,1} = 42.85\,\mu$M. Note that $\beta = 0$ refers to zero inhibition (i.e., $K_{I,1} = \infty$.). Notice that the pole loci in Fig. 16.15 are limited to a value of $\beta = -105.36$ for which the closed-loop poles are $[-359.4791 \quad - j \times 264.6368]$, $[-359.4791 \quad + j \times 264.6368]$, $[-823.5418 + 0j]$. Keeping the same initial concentration of $I_{0,1}$, inhibition constants which are lower than the latter value of $K_{I,1}$ correspond to the case where one or more enzyme acquire negative velocity. Since our model, in this simulation, involves only enzymes of the type considered in Sect. 16.2.1, where the product is not allowed to produce the substrate, negative velocities are not acceptable. By allowing the enzyme to catalyze the reverse reaction as in Sect. 16.2.2, the above result of negative velocity would be acceptable and probably would allow a larger range of inhibition constants. This topic is left for a further study.

Remark 16.5 We note that while the simulated enzyme velocities (not shown in Fig. 16.14) may be quite close in both cases, the product velocities, however, may deviate owing to the nonlinear nature of the MM kinetic equations. We also note that the latter deviation in less accentuated in the closed-loop system compared to the open-loop system, as one may expect.

16.5 Delayed Pathways

We consider the case where a transport delay is inherent to one or more of the steps included in the pathway. We use the first-order Pade [30] approximation

$$e^{-s\tau} = \frac{2 - \tau s}{2 + \tau s},$$

and consider the case where a transport delay occurs between the second and the third step in the above four-enzyme pathway. We thus have

$$\bar{P}_{2,del}/\bar{P}_2 = -1 + \frac{4/\tau}{s + 2/\tau}.$$

Taking

$$\xi(s)/\bar{P}_2(s) = \frac{4/\tau}{s + 2/\tau},$$

we obtain

$$\dot{\xi}(t) = \frac{-2}{\tau}\xi(t) + \frac{4}{\tau}\bar{P}_2(t).$$

Considering the latter, we obtain the following state-space representation for the delayed inhibited system:

$$\frac{d}{dt}\tilde{P} = \bar{A}\tilde{P} + \bar{B}\delta_{-1}(t) + B_1 w(t), \text{ where } \tilde{P} = col\{\bar{P}_1, \ldots, \bar{P}_4, \xi\}$$

with

$$\bar{A} = \begin{bmatrix} -\alpha_2 & 0 & -\beta & 0 & 0 \\ \alpha_2 & -\alpha_3 & 0 & 0 & 0 \\ 0 & -\alpha_3 & -\alpha_4 & 0 & \alpha_3 \\ 0 & 0 & \alpha_4 & 0 & 0 \\ 0 & 4/\tau & 0 & 0 & -2/\tau \end{bmatrix}, \quad \bar{B} = \begin{bmatrix} \tilde{V}_{12} \\ \tilde{V}_{23} \\ \tilde{V}_{34} \\ \tilde{V}_4 \\ 0 \end{bmatrix} \text{ and } B_1 = \begin{bmatrix} \alpha_1 \bar{S}_1 \\ 0 \\ 0 \\ 0 \\ 0 \end{bmatrix}.$$

$$(16.37)$$

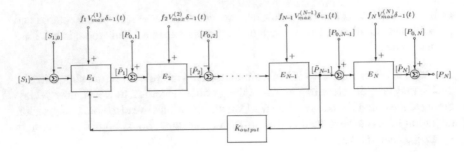

Fig. 16.16 The closed-loop pathway connection for the delayed system

Note that the above-inhibited system actually employs a static output-feedback control where the control signal is

$$u(t) = K_{output} C \hat{P}(t), \quad K_{output} = -\beta, \quad C = [0\ 0\ 1\ 0].$$

The following system is thus obtained:

$$\frac{d}{dt}\hat{P} = (\bar{A} + \tilde{B}_2 K_{output} C)\hat{P} + \bar{B}\delta_{-1}(t) + B_1 w(t),$$
$$Y(t) = C\hat{P}(t),$$

where

$$\hat{A} = \begin{bmatrix} -\alpha_2 & 0 & 0 & 0 \\ \alpha_2 & -\alpha_3 & 0 & 0 \\ 0 & \alpha_3 & -\alpha_4 & 0 \\ 0 & 0 & \alpha_4 & 0 \end{bmatrix}, \quad \tilde{B}_2 = \begin{bmatrix} 1 \\ 0 \\ 0 \\ 0 \end{bmatrix}, \quad \tilde{B} = \begin{bmatrix} \tilde{V}_{12} \\ \tilde{V}_{23} \\ \tilde{V}_{34} \\ \tilde{V}_4 \end{bmatrix}, \quad \text{and} \quad B_1 = \begin{bmatrix} \alpha_1 \bar{S}_1 \\ 0 \\ 0 \\ 0 \end{bmatrix},$$

(16.38)

and where K_{output} is the static output gain and $Y(t)$ is the measurement signal. It follows from (16.31) that a constant $\alpha_1 \bar{S}$ implies $w(t) = \delta_{-1}(t)$. The model of (16.34) allows, however, other kinds of "disturbances" such as an exponentially decreasing functions (Fig. 16.16).

16.6   Unbranched Enzyme Pathway—Linear Analysis via System Uncertainty

In Sects. 16.4.1 and 16.4.2, the open- and closed-loop systems, respectively, were analyzed using the linear models that were derived in Sects. 16.2.1 and 16.2.3 based on a simple Taylor expansion of the kinetic equations. Similarly to Sect. 16.3, where an alternative analysis has been applied to both single MM enzyme and inhibited enzyme, one can apply the same analysis to the four-block enzyme pathway. We recall that the alternative approach is based on transforming the nonlinearity embedded in the system to an uncertain system that can be determined in advance, thus tuning the "extent of nonlinearity". Considering the open-loop system of Fig. 16.10 where (16.28) holds, we obtain the following uncertain polytopic system:

$$\hat{\Omega} \stackrel{\triangle}{=} \{(\tilde{V}_1, \alpha_1, \tilde{V}_2, \alpha_2, \tilde{V}_3, \alpha_3, \tilde{V}_4, \alpha_4) | (\tilde{V}_1, \alpha_1, \tilde{V}_2, \alpha_2, \tilde{V}_3, \alpha_3, \tilde{V}_4, \alpha_4)$$

$$= \sum_{i=1}^{16} \tau_i (\tilde{V}_{1,i}, \alpha_{1,i}, \tilde{V}_{2,i}, \alpha_{2,i}, \tilde{V}_{3,i}, \alpha_{3,i}, \tilde{V}_{4,i}, \alpha_{4,i}); \ \tau_i \geq 0; \ \sum_{i=1}^{16} \tau_i = 1\}.$$

The latter stems from the fact that for each enzyme in the pathway, depicted by (16.28), the kinetic equation is affine in $(\tilde{V}_{1,i}, \ \alpha_{1,i})$. Thus, any convex combination of the 2^4 vertices (each enzyme is assigned an interval of two values) will reside in the polytope, and will thus cover a greater region of the nonlinear system.

Similarly to the above description, for the closed-loop system of Fig. 16.13, we have

$$\hat{\Psi} \stackrel{\triangle}{=} \{(\tilde{V}_1, \alpha_1, \beta, \tilde{V}_2, \alpha_2, \tilde{V}_3, \alpha_3, \tilde{V}_4, \alpha_4) | (\tilde{V}_1, \alpha_1, \beta, \tilde{V}_2, \alpha_2, \tilde{V}_3, \alpha_3, \tilde{V}_4, \alpha_4)$$

$$= \sum_{i=1}^{16} \tau_i (\tilde{V}_{1,i}, \alpha_{1,i}, \beta_i, \tilde{V}_{2,i}, \alpha_{2,i}, \tilde{V}_{3,i}, \alpha_{3,i}, \tilde{V}_{4,i}, \alpha_{4,i}); \ \tau_i \geq 0; \ \sum_{i=1}^{16} \tau_i = 1\},$$

and thus any set of $\{\tilde{V}_1, \alpha_1, \beta, \tilde{V}_2, \alpha_2, \tilde{V}_3, \alpha_3, \tilde{V}_4, \alpha_4\}$ that lies in $\hat{\Psi}$ will describe the enzyme pathway for a quadrupole of set points that lie in the corresponding interval.

16.7 Unbranched Enzyme Pathway—Sensitivity Analysis

Having shown that the closed-loop system obtained in the case of internal feedback inhibition utilizes, in fact, a static output-feedback control, we are able to apply various tools of both classical and optimal control theory to look into the feedback quality and cost in the system.

We begin by analyzing the open-loop four-block system of Sect. 16.4.1 where we explore the effect of variations in the total enzyme concentration (i.e., $[E]_t$) of the enzyme constituting the system. We then turn to the closed-loop configuration where, given a range of feedback gains, we seek the possible weights of a predetermine index of performance, thus clearing the way to asses the optimal aspects of the feedback-controlled system. We bring a numerical result that demonstrates the merit of the above research.

16.7.1   Open-Loop System: Sensitivity to $[E]_{total}$

We consider the case of the open-loop four-MM identical enzymes of Sect. 16.4.1
having, for simplicity, the same initial substrate concentrations but different enzyme
concentrations. Taking $K_M = S_0 \bar{f}$, we obtain from (16.31) and (16.32a, b) the following equivalent linearized rate equation:

$$v = \frac{[E]_t k_{catalytic}}{1+\bar{f}} + \frac{[E]_t k_{catalytic}}{S_0} \frac{\bar{f}}{(1+\bar{f})^2} \bar{S} = \frac{[E]_t k_{catalytic}}{1+\bar{f}} + [E]_t \bar{\alpha} \bar{S},$$

$$\bar{\alpha} \triangleq \frac{k_{catalytic} \bar{f}}{S_0 (1+\bar{f})^2},$$

where we recall from (1) that $k_{catalytic} = k_2$.

Applying the latter result to the open-loop four-block enzyme system of Sect.
16.4.1 we obtain, following the state-space representation of (16.30), the following
system matrices:

$$A = \begin{bmatrix} -[E_2]_t \bar{\alpha}_2 & 0 & 0 & 0 \\ [E_2]_t \bar{\alpha}_2 & -[E_3]_t \bar{\alpha}_3 & 0 & 0 \\ 0 & [E_3]_t \bar{\alpha}_3 & -[E_4]_t \bar{\alpha}_4 & 0 \\ 0 & 0 & [E_4]_t \bar{\alpha}_4 & 0 \end{bmatrix} \quad \text{and} \quad B = \begin{bmatrix} ([E_1]_t - [E_2]_t) \bar{V} \\ ([E_2]_t - [E_3]_t) \bar{V} \\ ([E_3]_t - [E_4]_t) \bar{V} \\ [E_4]_t \bar{V} \end{bmatrix},$$

$$\tag{16.39}$$

where $B_1 = \begin{bmatrix} [E_1]_t \bar{\alpha}_1 \bar{S}_1 \\ 0 \\ 0 \\ 0 \end{bmatrix}$, and $\bar{V} = \frac{k_{catalytic}}{1+\bar{f}}$.

Considering, for simplicity, the special case where we probe the system sensitivity
to changes in $[E_2]_t$, we obtain the following system:

$$\frac{d}{dt} \hat{P} = [A + \Delta A]\hat{P} + [B + \Delta B]\delta_{-1}(t) + B_1 w(t),$$

where

$$A = \begin{bmatrix} -\alpha_2 & 0 & 0 & 0 \\ \alpha_2 & -\alpha_3 & 0 & 0 \\ 0 & \alpha_3 & -\alpha_4 & 0 \\ 0 & 0 & \alpha_4 & 0 \end{bmatrix}, \quad \Delta A = \begin{bmatrix} -(\Delta[E_2]_t)\bar{\alpha}_2 & 0 & 0 & 0 \\ (\Delta[E_2]_t)\bar{\alpha}_2 & 0 & 0 & 0 \\ 0 & 0 & 0 & 0 \\ 0 & 0 & 0 & 0 \end{bmatrix}, \quad B = \begin{bmatrix} \bar{V}_{12} \\ \bar{V}_{23} \\ \bar{V}_{34} \\ \bar{V}_4 \end{bmatrix}, \quad \text{and} \quad B_1 = \begin{bmatrix} \alpha_1 \bar{S}_1 \\ 0 \\ 0 \\ 0 \end{bmatrix},$$

$$\tag{16.40}$$

where $\Delta B = \begin{bmatrix} -(\Delta[E_2]_t)\bar{V} & (\Delta[E_2]_t)\bar{V} & 0 & 0 \end{bmatrix}$, where $\Delta[E_2]_t$ denotes the uncertainty in the second enzyme and where in the above A and B, the nominal initial
enzymes concentrations are taken.

The latter case is readily cast into the following structure:

$$\Delta A = H_A \Delta[E_2]_t E_A, \quad H_A = [-1\ 1\ 0\ 0]^T, \quad E_A = \bar{\alpha}_2[1\ 0\ 0\ 0],$$

$$\Delta B = H_B \Delta[E_2]_t E_B, \quad H_B = [-1\ 1\ 0\ 0]^T, \quad E_B = \bar{V}.$$

This structure is amenable to standard technique of dealing with norm-bounded uncertainties (see [31]). We note that the above result is structurally similar to the one of the closed-loop systems, where $[E_2]_t$ is uncertain. In the latter case, the matrices A and B of (16.39) should be replaced by $\tilde{A}$ and $\tilde{B}$ of (16.32). In the following section, in the numerical example there, we consider the case of uncertainty in $[E_2]_t$.

16.8 Closed-Loop System: H_∞ Optimal Control Analysis

The issue of the optimal nature of biochemical pathways was highly developed in the last four decades and numerous studies have been published in this field, composed of different frameworks and computational techniques (for introductory aspects see [23, 25, 32] and for the issue of optimality see [26–29, 33–36]). To the best of our knowledge, the application of modern optimal control performance tests such as the H_∞ and the P2P [peak-to-peak] have never been applied in this field. These performance methods were highly explored within the traditional modern control community and were shown to represent important tools for both analysis and design of practical control engineering systems. The important and relevant issue here is that the latter optimal measures are very strong tools of **analysis** and one can apply them, in principle, to any dynamical control system in order to probe key issues like robust stability and robust disturbance analysis among other performances.

We now turn to inspect the nature of the closed-loop system as an optimal system via our approach. For such a quest, an index of performance should be formulated and its terms justified to be physically relevant to the type of biochemical systems considered here. We begin by considering the principle (general) problem of determining the weights of an index of performance (of a predetermined structure), in the case where static output gains are known to reside in a given interval. We then apply our result to the closed-loop configuration of the four-block system of Sect. 16.4.2.

We consider the following system:

$$\dot{x} = (A + B_2 K C)x + B_1 w, \quad x(0) = 0, \tag{16.41}$$

where the static output-feedback gain K is given. We consider the following performance index:

$$J = \int_0^\infty (x^T(Q + C^T K^T R K C)x - \gamma^2 w^T w)dt, \tag{16.42}$$

and we seek $0 \leq Q, 0 < R$ and $0 < \gamma$ such that the given gain matrix K achieves $J < 0$ for all nonzero $w \in L_2[0 \ \infty)$. We note that the value of γ represents the maximum singular value of the transference between the input and the output $z = col\{Q^{0.5}x, \ R^{0.5}u\}$.

Applying the standard BRL (for deterministic time-invariant linear systems [37]), we know that the latter requirement will be satisfied iff there exist $0 < P$ that satisfies

Table 16.3 Optimal analysis results of different cases of the closed-loop four-MM enzyme pathway of the irreversible type depicted in Table 16.2. Shown are the minimal attenuation level γ and the feasibility interval [0 q_{max}] denoted as feas. interval. The solution type refers to the various methods applied for obtaining the results. In Case 1, $A = \hat{A}$, $B_2 = \tilde{B}_2$ where $\hat{A}$ and $\tilde{B}_2$ are defined in (16.34c, d) and $K_{output} = -203$. The uncertain Cases 2–5 are similar to Case 1, except for an uncertainty in the total concentration of [E_2], according to (16.40a, b). Note that K_{output} is common to all cases and that the control cost is $R = 1$

Case no.	Uncertain parameter	Solution type	γ_{min}	Feas. interval
1	–	Nominal solution	8.818×10^{-4}	[0 12.00]
2	$(1 \pm 0.1)[E_2]_t$	Quadratic	8.899×10^{-4}	[0 12.95]
3	$(1 \pm 0.15)[E_2]_t$	Quadratic	8.990×10^{-4}	[0 15.58]
4	$(1 \pm 0.2)[E_2]_t$	Quadratic	9.110×10^{-4}	[0 16.55]
5	$(1 \pm 0.25)[E_2]_t$	Quadratic	9.262×10^{-4}	[0 14.16]

the following inequality:

$$\begin{bmatrix} P(A + B_2 KC) + (A^T + C^T K^T B_2^T)P + Q + C^T K^T RKC & PB_1 \\ * & -\gamma^2 I \end{bmatrix} < 0.$$

(16.43)

The latter is affine in the decision variables P, Q, R, and γ^2 and is therefore an LMI.

In our case, we seek a J that achieves $J < 0$ for the minimum γ^2 and we thus seek a solution for the latter LMI for the minimum possible γ^2.

There will probably exist many solutions for Q that will satisfy the above LMI for a given γ and, say, $R = I$. One can then impose some structure on Q, in order to provide some physical meaning for J. One such a structure, for the SISO case, where C is a row vector, is

$$Q = qC^T C, \quad q > 0.$$

The corresponding LMI will then be

$$\begin{bmatrix} P(A + kB_2C) + (A^T + kC^T B_2^T)P + (q + k^2)C^T C & PB_1 \\ * & -\gamma^2 I \end{bmatrix} < 0,$$

(16.44)

where k is the scalar version of K.

It is readily seen that the latter LMI always has a solution for large enough γ if the pair (A, C) is observable (we simply obtain then a Lyapunov inequality). For a prechosen γ and $R = I$, one may seek the range of positive values of q for which the given value of k achieves the required disturbance attenuation level γ. Obviously, the minimum value of q is zero and the problem thus becomes one of finding the maximum value of q for which (16.44) is satisfied.

Applying the latter results to the closed-loop system of the four enzymes of Sect. 16.4.2, where $x(t) = \hat{P}(t)$, we obtain the following results for different cases, where the feedback gain $K_{output} = k = -\beta = -203$ is kept constant:

Table 16.4 Optimal analysis results using the method of Sect. 16.3. The uncertainty interval (feas. interval) for each enzyme is indicated in the second column. The convex quadratic solution for each case is obtained by assembling the system as an uncertain polytopic one, around the predetermined set points f_1, f_2, f_3 of Table 16.2 and is denoted as Quad. Polytopic

Case no.	Uncertain parameter	Solution type	γ_{min}	Feas. interval
1	$(1 \pm 0.1) * \{f_1, f_2, f_3\}$	Quad. Polytopic	9.505×10^{-4}	[0 26.64]
2	$(1 \pm 0.2) * \{f_1, f_2, f_3\}$	Quad. Polytopic	10.405×10^{-4}	[0 31.64]
3	$(1 \pm 0.5) * \{f_1, f_2, f_3\}$	Quad. Polytopic	14.199×10^{-4}	[0 33.30]
4	$(1 \pm 0.1) * \{f_1, f_2, f_3\}, E_2$	Quad. Polytopic	10.030×10^{-4}	[0 33.41]
5	$(1 \pm 0.2) * \{f_1, f_2, f_3\}, E_2$	Quad. Polytopic	11.14×10^{-4}	[0 38.69]
6	$(1 \pm 0.5) * \{f_1, f_2, f_3\}, E_2$	Quad. Polytopic	15.70×10^{-4}	[0 26.97]

We note that for all cases listed in Table 16.3, the structure of the trajectory weight matrix is $Q = q I_3$, and in each case a maximum value of q is sought. The interval $[0 \; q_{max}]$ is thus obtained and is denoted as the feasibility interval. In Cases 2–5, an uncertainty in the concentration of $[E_2]$ is admitted to the system according to (16.40a, b). Note that, as expected, the minimum attenuation level γ increases (though slightly) as the uncertainty in $[E_2]$ increases, up to 25% tolerance in $[E_2]$. Note also that in Cases 2–5, the quadratic type of solution implies that the same LF (Lyapunov function) is assigned to both endpoints of the uncertainty interval [38].

The strikingly interesting results that stem from Table 16.3 is twofold: first, the values of the attenuation levels are relatively small in spite of the relatively large ratio between q_{max} and R. The second result is that the same feedback gain achieves almost the same attenuation levels and feasibility intervals for both the nominal and the uncertain cases. A similar result is also obtained (not shown) where given q_{max}, a search for minimal γ is performed. One obtains almost identical attenuation levels for the abovementioned uncertainty tolerance values.

The results of Table 16.3 were obtained for the case where the system, either nominal or uncertain, is linearized via the Taylor expansion. Thus, the uncertainties in $[E_2]$ (Cases 2–5) represent only **physical uncertainties** which are biochemically relevant. In Table 16.4, we bring the results of the H_∞ optimal analysis for both nominal systems (i.e., with no physical uncertainties) modelled as uncertain polytopic ones and for the latter systems with an additional physical uncertainty in $[E_2]$.

In Table 16.4, we apply the method of Sect. 16.3, where an advanced optimal analysis is proposed by modeling the nonlinear system (the pathway) as a linear uncertain one. In Cases 1–3, we apply our analysis where each set point of the enzymes (f_1, f_2, f_3) has a 10%, 20%, 50% tolerance, respectively; thus an eight-vertex polytope is obtained. Similarly to Table 16.1, relatively small attenuation levels are obtained, where in Cases 1, 2 these values are quite close. Moving to 50% uncertainty in all three set points, increases γ moderately in comparison to Case 2. In Cases 4–6, the previous three cases are considered where a 20% (physical) uncertainty in the initial concentration of E_2 is admitted to the system, thus a 16-vertex polytope is obtained. We note that the attenuation levels of each pair of cases having the same "virtual" uncertainty (for example, Cases 1, 4) are very close.

Remark 16.6 It is noted that the behavior of the uncertain closed-loop system is similar in the case with "virtual uncertainties" (Table 16.4) and the case without such uncertainties (Table 16.3). This implies that the linearized system around the selected set points is a good approximation to the nonlinear system.

In order to assess the importance of the results contained in both, Tables 16.3 and 16.4, we note that the pathway under study is stable in the open-loop configuration (excluding the last enzyme which acts as an integrator). Thus, the fact that in the closed-loop configuration, a stable system is obtained cannot be the principle importance of the feedback. The main question should be how robust is the closed-loop system performance, in face of system uncertainties or disturbances applied at the input of the system.

The results of the present section clearly demonstrate that using concentrations taken from typical biochemical enzymatic systems that the closed-loop configuration is **robust** in the H_∞ sense. The ability of the system in rejecting disturbances applied, either at the input of the system or in attenuating system uncertainties (for example in $[E_2]$), is clearly demonstrated. The results of Table 16.4, being the outcome of the "virtual" uncertainties that describe the linearization of the system, imply that the above robustness of the system is probably characteristic of the true nonlinear model.

At this point, we would like to stress the point that the simple pathway brought in this study is certainly not representative of a typical, even small, real biochemical pathway, nor there is an attempt here to deliberately imply that regulation is due mainly to the first enzyme of the pathway (for a comprehensive study of these crucial points, see [24, 39–42]). Also, the irreversibility of each of the enzyme pathway is taken here only to simplify the model. It is expected, however, that the main result of our study, namely, that Nature may adopt the same performance index for a large part of the nonlinearity and under significant physical parameter uncertainties will also prevail for real (more involved) pathway structures.

16.9   The Threonine Synthesis Pathway—H_∞ Optimal Control Analysis

In order to validate the applicability of our optimal analysis, we turn to the threonine synthesis pathway of the next section.

In Fig. 16.17, we bring the threonine synthesis pathway in E. Coli [11–13]. In [11], the system metabolites and modifiers are kept constant, save the concentrations of the four intermediates ASPP, ASA, HS, and HSP which are allowed to vary during the system dynamics toward steady state. The concentration of threonine, the end product of the pathway, was kept at a 3.49 mM. Since, in this study, we aim at assessing the quality of the three internal feedback loops involved in the system, we allow the threonine concentration to vary. Thus, we obtain a five-variable system on which we apply our analysis. In the sequel, we first bring the constant metabolites

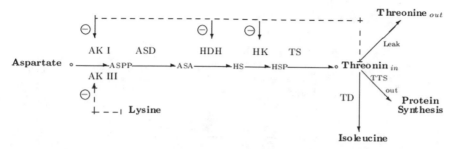

Fig. 16.17   The threonine synthesis pathway in E. coli. Shown are the three negative feedback loops exerted on the system by threonine (THR)—the end product of the pathway. The enzymes AKI and AKIII catalyze the formation of ASPP from the amino acid Aspartate (ASP)—the starting point of the pathway. The enzyme ASD catalyzes the formation of ASA from ASPP, HDH catalyzes the formation of HS from ASA and the formation of HSP is catalyzed by the enzyme HK. TS catalyzes the formation of THR from HSP. Threonine is removed by its conversion to Isoleucine (by the enzyme TD), by protein synthesis (by the enzyme TTS), and by leakage out of the cell. Note that lysine is an inhibitor of the AKIII enzyme reaction

concentrations (taken from [11]) and we then apply a simple Taylor expansion of the five enzymatic reactions that constitute the pathway.

The following metabolites are kept constants (concentrations are given in mM):

$$ATP = 1.31, \ ADP = 0.17, \ NADP^+ = 0.63, \ NADPH = 0.56, \ ASP = 1.34,$$
$$LYS = 0.46 \ Pi = 5.0, \ THR = 0.20, \ \text{and} \ [ASP] = [ASA] = [HS] = [HSP] = 0.005 \, \text{mM}.$$

We bring the following five consecutive kinetic equations of the pathway [11] and their corresponding Taylor expansions:

- AKI Reaction

$$v_{AKI} = \frac{V_{AKI}([ASP][ATP] - \frac{[ASPP][ADP]}{K_{eq}})}{\Upsilon_{ATP}(\Upsilon_{THR} + [ASP] + [ASPP]\frac{K_{ASP}}{K_{ASPP}})},$$

where

$$\Upsilon_{THR} \overset{\Delta}{=} K_{ASP} \frac{1 + (\frac{[THR]}{K_{iTHR}})^{h_{THR}}}{1 + (\frac{[THR]}{\alpha K_{iTHR}})^{h_{THR}}},$$

and

$$\Upsilon_{ATP} \overset{\Delta}{=} K_{ATP}[1 + \frac{[ADP]}{K_{ADP}}] + [ATP]. \tag{16.45}$$

The AKI parameters are

$$V_{AKI} = 0.68987, \ K_{ASP} = 0.97, \ K_{iTHR} = 0.167, \ \alpha = 2.47, \ h_{THR} = 4.09,$$

$$K_{ASPP} = 0.017, \ K_{ADP} = 0.25, \ K_{ATP} = 0.98, \ K_{eq} = 0.000639.$$

Following linearization, we obtain

$$v_{AKI} = V_{10} + \alpha_1 \overline{[ASPP]} + \alpha_2 \overline{[THR]},$$

$$\alpha_1 \stackrel{\Delta}{=} \frac{\partial v_{AKI}}{\partial ASPP} = \frac{-V_{AKI}\frac{ADP}{K_{eq}}\Delta_1}{\Delta_1^2} - \frac{\Upsilon_{ATP}\frac{K_{ASP}}{K_{ASPP}}\Psi_1}{\Delta_1^2} \qquad (16.46)$$

$$\alpha_2 \stackrel{\Delta}{=} \frac{\partial v_{AKI}}{\partial THR} = -\frac{\frac{\partial \Upsilon_{THR}}{\partial THR}\Upsilon_{ATP}\Psi_1}{\Delta_1^2},$$

where

$$\frac{d}{dt}(\Upsilon_{THR}) = \frac{K_{ASP}[h_{THR}(\frac{THR}{K_{iTHR}})^{h_{THR}-1}]}{K_{iTHR}DEN_1}$$

$$-\frac{K_{ASP}[h_{THR}(\frac{THR}{\alpha K_{iTHR}})^{h_{THR}-1}][1 + (\frac{THR}{K_{iTHR}})^{h_{THR}}]}{\alpha K_{iTHR}DEN_1^2}$$

with

$$DEN_1 = 1 + (\frac{THR}{\alpha K_{iTHR}})^{h_{THR}},$$

where Ψ_1 and Δ_1 are the numerator and the denominator of (16.45a), respectively, and where V_{10} is given in (16.45a) for the abovementioned constant initial conditions.

- AKIII Reaction

$$v_{AKIII} = \frac{V_{AKIII}([ASP][ATP]-\frac{[ASPP][ADP]}{K_{eq}})}{\Upsilon_{ATP}\Upsilon_{LYS}([ASP]+K_{ASP}(1+\frac{[ASPP]}{K_{ASPP}}))} \quad \text{where } \Upsilon_{LYS} \stackrel{\Delta}{=} [1 + (\frac{[LYS]}{K_{iLYS}})^{h_{LYS}}].$$
$$(16.47)$$

The AKIII parameters are

$$V_{AKIII} = 0.33488, \quad K_{ASP} = 0.323, \quad LYS = 0.46, \quad K_{ASPP} = 0.017, \quad K_{iLYS} = 0.391,$$

$$h_{LYS} = 2.78, \quad K_{ADP} = 0.25, \quad K_{ATP} = 0.225, \quad \text{and} \quad K_{eq} = 0.000639.$$

Following linearization, we obtain

$$v_{AKIII} = V_{10,b} + \alpha_3 \overline{[ASPP]}, \quad \alpha_3 \stackrel{\Delta}{=} \frac{\partial v_{AKIII}}{\partial ASPP} =$$
$$\frac{-V_{AKIII}\frac{ADP}{K_{eq}}\Delta_{1,b}}{\Delta_{1,b}^2} - \frac{\Upsilon_{III}\frac{K_{ASP}}{K_{ASPP}}\Psi_{1,b}}{\Delta_{1,b}^2}, \qquad (16.48)$$

where $\Upsilon_{III} = \Upsilon_{ATP}\Upsilon_{LYS}$ and Ψ_{1b}, Δ_{1b} are the numerator and the denominator of (16.47a), respectively, and where $V_{10,b}$ is (16.47a) for the abovementioned constant initial conditions.

- ASD Reaction

$$v_{ASD} = \frac{V_{ASD}([ASPP][NADPH] - \frac{[ASA][NADP^+][Pi]}{K_{iASD}})}{\Upsilon_{NADPH}[K_{ASPP}[1 + \frac{[ASA]}{K_{ASA}}\psi_{2b}] + [ASPP]]},$$

$$\Upsilon_{NADPH} \overset{\Delta}{=} K_{NADPH}[1 + \frac{[NADP^+]}{K_{NADP^+}}] + [NADPH],$$

(16.49)

$$\psi_{1b} \overset{\Delta}{=} [1 + \frac{[Pi]}{K_{Pi}}].$$

The ASD parameters are $V_{ASD} = 0.9203$, $K_{NADP} = 0.144$,

$K_{NADPH} = 0.0287$, $K_{ASPP} = 0.022$, $K_{Pi} = 10$, $K_{eq} = 56.415$, $K_{ASA} = 0.11$.

Following linearization, we obtain

$$v_{ASD} = V_{20} + \beta_1\overline{[ASPP]} + \beta_2\overline{[ASA]}, \quad \beta_1 \overset{\Delta}{=} \frac{\partial v_{ASD}}{\partial ASPP} =$$
$$\frac{V_{ASD}[NADPH]\Delta_2 - \Upsilon_{NADPH}\Psi_2}{\Delta_2^2}$$

(16.50)

$$\beta_2 \overset{\Delta}{=} \frac{\partial v_{ASD}}{\partial ASA} = \frac{-V_{ASD}[NADP^+][Pi]}{K_{eq}\Delta_2} - \frac{\Upsilon_{NADPH}\psi_{2b}}{K_{ASA}\Delta_2^2}\Psi_2,$$

where Ψ_2, Δ_2 are the numerator and the denominator of (16.49a), respectively, and where V_{20} is (16.49a) for the abovementioned constant initial conditions.

- HDH Reaction

$$v_{HDH} = \frac{V_{HDH}([NADPH][ASA] - \frac{[HS][NADP^+]}{K_{eq}})}{\Upsilon_{NADPH}\Upsilon_{THR_b}(K_{ASA} + [ASA] + [HS]\frac{K_{ASA}}{K_{HS}})}, \quad \text{where } \Upsilon_{THR_b} \overset{\Delta}{=} \frac{1 + (\frac{[THR]}{K_{iTHR}})^h}{1 + (\frac{[THR]}{\alpha K_{iTHR}})^h},$$

(16.51)

and where Υ_{NADPH} is given in (16.49b) with the following HDH parameters:

$V_{HDH} = 0.9203$, $K_{NADP} = 0.067$, $K_{NADPH} = 0.037$, $K_{ASP} = 10$, $h = 1.41$,

$K_{eq} = 3162.28$, $K_{ASA} = 0.245$, $K_{HS} = 3.39$, $K_{iTHR} = 0.094$, $\alpha = 3.93$.

Following linearization, we obtain

$$v_{HDH} = V_{30} + \delta_1\overline{[ASA]} + \delta_2\overline{[HS]} + \delta_3\overline{[THR]}, \quad \delta_1 =$$
$$\frac{V_{HDH}[NADPH]\Delta_3 - \Upsilon_{NADPH}\Upsilon_{THR_b}\Psi_3}{\Delta_3^2}$$

(16.52)

$$\delta_2 = \frac{-V_{HDH}\frac{NADP^+}{K_{eq}}}{\Delta_3} - \frac{\Upsilon_{NADPH}\Upsilon_{THR_b}K_{ASA}\Psi_3}{K_{HS}\Delta_3^2},$$

$$\delta_3 = -V_{HDH}\frac{\Upsilon_{NADPH_b}\frac{\partial \Upsilon_{THR_b}}{\partial THR}\Psi_3}{\Delta_3^2},$$

$$\Upsilon_{NADPH_b} = \Upsilon_{NADPH}(K_{ASA} + [ASA] + [HS]\frac{K_{ASA}}{K_{HS}}), \; \delta_1 \stackrel{\Delta}{=} \frac{\partial v_{HDH}}{\partial ASA}, \; \delta_2 \stackrel{\Delta}{=} \frac{\partial v_{HDH}}{\partial HS},$$

$\delta_3 \stackrel{\Delta}{=} \dfrac{\partial v_{HDH}}{\partial THR}$, where Ψ_3 and Δ_3 are the numerator and the denominator of (16.51a), respectively, and where V_{30} is (16.51a) for the above given initial conditions.

- HK Reaction

$$v_{HK} = \frac{V_{HK}[HS][ATP]}{\Upsilon_{ATP,b}\Upsilon_{LYS}(\Upsilon_{HS}(1+\frac{[THR]}{K_{iTHR}})+[HS])}, \; \Upsilon_{HS} \stackrel{\Delta}{=} K_{HS}(1 + \frac{ATP}{K_{iATP}}), \quad \text{where} \tag{16.53}$$

$$\Upsilon_{ATP,b} \stackrel{\Delta}{=} K_{ATP}[1 + \frac{[HS]}{K_{iHS}}] + [ATP].$$

The HK parameters are

$$V_{HK} = 0.5457, \; K_{HS} = 0.11, \; K_{iTHR} = 1.09, \; K_{iLYS} = 9.45, \; K_{iHS} = 4.7,$$

$$K_{iATP} = 4.35, \; K_{ATP} = 0.072.$$

Following linearization, we obtain

$$v_{HK} = V_{40} + \epsilon_1\overline{[HS]} + \epsilon_2\overline{[THR]}, \; \epsilon_1 \stackrel{\Delta}{=} \frac{\partial v_{HK}}{\partial HS} = \frac{V_{HK}[ATP]}{\Delta_4} - \frac{\Upsilon_{HS,b}\Psi_4}{\Delta_4^2} \tag{16.54}$$

$$\epsilon_2 \stackrel{\Delta}{=} \frac{\partial v_{HK}}{\partial THR} = -\frac{\Upsilon_{ATP,b}\Upsilon_{LYS}\frac{\Upsilon_{HS}}{K_{iTHR}}\Psi_4}{\Delta_4^2},$$

where

$$\Upsilon_{HS,b} = [2\frac{K_{ATP}}{K_{iHS}}[HS] + K_{ATP} + [ATP] + \frac{K_{ATP}}{K_{iHS}}\Upsilon_{HS,b}[1 + \frac{THR}{K_{iTHR}}]]\Upsilon_{LYS},$$

where Ψ_4, Δ_4 are the numerator and the denominator of (16.53a), respectively, and where V_{40} is (16.53a) given the constant initial conditions.

- TS Reaction

$$v_{TS} = \frac{V_{TS}[HSP]}{K_{HSP}+[HSP]}. \tag{16.55}$$

The TS parameters are $V_{TS} = 0.2392$, $K_{HSP} = 0.307$. Following linearization, we obtain

$$v_{TS} = V_{50} + \xi_1 \overline{[HSP]} \text{ where } \xi_1 \triangleq \frac{\partial v_{TS}}{\partial HSP} = \frac{V_{TS}[(K_{HSP} + [HSP]) - [HSP]]}{(K_{HSP} + [HSP])^2},$$

$$(16.56)$$

where V_{50} is obtained by placing the above TS parameters and the constant initial condition in (16.55).

- Sink Reaction

  In order to provide a sink for the threonine concentration, which accumulates via the TS step, three sink routs were considered by [11] where the major one involves association of threonine with tRNA. The rate equation for this reaction is

$$v_{SINK} = \frac{V_{SINK} \frac{[THR]}{K_{THR}} \frac{[tRNA]}{K_{tRNA}}}{(1 + \frac{[THR]}{K_{THR}}) \frac{[tRNA]}{K_{tRNA}}}, \tag{16.57}$$

where $V_{SINK} = 0.3$, $[tRNA] = 0.01$ mM, $K_{THR} = 0.5$ and $K_{tRNA} = 0.12$. Following linearization, we obtain $v_{SINK} = V_{60} + \zeta_1 \overline{THR}$ and

$$\zeta_1 \triangleq \frac{\partial v_{SINK}}{\partial THR} = \frac{V_{SINK} K_{THR}[tRNA]}{(K_{THR} + [THR])^2([tRNA] + K_{tRNA})^2}.$$

16.9.1 The Complete Linearized Threonine Synthesis Pathway

Following the derivations in Sect. 16.9, we are able to assemble the complete pathway, where the feedback loops are explicitly exerted in the system via the dependence of the AKI, HDH, and HK reactions on the threonine. We obtain the following state-space realization for the linearized pathway:

$$\frac{d}{dt}\bar{P} = A\bar{P} + B\delta_{-1}(t), \quad \bar{P} = col\{\bar{P}_1, \ldots, \bar{P}_5\}, \tag{16.58}$$

where δ_{-1} denotes a unity step function,

$$
\begin{matrix}
\bar{P}_1 = \overline{ASPP} \\
\bar{P}_2 = \overline{ASA} \\
\bar{P}_3 = \overline{HS} \\
\bar{P}_4 = \overline{HSP} \\
\bar{P}_5 = \overline{THR}
\end{matrix}
, \quad
A = \begin{bmatrix}
\alpha_1 + \alpha_3 - \beta_1 & -\beta_2 & 0 & 0 & \alpha_2 \\
\beta_1 & \beta_2 - \delta_1 & -\delta_2 & 0 & -\delta_3 \\
0 & \delta_1 & \delta_2 - \epsilon_1 & 0 & \delta_3 - \epsilon_2 \\
0 & 0 & \epsilon_1 & -\xi_1 & \epsilon_2 \\
0 & 0 & 0 & \xi_1 & -\zeta_1
\end{bmatrix}
, \quad
B = \begin{bmatrix}
V_{10} + V_{10b} - V_{20} \\
V_{20} - V_{30} \\
V_{30} - V_{40} \\
V_{40} - V_{50} \\
V_{50} - V_{60}
\end{bmatrix},
$$

$$(16.59)$$

and where $\overline{ASPP}$, $\overline{ASA}$, $\overline{HS}$, $\overline{HSP}$, $\overline{THR}$, denote the deviations of these metabolites from their initial values. Thus, $\overline{ASPP} = ASPP(t = 0) - ASPP(t)$ and so on.

Adding a disturbance w in the concentration of the metabolites, it is readily seen that the model of (16.58) can be represented by the following static output-feedback model.

$$\frac{d}{dt}\bar{P} = (\tilde{A} + B_2 K C)\bar{P} + B\delta_{-1}(t) + B_1 w(t), \tag{16.60}$$

where

$$\tilde{A} = \begin{bmatrix} \alpha_1 + \alpha_3 - \beta_1 & -\beta_2 & 0 & 0 & 0 \\ \beta_1 & \beta_2 - \delta_1 & -\delta_2 & 0 & 0 \\ 0 & \delta_1 & \delta_2 - \epsilon_1 & 0 & 0 \\ 0 & 0 & \epsilon_1 & -\xi_1 & 0 \\ 0 & 0 & 0 & \xi_1 & -\zeta_1 \end{bmatrix}, \quad B_2 = \begin{bmatrix} 1 & 0 & 0 \\ 0 & -1 & 0 \\ 0 & 1 & -1 \\ 0 & 0 & 1 \\ 0 & 0 & 0 \end{bmatrix},$$

$$C^T = \begin{bmatrix} 0 \\ 0 \\ 0 \\ 0 \\ 1 \end{bmatrix}, \quad \text{and} \quad K = \begin{bmatrix} \alpha_2 \\ \delta_3 \\ \epsilon_2 \end{bmatrix}.$$

Applying, for example, a step disturbance of magnitude $\widetilde{ASPP}$ at the entry point of the ASPP, the following disturbance matrix is obtained:

$$B_1 = \begin{bmatrix} (\alpha_1 + \alpha_3)\widetilde{ASPP} & 0 & 0 & 0 & 0 \end{bmatrix}^T.$$

The model of (16.60) reveals three static output-feedback loops in the system. In order to asses the effect of these loops, we consider a step disturbance at either the ASP, which is the natural entry to the threonine synthesis system, or the ASPP. In fact, a disturbance signal can be applied to any one of the system metabolites. We note that in the case of a disturbance in ASP the following input matrix is obtained:

$$B_1 = \begin{bmatrix} \alpha_0 \widetilde{ASP} & 0 & 0 & 0 & 0 \end{bmatrix}^T,$$

where $\widetilde{ASP}$ is a step disturbance in ASP and where $\alpha_0 \triangleq \dfrac{\partial v_{AKI}}{\partial ASP}$.

16.9.2   The Threonine Synthesis Pathway—Research Procedure

We first linearize the system around the given set points and obtain the system coefficients (i.e., $\alpha_1, \alpha_2 \ldots \beta_1$ and so on). We thus derive also the control gain K of (16.60e). Performing simulations of the response of the linearized system and comparing the results with those obtained for the original nonlinear system show that the approximation by the linear model is remarkably good. We then apply the optimality measure of Sect. 16.8. That is, we seek, for a given interval for q and $R = I_3$, a minimal value for γ. We apply this measure to probe the closed-loop system sensitivity to uncertainties in several enzymes of the pathway. For each uncertain concentration of an enzyme, a set of two LMIs should be solved, each corresponds to edge point of the uncertainty interval. An uncertainty of $\pm 40\%$ in HDH, for example, is taken as an interval of $[0.6V_{HDH} \quad 1.4V_{HDH}]$, where V_{HDH} is the maximal velocity and where we note that this term is proportional to the concentration of HDH [11].

In addition to the above treatment, we explore the effect of each individual feedback loop on the system sensitivity to enzyme uncertainties. "Detaching" a feedback loop is justified providing that it is completely identified in the system. This task, however, is quite easy in the threonine synthesis pathway case, since the actual negative feedback inhibition loops are completely evident. The theoretical possibility of detaching a feedback loop in a given pathway can be very meaningful in trying to understand its importance in the intact pathway.

16.9.3 The Threonine Synthesis Pathway—Numerical Results

We have investigated the system behavior by looking at its time-domain dynamics and by considering its optimal behavior for different set points. Nearly, identical results were obtained for all the cases examined. In this section, we bring various results which correspond to the initial concentration given in Sect. 16.9 and for 90% disturbance, namely, of magnitude of 90% of its initial concentration, in either ASP or ASPP. In both tests, we obtain the system of (16.60a–e) with the following matrices:

$$\tilde{A} = \begin{bmatrix} -41.1280 & 2.4459 & 0 & 0 & \alpha_2 \\ 16.1855 & -5.4604 & 0.0055 & 0 & -\delta_3 \\ 0 & 3.0145 & -2.7343 & 0 & \delta_2 - \epsilon_2 \\ 0 & 0 & 2.7288 & -0.7544 & \epsilon_2 \\ 0 & 0 & 0 & 0.7544 & -.027 \end{bmatrix}, \quad \text{and} \quad B_2 = \begin{bmatrix} 1 & 0 & 0 \\ 0 & -1 & 0 \\ 0 & 1 & -1 \\ 0 & 0 & 1 \\ 0 & 0 & 0 \end{bmatrix},$$

where

$$C^T = \begin{bmatrix} 0 \\ 0 \\ 0 \\ 0 \\ 1 \end{bmatrix} \quad \text{and} \quad K^T = \begin{bmatrix} -0.1748 & -0.0461 & -0.0105 \end{bmatrix},$$

and where $\alpha_2 = -0.1748$, $\delta_2 = -0.0055$, $\delta_3 = -0.0461$ and $\epsilon_2 = -0.0105$. Applying a 90% step disturbance at the entry of ASP (i.e., 90% of the initial concentration of ASP), we obtain the following disturbance matrix: $B_1^T = \begin{bmatrix} 0.0710 & 0 & 0 & 0 & 0 \end{bmatrix}$. Applying a 90% step disturbance at the entry of ASPP, we obtain the following disturbance matrix: $B_1^T = \begin{bmatrix} -0.1122 & 0 & 0 & 0 & 0 \end{bmatrix}$. We start by taking $Q = q I_5$, where q is a scalar in the feasibility interval $\in [0\ 1]$ and a control effort weight of $R = I_3$. For the nominal system, we obtain a minimum disturbance attenuation level of $\gamma = 0.2149$. Next, we keep the same nominal gain vector of $K^T = \begin{bmatrix} -0.1748 & -0.0461 & -0.0105 \end{bmatrix}$

Table 16.5 Optimal analysis results for uncertainty in AKI, AKIII, ASD, HDH. Shown are the minimal attenuation levels γ for the feasibility interval of [0 1]

Case	Uncertainty interval	$\gamma_{min} - AKI$	$\gamma_{min} - AKIII$	$\gamma_{min} - ASD$	$\gamma_{min} - HDH$
1	–	0.214	0.214	0.214	0.214
2	[0.9 1.1]	0.220	0.2190	0.219	0.220
3	[0.7 1.3]	0.229	0.227	0.227	0.229
4	[0.5 1.5]	0.241	0.237	0.233	0.236
5	[0.4 1.6]	0.250	0.240	0.235	0.238

Table 16.6 Optimal analysis results for detached systems. Shown are the minimal attenuation level γ obtained for the nominal system

Case	Detached loop	γ_{min}
1	AKI	0.5549
2	HDH	0.284
3	HK	0.2149.

and introduce various uncertainty levels in the enzymes AKI, AKII, ASD, HDH, and HK. In Table 16.5, we bring the results of the first four tests.

Similar results to the above ones are obtained for uncertainties in HK. Decreasing the control effort weight for the nominal system results in weakly decreased values of γ as follows: for $R = [0.01I_3 \ 0.1I_3 \ 1.0I_3 \ 10I_3]$, we obtain $\gamma = [0.211 \ 0.211 \ 0.214 \ 0.243]$, respectively. The same phenomenon is observed also for the uncertain cases.

"Detaching" each of the three feedback loops (setting its gain to zero), and correspondingly calculating the new coefficient values of the detached enzyme (by taking $THR = 0mM$ in its linearized model), a new feedback system emerges. On this new system, we apply the above H_∞ optimality tests. Detaching, for example, the loop of AKI, we set $\alpha_2 = 0$ and calculate α_1, α_0 taking THR=0 in their derivations. Note that although the "detachment" removes the term of THR in the detached enzyme, it leaves the maximal velocity unchanged. The results obtained for the detached systems are given in Table 16.6.

We note that in agreement with the results of [11], the detachment effect of AKI has the largest impact.

The above tests are readily repeated for a 90% initial concentration step disturbance at the entry of ASPP. We obtain the following results (Table 16.7), which are similar to those obtained for the ASP disturbance (note that HDH was replaced by HK).

Our next step is based on transforming the linearized pathway that we have we obtained to an uncertain polytopic system, as explained in Sect. 16.3. It is expected that since the linear simulation is very close to the simulation of the nonlinear system, the attenuation levels for the polytopic case will not deviate much compared to the simply linearized system which is indeed the case. In Table 16.8, we bring the results of the optimality test for the nominal polytopic case (i.e., with no actual (physical) uncertainties), driven by a disturbance signal on ASP. The fictitious uncertainty interval is denoted as "virtual interval."

Table 16.7 Optimal analysis results for uncertainties in AKI, AKII, ASD, and HK for a 90% step disturbance in ASPP. Shown are the minimal attenuation levels γ for the feasibility interval of [0 1]

Case	Uncertainty interval	$\gamma_{min} - AKI$	$\gamma_{min} - AKIII$	$\gamma_{min} - ASD$	$\gamma_{min} - HK$
1	–	0.339	0.339	0.339	0.339
2	[0.9 1.1]	0.346	0.348	0.347	0.348
3	[0.7 1.3]	0.365	0.366	0.359	0.362
4	[0.5 1.5]	0.393	0.383	0.368	0.374
5	[0.4 1.6]	0.413	0.391	0.375	0.378

Table 16.8 Optimal analysis results for the polytopic system. Shown are the minimal attenuation level γ obtained for three "virtual intervals"

Case	"Virtual interval"	γ_{min}
1	0	0.2149
2	0.1	0.223
3	0.3	0.24
4	0.8	0.40

Note that even the result that is achieved for the very large fictitious interval of 0.8 does not deviate much from the one obtained for the nominal linear system (i.e., for a 0 interval). Similar results are obtained where to the "virtual uncertainty" an actual uncertainty in enzyme concentration is added.

16.10   Conclusions

In this chapter, an engineering feedback control-oriented framework is introduced for the study of the control and regulation of biochemical systems. We adopt an approach by which we first linearize each unit of a given biochemical system using a simple Taylor expansion. The linearized system is then assembled into a linear polytopic uncertain system. By "tuning" the range of these "fictitious" uncertainty intervals, we thus account for the nonlinearity that is inherent to all biochemical systems. The principle advantage of this approach is manifested by the applicability of linear classical and modern control theory tools to biochemical systems.

The proximity of the linearized models with and without the "fictitious" uncertainties to the original nonlinear system is assessed by simulating the system response (the linearized and nonlinear systems) to various biochemical signals (such as concentration jumps) and comparing the results. In cases where the linearized model and the nonlinear model achieve similar results by way of simulation, one may conclude that the linearized model is a good approximation to the original nonlinear system (for example, the optimality results of the threonine synthesis pathway in Sect. 16.9). Once a good approximated nonlinear system is at hand, the possible application of robust linear control theory is a natural step to explore the possible optimality of biochemical systems, in face of common parameter uncertainties. These uncertainties include, for example, uncertainties in the concentration of the various enzymes that constitute a given biochemical pathway.

By adopting a control engineering framework to biochemical systems, three basic points should be bear in mind. The first is that in both the theory and practice of traditional control theory, the controller and the controlled system(s) are different physical entities. Thus, one can apply different controllers to the same plant resulting in different performance objectives. In biochemical system (and indeed in chemical reactions), however, the controller and the controlled system are typically integrated in the same molecular structure and cannot be **physically** separated. In enzymes, the inhibitory site may be the same site where the catalysis of the substrate occurs (the "catalytic site") or a different (modulatory) site residing in a close or remote site to the catalytic one.

The second point is that in spite of the latter fact, one can distinguish between the case where the enzyme is not subjected to an inhibitory substance or is subjected to different levels (concentrations and inhibition constant) of the inhibitor. This allows us, in principle, to view the system in both configurations: On one hand as "open-loop" configuration and on the other hand as a "closed-loop" one, where the system is inhibited by a remote substance of the same pathway (say a metabolic pathway) thus forming the "closed-loop" structure. This reasoning must be adopted a priori before approaching these systems. Conceptually, as long as a feedback is evident in the system, one needs only to adjust the tools of linear (or nonlinear) control theory as practiced in engineering.

The third, not less important, basic point is that all biochemical enzyme systems are evidently **nonlinear in nature**. Hence, the question arises why should one linearize these systems. The answer depends on what one seeks to achieve. If one aims at probing the kinetic behavior of a given enzyme pathway in both, the transitory and the steady-state phase, then one can apply strong tools of nonlinear analysis (both theoretically and numerically), and achieves a complete simulated system (depending on the experimental data at hand and the complexity of the system). However, if one aims at studying the control strategy in biochemical pathways (and indeed biochemical systems by large), then, by either linearizing the system, **as a starting point**, or model it as a linear uncertain system, one can apply the strong tools of engineering control theory which is mainly centered around the concepts of feedback, robustness of the system, and the type of feedback strategy that Nature applies. In biochemistry, robustness relates to keeping the system performance in face of variation in the enzyme concentration and activities as well as variation in the substrate concentrations. These concepts can be studied once either linear or nonlinear control models are adopted. It is quite obvious that linear models are likely to provide an insight into the control strategy of the system, thus paving the way to the more involved nonlinear control analysis, a subject which is left for a further study.

Following the above lines of reasoning for the study of biochemical pathways, a complete framework is presented in this chapter which enables us to explore the possible optimality of biochemical systems in the H_∞-norm sense. We have begun by constructing a simple four-MM enzyme reaction blocks where, in the closed-loop configuration, the first enzyme is inhibited by the product of the third enzyme. This inhibition is achieved by applying a static output-feedback control strategy. In order to probe the qualities of the feedback control of such a constructed pathway, we have brought a numerical example that adopts kinetic and thermodynamic constants as well as substances concentrations having typical values in Biochemistry. We consider in this example a special case where all the substrates of the enzymes involved are "supplied" as step functions, thus enabling us to probe the transients in their perturbed counterparts and their steady-state values. Comparing the simulation results of both the open- and the closed-loop configurations of the linearized pathway to those obtained for the nonlinear one, it is clearly shown that the linear approximation is sound.

We were therefore confident in applying the H_∞ optimality test to the static output feedback that is inherent to the inhibition process. It is shown that one can apply principles of optimal control theory to probe one of the most important question in the study of biochemical enzyme feedback pathways which is: How optimal is the system performance, given the feedback inhibition dynamics of the system.

The numerical results that have been obtained for the closed-loop system of Sect. 16.8 clearly show that the feedback inhibition of this system achieves optimal H_∞ performance, for a wide range of the weighting assigned to the deviations of the enzyme products. It is shown that the feedback gain allows similar small attenuation levels, for essentially the same performance index, for both nominal and physically uncertain systems. The latter property is also manifested by assembling the system as a linear uncertain one, thus providing better account for its nonlinearity. It is shown in Table 16.4 that a greater portion of the nonlinear system can be captured at the expense of relatively small increase in the attenuation levels, while still preserving a robust performance in face of real physical uncertainties. The optimality test of the hypothetical four-block pathways could have been readily extended to a much more involved structure containing more enzymes (of several types) and a multiple of regulation points. However, in order to simplify the mathematical burden of the results, we have chosen a simple structure containing only irreversible MM enzyme reactions.

In order to assess the possible validity of our results in the biochemical realm, we turned to the threonine synthesis pathway. Similarly to the four-block system, we adopt a two-phase approach:

In phase I, following a simple Taylor expansion of the system kinetics, we have obtained a linear model for the system from which we derived the nominal controller gain matrix. We then applied the optimality test to various uncertainties in the system. In this phase, we have also "detached" each of the negative feedback loops and examined the performance of the resulting controller in face of the above system uncertainties. The numerical results that were obtained clearly suggest that the nominal controlled system (i.e., with the nominal controller) is robust to the various enzyme uncertainties that were examined. In particular, it is shown that the sensitivity of the system to uncertainty is very similar for all the enzymes that were tested. A similar behavior is observed where the disturbance is shifted to the ASPP concentration.

In phase II, the various tests of phase I have been repeated, where the system has been transformed into one with polytopic uncertainties, by introducing a "fictitious" uncertainty interval centered around each coefficient that describes the linearized kinetic equation. Applying this approach the accuracy of our linearized model has been assessed. In the case where the linearized system is a good approximation to the nonlinear one, the attenuation levels do not deviate much, compared to the system without these "fictitious" uncertainties. Certainly, the attenuation levels must increase due to the later uncertainties, but the extent of this increase depends on the quality of the approximation. It should, therefore, be expected that since the time behavior of the linearized system that we have tested is almost the same as the nonlinear one, the attenuation levels obtained in phase I or II indicate the levels that can be achieved by the actual nonlinear system. Our results clearly show that this is indeed the case.

The results that are described in this chapter, going from the simple block models to a hypothetical (simple) pathway and culminating in the threonine synthesis pathway, clearly indicate that a further research should be conducted in order to understand the control strategy of biochemical systems via our approach. Specifically, the results obtained so far in the study of the threonine synthesis pathway clearly emphasizes the need for a further research of additional important issues such as the relative significance of the various negative loops in the system. Moreover, it is not clear, at this point, whether the H_∞ control framework significantly accounts for the robustness of the pathway under study. In Chap. 17, we extend the approach brought in Sect. 16.9 to other optimal measures and we demonstrate the impact of these measures on the robustness of the threonine synthesis pathway to similar uncertainties and disturbances as in Sect. 16.9. In Chap. 17, we also consider the glycolytic pathway as an additional case study.

References

1. Gershon, E., Hiler, R., Shaked, U.: Classical control theory approach to enzymatic reactions. In: Proceedings of the European Control Conference (ECC03), Cambridge, England (2003)
2. Gershon, E., Shaked, U.: H_∞ feedback control of biochemical pathways via system uncertainty. In: Proceedings of the 5rd IFAC Symposium on Robust Control Design (ROCOND), Toulouse, France, July 2006
3. Gershon, E., Shaked, U.: H_∞ feedback-control theory in biochemical systems. Int. J. Robust Nonlinear Control **18**, 18–50 (2008)
4. Gershon, E., Yokev, O., Shaked, U.: H_∞ feedback-control of the Threonine Synthesis pathway via system uncertainty. In: Proceedings of the European Control Conference (ECC07), Kos, Greece, June 2007
5. Gershon, E., Shaked, U.: Robust polytopic analysis of the feedback-control of Glycolysis in Yeasts via some system norms. In: Proceedings of the 20th Mediterranean Conference on Control and Automation (MED12), Barcelona, Spain, July 2012
6. Gershon, E., Navon, M.: Robust feedback-control analysis of the Threonine Synthesis Pathway via various system norms. In: Proceedings of the 22nd Mediterranean Conference on Control and Automation (MED14), Palermo, Sicily, June 2014

7. Gershon, E., Navon, M., Shaked, U.: Robust peak-to-peak and H_∞ static output-feedback control of the Threonine Synthesis Pathway. In: Proceedings of the European Control Conference (ECC15), Linz, Austria, July 2015

8. Lehninger, A.L.: Principles of Biochemistry. Worth Publishers (1982)

9. Segel, I.R.: Enzyme Kinetics—Behavior and Analysis of Rapid Equilibrium and Steady-state Enzyme Systems. Wiley (1975)

10. Wilkinson, F.: Chemical Kinetics and Reaction Mechanisms. Van Nostrand Reinhold Company (1980)

11. Chassagnole, C., Rais, B., Quentin, E., Fell, D., Mazat, J.P.: An integrated study of Threonine-pathway enzyme kinetics in Echerichia coli. Biochem. J. **356**, 415–423 (2001)

12. Chassagnole, C., Rais, B., Quentin, E., Fell, D., Mazat, J.P.: Threonine synthesis from aspartate in Echerichia coli cell-free extract: pathway dynamics. Biochem. J. **356**, 425–432 (2001)

13. Chassagnole, C., Rais, B., Quentin, E., Fell, D., Mazat, J.P.: Control of Threonine-synthesis pathway in Echerichia coli: theoretical and experimental approach. Biochem. J. **356**, 433–444 (2001)

14. Franklin, G.F., Fowell, J.D., Emami-Naeini, A.: Feedback Control of Dynamic Sysytems, 2nd edn. Addison-Wiesley, New York (1991)

15. Atkinson, D.E.: Enzymes as control elements in metabolic regulation. In: Boyer, P. (ed.) The Enzymes, vol. 1, 3rd edn., p. 461. Academic Press (1970)

16. Ellison, W.R., Lueck, J.D., Fromm, H.J.: Studies on the mechanism of orthophosphate regulation of Bovine Brain Hexokinase. J. Biol. Chem. **250**(5), 1864–1871 (1975)

17. Ardehali, H., Printz, R.L., Whitesell, R.R., May, J.M., Granner, D.K.: Functional interaction between the N- and C-terminal halves of Human Hexokinase II. J. Biol. Chem. **274**, 15986–15989 (1999)

18. Aleshin, A.E., Zeng, C., Bourenkov, G.P., Bartunik, H.D., Fromm, H.J., Honzatko, R.B.: The mechanism of regulation of Hexokinase: new insight from the crystal structure of recombinant human brain hexokinase comlexed with glucose and glucose-6-phospahte. Structure **6**(1), 39–50 (1998)

19. Aleshin, A.E., Kirby, C., Xiaofeng, L., Bourenkov, G.P., Bartunik, H.D., Fromm, H.J., Honzatko, R.B.: Crystal structures of motant monomeric Hexokinase I reveal multiple ADP binding sites and conformational changes relevant to allosteric regulation. J. Mol. Biol. **296**, 1001–1015 (2000)

20. Liu, X., Sup Kim, C., Kurbanov, F.T., Honzatko, R.B., Fromm, H.J.: Dual mechanism for Glucose 6-Phosphate inhibition of human Hexokinase. J. Biol. Chem. **274**, 31155–31159 (1999)

21. de Cerqueira Cesar, M., Wilson, J.E.: Functional characteristics of hexokinase bound to the type a and type B sites of bovine brain mitochondria. Arch. Biochem. Biophys. **397**(1), 106–112 (2002)

22. Arora, K.K., Filburn, C.R., Pederson, P.I.: Structure/function relationships in hexokinase. Site-directed mutational analyses and characterization of overexpressed fragments implicate different functions for the N- and C-terminal halves of the enzyme. J. Biol. Chem. **268**, 18259–18266 (1993)

23. Wolkenhauer, O., Ghosh, B.K., Cho, K.H.: Control and coordination in biochemical networks. IEEE Control Syst. **24**(4), 30–34 (2004)

24. Fell, D.: Metabolic control analysis: a survey of it's theoretical and experimental development. Biochem. J. **286**, 313–330 (1992)

25. Fell, D.: Frontiers in Metabolism: Understanding the Control of Metabolism. Portland Press (1997)

26. Voit, E.O.: Computational Analysis of Biochemical Systems: A Practical Guide for Biochemists and Molecular Biologists. Cambridge University Press, Cambridge, UK (2000)

27. Goel, G.: Reconstructing Biochemical Systems. Systems Modeling and Analysis Tools for Decoding Biological Designs. VDM Verlag Dr. Mller, Saarbrcken, Germany (2008)

28. Torres, N.V., Voit, E.O.: Pathway Analysis and Optimization in Metabolic Engineering. Cambridge University Press, Cambridge, UK (2005)

29. Voit, E.O.: Biochemical systems theory: a review. ISRN Biomath. **2013**(Article ID 897658), 53 (2013). https://doi.org/10.1155/2013/897658
30. Baker, G.A., Graves-Morris, P.: Pade Approximation. Cambridge University Press (2009)
31. Xie, L., Fu, M., de Souza, C.E.: H_∞ control and quadratic stabilization of systems with parameter uncertainty via output feedback. IEEE Trans. Autom. Control **37**, 1253–1256 (1992)
32. Alves, R., Savageau, M.A.: Effect of overall feedback inhibition in unbranched biosynthetic pathways. Biophys. J. **79**, 2290–2304 (2000)
33. Schuster, S., Heinrich, R.: Minimization of intermediate concentrations as a suggested optimality principle for biochemical networks, I. Theoretical analysis. J. Math. Biol. **29**, 425–442 (1991)
34. Savinell, J.M., Palsson, B.O.: Network analysis of intermediary metabolizm using linear optimization, I. Developmenmt of mathematical formalism. J. Theor. Biol. **154**, 421–454 (1992)
35. Mends, P., Kell, D.B.: Non-linear optimaization of biochemical pathways: applications to metabolic engineering and parameter estimation. Bioinform. **14**(10), 869–883 (1998)
36. Morohashi, M., Winn, A.E., Borisuk, M.T., Bolouri, H., Doyle, L., Kitano, H.: Robustness as a measure of plausibilty in models of biochemical netwroks. J. Theor. Biol. **216**, 19–30 (2002)
37. Boyd, S., El Ghaoui, L., Feron, E., Balakrishnan, V.: Linear Matrix Inequalities in System and Control Theory. SIAM, Philadelphia (1994)
38. Petersen, I.R.: Quadratic stabilizability of uncertain linear systems containing both constant and time varying uncertain parameters. J. Optim. Theory Appl. **57**(3), 439–461 (1988)
39. Savageau, M.A.: Biochemical systems analysis. I. Some mathematical properties of the rate law for the component enzymatic reactions. J. Theor. Biol. **25**, 365–369 (1969)
40. Savageau, M.A.: Biochemical systems analysis. II. The steady-state solutions for an n-pool system using a power-law approximation. J. Theor. Biol. **25**, 370–379 (1969)
41. Savageau, M.A.: Biochemical systems analysis. 3. Dynamic solutions using a power-law approximation. J. Theor. Biol. **26**, 215–226 (1970)
42. Savageau, M.: Biochemical Systems Analysis: A Study of Function and Design in Molecular Biology. Addison Wesley, Reading, MA (1976)

Chapter 17
Feedback-Control Theory in Biochemical Systems—Various System Norms

Abstract The theories of optimal H_∞ control and three other optimal measures are applied to the study of biochemical pathways. The additional optimal measures include the H_2, energy-to-peak and the peak-to-peak optimal measures. Based on a static output-feedback configuration, the sensitivity of the pathways to various system uncertainties that include uncertainties in the rate constants and the concentrations of the enzyme involved is addressed. The results are applied to the study of the possible optimality in the above measures of both the threonine synthesis pathway and the glycolytic pathway. It is shown that the H_∞ and the peak-to-peak optimal measures are better suited to describe the sensitivity of these pathways to various parameter uncertainties.

17.1 Introduction

The issue of robustness of biochemical pathways with respect to changes in the concentration and the rate constants of the enzymes and the substrates involved, has been introduced in Chap. 15. In short, the main two major frameworks that deal with the control and regulation of biochemical pathways [among other cellular processes] are the BST [Biochemical System Theory] framework (see [1–6] and the references therein) and the MCA [Metabolic Control Analysis] approach (see [7, 8] for a comprehensive account of this topic). We note that numerous studies that are based on different approaches and aided by a host of mathematical techniques and sophisticated software packages were conducted in the last four decades in the said research subject (see, for example, [9–11]).

In Chap. 16, the issue of robust performance of biochemical pathways in the H_∞ sense has been introduced and demonstrated via the threonine synthesis pathway. The basic approach in Chap. 16 relies on the linearization of the pathway under study. Following verification that its response to an external stimulus is a good approximation to the original nonlinear model of the system, we are able to apply the traditional tools of modern control theory and especially the linear H_∞ control theory approach.

Concerning metabolic pathways, the glycolytic pathway is a natural candidate to feedback-control theory analysis, being a major metabolic pathway in almost all

© Springer Nature Switzerland AG 2019
E. Gershon and U. Shaked, *Advances in H_∞ Control Theory*,
Lecture Notes in Control and Information Sciences 481,
https://doi.org/10.1007/978-3-030-16008-1_17

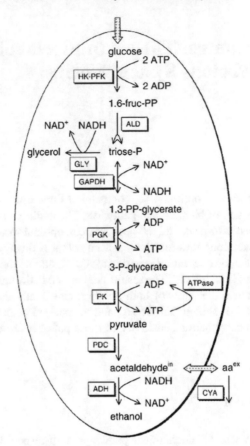

Fig. 17.1 Schematic representation of the anaerobic glycolytic pathway.The reactions are: HK-PFK, lumped reaction of hexokinase and phosphofructokinase; ALD, aldolase reaction; GLY, glycerol-producing branch; GAPDH, reaction of glyceraldehyde-3-phosphate dehydrogenase; PGK, reaction catalyzed by phosphoglycerate kinase; PK, lumped reactions of phosphoglycerate mutase, enolase, and pyruvatekinase; PDC, reaction catalyzed by pyruvate decarboxylase; ADH, alcohol dehydrogenase reaction; ATPase, total cellular ATP consumption; CYA, degradation of acetaldehyde by cyanide. Fluxes: glucose influx, exchange of acetaldehyde between the cell and the external medium

forms of organic life. The glycolytic pathway has been studied extensively in several organisms both in vitro and in vivo (see [12] for the general introduction of the pathway, see also [13] and the references therein). The kinetics and regulation of the enzymes involved in the pathway were studied by numerous research groups due to its central role in human metabolism [12]. In Figs. 17.1, 17.2 we bring a simple model of the glycolytic pathway in Yeasts [13]. This model was originally proposed in order to account for the possible oscillatory aspect of the pathway, as was experimentally measured. The model is composed of nine steps and it was shown to mimic successfully the measured oscillations. In the present chapter we

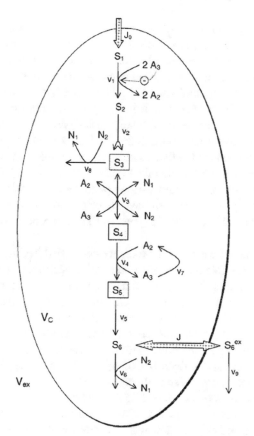

Fig. 17.2 Schematic representation of the models. The 9-variable model includes all shown metabolites as variables, whereas in the 6-variable model, S3, S4, and S5 are treated as parameters. The variables are: S1, glucose; S2, fructose-1,6-bisphosphate, S3, pool of the triosephosphates, glyceraldehyde- 3-phosphate, and dihydroxyacetone phosphate; S4, 3-phosphoglycerate; S5, pyruvate; S6, acetaldehyde in the cell; S6 ex, extracellular acetaldehyde; A3, ATP, and N2, NADH

concentrated on the non-oscillatory behavior of the glycolytic model which was also shown to yield similar results to those obtained experimentally. The non-oscillatory behavior follows a small change in one of the rate constants of the original oscillatory system.

The theory developed in Chap. 16 (see also [14, 15]) has been applied to the study of the threonine synthesis pathway [16–18]. Following a simple Taylor expansion of each block of the latter pathway around a pre-chosen set point, the resulting linearized system was assembled as a linear one with polytopic uncertainties. Thus, we were able to better account for the nonlinearity of the system by tuning the uncertainty interval of the obtained polytope. In Chap. 16, Sect. 16.9, it was shown that the optimality of the feedback loops constituting the **threonine synthesis** system is amenable to the H_∞ control theory measure of optimality. To the best of our knowledge, no previous attempt was carried out in order to evaluate the optimality of any enzymatic pathway, either (physically realizable) theoretical one or real one, in the H_∞ sense, as presented in Chap. 16, or some related control-oriented optimal measures which are key elements in the optimal control field. The motivation for such a study, as noted already in the Chap. 15, stems from the powerful physical meanings of the optimal control measures such as the (already treated) H_∞, or the P2P (Peak to Peak) norm measures (to name a few), within the modern control engineering field. It is conceivable that biochemical feedback systems may fall into one of these categories. The question whether a real biochemical pathway "applies" a worst-case strategy in the face of the disturbances and various parameter uncertainties or whether it "applies" a peak to peak strategy, or say, H_2 or energy to peak strategy, is a key question that can be addressed by the powerful and deep fields of modern control techniques.

In the present chapter we apply the tools we developed in Chap. 16 (see also [14, 15]) to explore the possible optimality of both: the glycolytic and the threonine synthesis pathways, being regulated by feedback-control, via the four indicated system norms. These norms are given Sect. 17.2. Our aim is to "track" the adequate system norm that better describes the system sensitivity to parameter uncertainties which may be caused by varying enzyme concentrations and uncertain kinetic rate constants (and possibly by varying temperature, pH environment, etc.).

In Sect. 17.3, following the linearization of each block of the glycolytic pathway, we look at the complete linearized pathway and extract the feedback gains involved in the system. We then apply, in Sect. 17.5, four optimality tests to explore the sensitivity of the system to uncertainties in the kinetic rate constants of the enzymes involved. Our results clearly show that the sensitivity of the system, in the face of relatively large uncertainties, depends on the system norm that is chosen and that, probably, the H_∞ and the peak to peak norms better represent the system sensitivity in the biochemical realm [19]. Similar results are obtained for the threonine synthesis pathway in Sect. 17.6, complementing the results already obtained for this pathway in Chap. 16, Sect. 16.9 [20]. In Sect. 17.7, we look at the threonine synthesis pathway in the face of uncertainties in the controller gain matrix [21].

17.2   Optimal Control of Biochemical Systems via some System Norms

In order to asses the performance of the pathways under study, we apply four optimality measures which include the H_2, H_∞ (the induced L_2 norm) and the $L_2 - L_\infty$ (energy to peak) and $L_\infty - L_\infty$ (peak-to-peak, L_1) system norms. We repeat here, for the sake of clarity, the H_∞ optimal criterion that was already developed in Chap. 16, Sect. 16.9 (see also [14, 15]). We then bring three propositions that deal with the other system norms used in our study.

17.2.1 H_∞ Optimal Control

This measure relates to the attenuation level of an arbitrary (energy bounded) disturbance in the system. In this context, we consider the following H_∞ performance index:

$$J = \int_0^\infty [x^T (Q + C^T K^T R K C)x - \gamma^2 w^T w]dt, \tag{17.1}$$

applied to a dynamic system described by the following state-space model.

$$\dot{x} = \tilde{A}x + B_1 w, \quad \tilde{A} = A + B_2 K C, \quad x(0) = 0. \tag{17.2}$$

The latter model describes a closed-loop system with a static output-feedback gain matrix K. It will be shown below that nature indeed applies such a feedback in the given glycolytic pathway.

In our case, the vector x represents the concentrations of the various metabolites in the pathway, the matrix C is a two-row matrix and B_2 is composed of two columns. The matrices Q and R in (17.1) represent the different weighting we put on the metabolic concentration and the control effort. Choosing $Q = q I_n$, where q is a scalar in the interval [0 0.001] (if not otherwise stated), and $R = I_9$, the corresponding H_∞-norm of the pathway is calculated for the known K.

For the above choice of Q and R, the H_∞-norm from the disturbance input w to the system output is obtained by seeking a positive definite matrix P that satisfies the following Linear Matrix Inequality (LMI) for a minimum value of $0 < \gamma$,

$$\begin{bmatrix} \Upsilon + \Upsilon^T + q I_n + C^T K^T R K C & P B_1 \\ * & -\gamma^2 I \end{bmatrix} < 0, \tag{17.3}$$

where $\Upsilon = P(A + B_2 K C)$.

It is readily seen that the latter LMI always has a solution, in the case where the system with the given static output feedback is stable, for large enough γ if the pair (A, C) is observable (we then simply obtain a Lyapunov inequality).

The LMI of (17.3) refers to the nominal case where the various enzyme concentrations are known constants. In the uncertain case, which stems either from the above "virtual uncertainty" that better fits the nonlinear nature of the problem, or from actual uncertainty in the enzyme's kinetic constants and concentrations, the matrix P is sought such that the following LMIs are satisfied for the minimum γ.

$$\begin{bmatrix} \tilde{\Upsilon}_j + \tilde{\Upsilon}_j^T + q I + C^T K^T K C & P B_{1j} \\ * & -\gamma^2 I \end{bmatrix} < 0, \quad j = 1, \ldots, L. \tag{17.4}$$

In these LMIs, the index j refers to the jth vertex of the uncertainty polytope (all together we have L vertices) and $\tilde{\Upsilon}_j = P(A_j + B_2 K C)$. In our case the uncertain parameters appear in the matrices A and B_1 and for simplicity we seek a single

$0 < P$ (the quadratic solution) which satisfies all the L LMIs for the minimum value of γ. The latter solution may be conservative and some non-quadratic solutions are suggested in the literature that reduce the entailed conservatism.

17.2.2 H_2 Optimal Control

The following result gives a sufficient condition for the H_2-norm of the system (17.2) to be finite [22].

Proposition 17.1 *Consider (17.2) where $\tilde{A}$ is Hurwitz. If there exists $P > 0$, and Z such that*

$$\begin{bmatrix} P\tilde{A} + \tilde{A}^T P & P B_1 \\ * & -I \end{bmatrix} < 0, \quad \begin{bmatrix} P & C^T \\ * & Z \end{bmatrix} > 0, \text{ and } Trace(Z) < \gamma^2, \qquad (17.5)$$

then the H_2−norm of the system is less than γ.

17.2.3   Energy-to-Peak ($L_2 - L_\infty$) Optimal Control

The following result gives a sufficient condition for an upper-bound γ of the energy-to-peak gain of the system (17.2) [22].

Proposition 17.2 *Consider (17.2) where $\tilde{A}$ is Hurwitz. If there exists $P > 0$ such that*

$$\begin{bmatrix} P\tilde{A} + \tilde{A}^T P & P B_1 \\ * & -I \end{bmatrix} < 0 \quad and \quad \begin{bmatrix} P & C^T \\ * & \gamma^2 I \end{bmatrix} > 0, \qquad (17.6)$$

then the $L_2 - L_\infty$ norm of the system is less than γ.

17.2.4   Peak-to-Peak ($L_\infty - L_\infty$) Optimal Control

The following result gives a sufficient condition for an upper-bound γ of the peak-to-peak gain of the system (17.2) [22].

Proposition 17.3 *Consider (17.2) where $\tilde{A}$ is Hurwitz. If there exists $P > 0$, $\lambda > 0$ and $\mu > 0$ such that*

$$\begin{bmatrix} P\tilde{A} + \tilde{A}^T P + \lambda P & PB_1 \\ * & -\mu I \end{bmatrix} < 0, \quad and \quad \begin{bmatrix} \lambda P & 0 & C^T \\ * & (\gamma - \mu)I & 0 \\ * & * & \gamma I \end{bmatrix} > 0, \quad (17.7)$$

then the peak-to-peak (or L_∞) induced norm of the system is smaller than γ.

We note that similarly to the uncertain polytopic case of the H_∞ system norm, the uncertain cases for the other systems norms are treated as in (17.4).

17.3 The Linearized Minimal Model of Yeast Glycolytic Pathway

In [13], a minimal model description of the yeast glycolytic pathway is developed. There, the nonlinear system rate equations are:

$$\begin{aligned} \dot{S}_1 &= J_0 - v_1, \\ \dot{S}_2 &= v_2 - v_1, \\ \dot{S}_3 &= 2v_2 - v_3 - v_8, \\ \dot{S}_4 &= v_3 - v_4, \\ \dot{S}_5 &= v_4 - v_5, \\ \dot{S}_6 &= v_5 - v_6 - J, \\ \dot{S}_6^{ex} &= \phi J - v_9, \\ \dot{A}_3 &= -2v_1 + v_3 + v_4 - v_7, \\ \dot{N}_2 &= v_3 - v_6 - v_8, \end{aligned} \qquad (17.8)$$

with the following system velocities:

$$\begin{aligned} J_0 &= \text{constant}, \\ v_1 &= k_1 S_1 A_3 f(A_3), \\ v_2 &= k_2 S_2, \\ v_4 &= k_4 S_4 (A - A_3), \\ v_5 &= k_5 S_5, \\ v_6 &= k_6 S_6 N_2, \\ v_7 &= k_7 S_3, \\ v_8 &= k_8 S_3 N_2, \\ v_9 &= k_9 S_6^{ex}, \end{aligned} \qquad (17.9)$$

where the kinetic constants of the system are given in the following table.

Following linearization of each of the rate equations of (17.8), we are able to assemble the complete linearized pathway, where the feedback loop around the first reaction of the system is explicitly exerted in the system via its dependence on ATP. We obtain the following state-space realization for the linearized pathway:

$$\tfrac{d}{dt}\bar{P} = A\bar{P} + B\delta_{-1}(t), \quad \bar{P} = col\{\bar{P}_1, \dots, \bar{P}_9\}, \tag{17.10}$$

where $\overline{M}$ denotes the deviation of any metabolite M from its initial value (for example, $\overline{Glucose} = Glucose(t = 0) - Glucose(t))$, δ_{-1} denotes a unity step function and where

$$\bar{P}_1 = \overline{Glucose}$$
$$\bar{P}_2 = \overline{Fructose - 1, 6\ biphosphate}$$
$$\bar{P}_3 = \overline{Triophosphates\ pool}$$
$$\bar{P}_4 = \overline{3 - Phosphoglycerate}$$
$$\bar{P}_5 = \overline{Pyruvate}$$
$$\bar{P}_6 = \overline{acetaldehyde\ in\ the\ cell};$$
$$\bar{P}_7 = \overline{extracellular; acetaldehyde}$$
$$\bar{P}_8 = \overline{ATP}$$
$$\bar{P}_9 = \overline{NADH},$$

and where

$$A = \begin{bmatrix} \alpha_1 & 0 & 0 & 0 & 0 & 0 & 0 & \alpha_2 & 0 \\ -\alpha_1 & \beta_1 & 0 & 0 & 0 & 0 & 0 & -\alpha_2 & 0 \\ 0 & -2\beta_1 & \chi_1 & \chi_2 & 0 & 0 & 0 & \chi_3 & \chi_4 \\ 0 & 0 & \gamma_1 & \gamma_2 & 0 & 0 & 0 & \gamma_3 & \gamma_4 \\ 0 & 0 & 0 & \delta_1 & \delta_2 & 0 & 0 & \delta_3 & 0 \\ 0 & 0 & 0 & 0 & -\delta_2 & \epsilon_1 & \epsilon_2 & 0 & \epsilon_3 \\ 0 & 0 & 0 & 0 & 0 & \varphi_1 & \varphi_2 & 0 & 0 \\ 2\alpha_1 & 0 & \gamma_1 & \gamma_2 & 0 & 0 & 0 & \eta_1 & \gamma_4 \\ 0 & 0 & -\chi_1 & -\chi_2 & 0 & \iota_1 & 0 & -\chi_3 & \iota_2 \end{bmatrix},$$

$$B = \begin{bmatrix} J_0 - V_{1,0} \\ V_{1,0} - V_{2,0} \\ 2V_{2,0} - V_{3,0} - V_{8,0} \\ V_{3,0} - V_{4,0} \\ V_{4,0} - V_{5,0} \\ V_{5,0} - V_{6,0} - J \\ \varphi J - V_{9,0} \\ -2V_{2,0} + V_{3,0} + V_{4,0} - V_{7,0} \\ V_{3,0} - V_{6,0} - V_{8,0} \end{bmatrix}.$$

The various entries (i.e., $\alpha_1, \alpha_2, \dots \iota_1..$) in (17.10) were obtained by the linearization (omitted here due to space limitations) and $V_{i,0}$ denotes the given initial velocity of the rate equation of metabolite i of the system based on the initial values of the system metabolites given in Table 17.2 in the sequel.

Adding a disturbance w in the concentration of the metabolites, it is readily seen that the model of (17.10) can be represented by the following static output-feedback model:

$$\tfrac{d}{dt}\bar{P} = (\tilde{A} + B_2 K C)\bar{P} + B\delta_{-1}(t) + B_1 w(t), \qquad (17.11)$$

where

$$\tilde{A} = \begin{bmatrix} 0 & 0 & 0 & 0 & 0 & 0 & 0 & 0 & 0 \\ 0 & \beta_1 & 0 & 0 & 0 & 0 & 0 & 0 & 0 \\ 0 & -2\beta_1 & \chi_1 & \chi_2 & 0 & 0 & 0 & \chi_3 & \chi_4 \\ 0 & 0 & \gamma_1 & \gamma_2 & 0 & 0 & 0 & \gamma_3 & \gamma_4 \\ 0 & 0 & 0 & \delta_1 & \delta_2 & 0 & 0 & \delta_3 & 0 \\ 0 & 0 & 0 & 0 & -\delta_2 & \epsilon_1 & \epsilon_2 & 0 & \epsilon_3 \\ 0 & 0 & 0 & 0 & 0 & \varphi_1 & \varphi_2 & 0 & 0 \\ 0 & 0 & \gamma_1 & \gamma_5 & 0 & 0 & 0 & \eta_1 & \gamma_4 \\ 0 & 0 & -\chi_5 & -\chi_2 & 0 & \iota_1 & 0 & -\chi_3 & \iota_2 \end{bmatrix}$$

$$B_2 = \begin{bmatrix} 1 & 1 \\ -1 & -1 \\ 0 & 0 \\ 0 & 0 \\ 0 & 0 \\ 0 & 0 \\ 0 & 0 \\ 4 & 0 \\ 0 & 0 \end{bmatrix}, \quad C^T = \begin{bmatrix} 1 & 0 \\ 0 & 0 \\ 0 & 0 \\ 0 & 0 \\ 0 & 0 \\ 0 & 0 \\ 0 & 0 \\ 0 & 1 \\ 0 & 0 \end{bmatrix}, \quad \text{and} \quad K = \begin{bmatrix} 0.5\alpha_1 & 0 \\ 0.5\alpha_1 & \alpha_2 \end{bmatrix}.$$

The model of (17.11) reveals a feedback loop which is exerted on the first step of the pathway via ATP. In order to asses the effect of this loop, we consider the above step disturbance at the entry to the Glycolysis system. In fact, a disturbance signal can be applied to any of the system metabolites.

17.4 The Research Procedure

We repeat here, in short, the research procedure brought in Chap. 16 for conveniens. We first linearize the system around the steady-state level values given in [13] and obtain the system coefficients (i.e., $\alpha_1, \ldots \beta_1, \gamma_1 \ldots \delta_1$ and so on). We thus derive also the control gain K of (17.11e). Applying simulations of the response of the linearized system and comparing these results with those obtained for the original nonlinear system, it is found that the approximation by the linear model is very good. We then apply the optimality measures of Sect. 17.2. That is, we seek, for a given interval, say $q \in [0\ \ 0.001]$ and $R = I_9$, a minimal value for γ for the cases of

Table 17.1 Values of the enzymatic kinetic rate constants and additional parameters of the 9-variable model of yeast glycolysis taken from [13]

Parameter	Value
J_0	50 mM min^{-1}
k_1	550.0 mM^{-1} min^{-1}
K_i	1.0 mM
k_2	9.8 min^{-1}
k_{GAPDH+}	323.8 mM^{-1} min^{-1}
k_{GAPDH-}	57823.1 mM^{-1} min^{-1}
k_{PGK+}	76411.1 mM^{-1} min^{-1}
k_{PGK-}	23.7 mM^{-1} min^{-1}
k_4	80.0 mM^{-1} min^{-1}
k_5	9.7 min^{-1}
k_6	2000.0 mM^{-1} min^{-1}
k_7	28.0 min^{-1}
k_8	85.7 mM^{-1} min^{-1}
κ	375.0 min^{-1}
ϕ	0.1
A	4 mM
N	1.0 mM
n	4

H_∞, $L_2 - L_\infty$ and $L_\infty - L_\infty$ system norms and an upper-bound for the H_2−norm. We apply these measures to probe the closed-loop system sensitivity to uncertainties in several enzymes of the pathway. For each uncertain kinetic rate constant of an enzyme, a set of two LMIs should be solved, each corresponds to edge point of the uncertainty interval. An uncertainty of ±40% in k_2 (which leads from S_1 to S_2 in Fig. 17.2), for example, is taken as an interval of $[0.6 V_{k_2} \quad 1.4 V_{k_2}]$ where its nominal value is given in Table 17.1 [13].

17.5 The Glycolytic Pathway—Numerical Results

We have probed the system behavior by looking at its time domain dynamics and by considering its optimal behavior for different set points. Nearly identical results were obtained for all the cases examined. In this section, we bring various results which correspond to the initial concentration given in Table 17.2 and for 90% disturbance, namely change of magnitude of 90% in the initial concentration of the Glucose flux. In all the tests, we obtain the system of (17.11) with the following matrices:

Table 17.2 Initial values of the glycolytic metabolites

Metabolite	Steady-State Concentration (mM)
S_1	1.09
S_2	5.10
S_3	0.55
S_4	0.66
S_5	8.31
S_6	0.08
S_6^{ex}	0.02
A_3	2.19
N_2	0.41

$$\tilde{A} = \begin{bmatrix} 0 & 0 & 0 & 0 & 0 & 0 & 0 & 0 & 0 \\ 0 & -0.1633 & 0 & 0 & 0 & 0 & 0 & 0 & 0 \\ 0 & 0.3267 & -3.3037 & 0.1266 & 0 & 0 & 0 & 0.1984 & 2.4558 \\ 0 & 0 & 2.7181 & -2.5399 & 0 & 0 & 0 & 0.6816 & -3.2413 \\ 0 & 0 & 0 & 2.4133 & -0.1617 & 0 & 0 & -0.8800 & 0 \\ 0 & 0 & 0 & 0 & 0.1617 & -19.9167 & 6.2500 & 0 & -2.6667 \\ 0 & 0 & 0 & 0 & 0 & 0.6250 & -1.9583 & 0 & 0 \\ 0 & 0 & 2.7181 & 2.2867 & 0 & 0 & 0 & -3.3397 & -3.2413 \\ 0 & 0 & 2.1325 & -0.1266 & 0 & -13.6667 & 0 & -0.1984 & -6.6936 \end{bmatrix}, \quad (17.12)$$

where B_2 and C are given in (17.11) and where the gain matrix is
$$K = \begin{bmatrix} -7.8175 & 0 \\ -7.8175 & -0.8973 \end{bmatrix}.$$
The initial values of the system metabolites are given in Table 17.2.

Applying a disturbance at the entry of the Glucose, we obtain the following disturbance matrix:
$$B_1^T = \begin{bmatrix} 2.5 & 0 & 0 & 0 & 0 & 0 & 0 & 0 & 0 \end{bmatrix}.$$

We start by taking $Q = qI_9$, where q is a positive scalar in the feasibility interval [0 0.001] and a control effort weight of $R = I_9$. For the nominal system, we obtain an H_∞ minimum disturbance attenuation level of $\gamma = 5.69$. Similarly we obtain for the $L_2 - L_\infty$ and $L_\infty - L_\infty$ minimal values of $\gamma = 4.75$ and $\gamma = 13.76$,, respectively. The obtained upper-bound of the H_2 norm is 5.33. Next, we keep the same nominal gain matrix of $K = \begin{bmatrix} -7.8175 & 0 \\ -7.8175 & -0.8973 \end{bmatrix}$ and introduce various uncertainty levels in several of the enzymes kinetic constants: k_1, k_2, k_5, k_6 and k_7. In Tables 17.3 and 17.4, we bring the results of these tests.

In Table 17.3, for each row (test), we keep the same feedback controller K and the same system kinetic constants, save those indicated in the table. Thus, for case 1 in Table 17.3, only k_1 was changed in the interval $[0.9k_1 \ 1.1k_1]$. The results were obtained for the feasibility interval of $q \in [0 \ 0.001]$ and the given values of Tables 17.1 and 17.2, where in each test, the response of the linearized system was compared to

Table 17.3 Optimal analysis results for various uncertainties in the system kinetic constants. Shown are the minimal attenuation levels of γ for the feasibility interval of $q \in [0\ 0.001]$ for the H_∞, $L_2 - L_\infty$ and the $L_\infty - L_\infty$ system norms, and for the upper-bound on the H_2 norm. In all these tests, the same feedback gain and initial values are kept constants. Uncer. Par. stands for uncertain parameter and n.f for non-feasible

Case	Uncer. Par.	H_2	H_∞	$L_2 - L_\infty$	$L_\infty - L_\infty$
0	Nominal	5.33	5.69	4.75	13.76
1	$(k_1 \pm 10\%)$	5.60	8.22	4.75	17.16
2	$(k_1 \pm 25\%)$	8.0	45.29	6.45	63.6
3	$(k_1 \pm 80\%)$	n.f	n.f	n.f	n.f
4	$(k_5 \pm 50\%)$	5.43	6.21	4.73	17.12
5	$(k_5 \pm 80\%)$	5.64	7.39	4.74	27.66
6	$(k_6 \pm 50\%)$	5.35	5.81	4.73	13.89
7	$(k_6 \pm 80\%)$	5.35	5.81	4.73	13.89
8	$(k_7 \pm 25\%)$	5.48	7.08	4.74	15.49
9	$(k_7 \pm 50\%)$	5.74	9.61	4.75	18.98
10	$(k_7 \pm 80\%)$	6.51	18.36	4.87	31.04
11	$(J_0 \pm 25\%)$	5.33	5.69	4.75	13.76
12	$(J_0 \pm 80\%)$	5.33	5.69	4.75	13.76

the one of the original nonlinearized systems. Only cases where the two responses were identical within a tolerance 5% have been considered. Case 0 in the table brings the various attenuation levels for the nominal case (i.e., without uncertainties). It is shown that in all the cases relating to the various uncertainties (Cases 1–12), the H_2 norm is insensitive to the system parameter uncertainties. The same behavior is also observed for the $L_2 - L_\infty$ measure. On the other hand, the H_∞ and the peak to peak ($L_\infty - L_\infty$) attenuation levels were sensitive to some of the system uncertainties even for mild changes (see for example, Cases 1–3 and 8–10). We note that some of the system kinetic rate constants were actually estimated based on relevant kinetic data that are found in the literature [13]. A 50% uncertainty in an estimated kinetic constant of an enzyme is considered, in some cases, to be a good approximation. Note that in some cases, the uncertainty in the value of an enzymatic kinetic constant may amount to several folds [12, 23]. It is interesting to try to track the **adequate system norm** which is suitable for probing the sensitivity of the system. In biochemical systems, which are the backbone of life, metabolism is confined within cells and within various compartments in the cells (for example, the nucleus and other internal organelles). Metabolic pathways are always intertwined via the energy sources (i.e., ATP and similar molecules) and their redox capacity. The prevalent case is that some metabolite that 'belongs' to a certain pathway (say glycolysis) is also an intermediate in other pathways. The inflow of such a metabolite from one pathway to a second one affects the transient and steady-state behaviors of both pathways, therefore the peak to peak norm may be—probably in some of the cases—a proper norm to probe

Table 17.4 Optimal analysis results for various uncertainties in different pairs of the system kinetic constants. Shown are the minimal attenuation levels of γ for the feasibility interval of $q \in [0\ 0.001]$ for the H_∞, $L_2 - L_\infty$ and the $L_\infty - L_\infty$ system norms. In all these tests the same feedback gain and initial steady-state values are kept constants. Uncer. Par. stands for uncertain parameter and n.f for non-feasible

Case	Uncer. Par.	$L_2 - L_\infty$	H_∞	$L_\infty - L_\infty$
0	Nominal	4.75	5.69	13.76
1	$(k_2, k_6) \pm 25\%$	4.74	6.88	15.59
2	$(k_2, k_6) \pm 50\%$	4.75	8.89	20.24
3	$(k_2, k_6) \pm 80\%$	n.f	n.f	n.f
4	$(k_2, k_7) \pm 25\%$	4.75	8.37	17.91
5	$(k_2, k_7) \pm 50\%$	4.90	14.79	29.92
6	$(k_2, k_7) \pm 80\%$	n.f	n.f	n.f
7	$(k_5, k_7) \pm 25\%$	4.74	7.36	16.74
8	$(k_5, k_7) \pm 50\%$	4.77	10.65	23.52
9	$(k_5, k_7) \pm 80\%$	5.86	24.56	57.04

the robustness of the given feedback system. It is also possible that the H_∞ optimal criteria may be suitable for sensitivity probing of the system as was shown in the case of threonine synthesis system in Chap. 16, Sect. 16.9.3.

In order to further investigate the latter behaviors of the system norms, we applied the procedure already that lead to Table 17.3, to pairs of kinetic constants that were chosen according to their biochemical role. Again, in Table 17.4 the same picture emerges as in Table 17.3 (note that the missing H_2 test provides result that are similar to the one obtained for the $L_2 - L_\infty$). Note that in cases 3 and 6, the linearized system behavior deviated from the nonlinear one and therefore these cases were ruled out. Similar results to those reported in Tables 17.3 and 17.4 were obtained for different values of the control effort R and the feasibility interval of q.

17.6 The Threonine Synthesis Pathway

Similarly to Sect. 16.9, we have investigated the behavior of the threonine synthesis system by looking at its time domain dynamics and by considering its optimal behavior for **different set points**. Almost identical results were obtained for all the cases examined. In the present section, we bring various results which correspond to the initial concentration given in Sect. 16.9 and for 90% concentration disturbance, namely of magnitude of 90% of its initial concentration, in either ASP or ASPP. In both tests, we obtain the following system (See Sect. 16.9 for further details):

$$\tfrac{d}{dt}\bar{P} = (\tilde{A} + B_2 K C)\bar{P} + B\delta_{-1}(t) + B_1 w(t), \qquad (17.13)$$

$$\tilde{A} = \begin{bmatrix} -41.1280 & 2.4459 & 0 & 0 & \alpha_2 \\ 16.1855 & -5.4604 & 0.0055 & 0 & -\delta_3 \\ 0 & 3.0145 & -2.7343 & 0 & \delta_2 - \epsilon_2 \\ 0 & 0 & 2.7288 & -0.7544 & \epsilon_2 \\ 0 & 0 & 0 & 0.7544 & -0.0229 \end{bmatrix},$$

where $\alpha_2 = -0.1748$, $\delta_2 = -0.0055$, $\delta_3 = -0.0461$, $\epsilon_2 = -0.0105$,

$$B_2 = \begin{bmatrix} 1 & 0 & 0 \\ 0 & -1 & 0 \\ 0 & 1 & -1 \\ 0 & 0 & 1 \\ 0 & 0 & 0 \end{bmatrix}, \quad C^T = \begin{bmatrix} 0 \\ 0 \\ 0 \\ 0 \\ 1 \end{bmatrix}, \quad \text{and} \quad K = \begin{bmatrix} -0.1748 \\ -0.0461 \\ -0.0105 \end{bmatrix}. \quad (17.14)$$

Applying a 90% step disturbance at the entry of ASP (i.e., 90% of the initial concentration of ASP), the following disturbance matrix is obtained:

$$B_1^T = \begin{bmatrix} 0.0710 & 0 & 0 & 0 & 0 \end{bmatrix}.$$

Applying a 90% step disturbance at the entry of ASPP, we obtain the following disturbance matrix:

$$B_1^T = \begin{bmatrix} -0.1122 & 0 & 0 & 0 & 0 \end{bmatrix}.$$

We begin by taking $Q = q I_5$, where q is a scalar in the feasibility interval $\in [0\ 1]$ and a control effort weight of $R = 1$. For the nominal system, we obtain, for example, a minimum disturbance H_∞ attenuation level of $\gamma = 0.357$, where a 90% ASPP step disturbance is applied. Next, we keep the same nominal gain vector of $K^T = \begin{bmatrix} -0.1748 & -0.0461 & -0.0105 \end{bmatrix}$ we introduce three uncertainty levels in the enzymes AKI, AKII, ASD, HDH and HK. In each uncertain case, we probe the four system norms. In Table 17.5 we bring the results of this test [excluding the case of HK], where f_1, f_2, f_3 represents 10, 50 and 90% uncertainty level, respectively, in the enzyme indicated next to these measures (thus $f_1[AKI]$ stands for 10% uncertainty in the concentration of the enzyme AKI, see Fig. 16.17).

The above procedure is repeated for a 90% step disturbance in the initial concentration of ASP and the obtained results are similar to those that are given in Table 17.5.

It can be seen in both tests that the H_∞ and the peak to peak norms are better suited to describe the system sensitivity to variations of its enzyme concentrations, since they match the reported experimental results concerning the system sensitivity to uncertainties in the various enzymes. It is also quite evident that the various systems norms are especially sensitive to variations in the concentration of AKI (compare, for example, the $L_\infty - L_\infty$ of Cases 1–3 in Table 17.5 relating to AKI to Cases 10–12 relating to HDH).

Table 17.5 Optimal analysis results for various uncertainties in the system enzyme concentrations following a step disturbance in the entry point of ASPP. Shown are the minimal attenuation levels γ for the feasibility interval of [0 1] for the H_∞, $L_2 - L_\infty$, and the $L_\infty - L_\infty$ system norms, and for the upper-bound on the H_2 norm. In all these tests the same feedback gain and initial steady-state values are kept constants

Case	Uncer. Par.	H_2	H_∞	$L_2 - L_\infty$	$L_\infty - L_\infty$
0	Nominal	0.078	0.357	0.072	0.377
1	$f_1[AKI]$	0.080	0.368	0.074	0.389
2	$f_2[AKI]$	0.086	0.409	0.080	0.432
3	$f_3[AKI]$	0.091	0.445	0.084	0.470
4	$f_1[AKIII]$	0.080	0.367	0.074	0.388
5	$f_2[AKIII]$	0.085	0.405	0.079	0.427
6	$f_3[AKIII]$	0.090	0.438	0.084	0.464
7	$f_1[ASD]$	0.080	0.360	0.074	0.383
8	$f_2[ASD]$	0.087	0.372	0.080	0.411
9	$f_3[ASD]$	0.096	0.436	0.086	0.497
10	$f_1[HDH]$	0.080	0.366	0.074	0.387
11	$f_2[HDH]$	0.088	0.392	0.080	0.415
12	$f_3[HDH]$	0.095	0.409	0.085	0.436

Decreasing the control effort weight for the nominal system leads, in the H_∞ case, in weakly decreased values of γ as follows: for $R = [0.01\ 0.1\ 1.0\ 10]$ we obtain $\gamma = [0.351\ 0.352\ 0.357\ 0.405]$, respectively. The same phenomenon is observed also for the other systems norms and for various measures of uncertainty.

"Detaching" each of the three feedback loops (setting its gain to zero), and correspondingly calculating the new coefficient values of the detached enzyme (by taking THR $= 0$ mM in its linearized model), a new feedback system emerges. On this new system we apply the above four optimality tests (that is the H_∞, H_2, $L_2 - L_\infty$ and the $L_\infty - L_\infty$ system norms). Detaching for example the loop of AKI, we set $\alpha_2 = 0$ and calculate α_1 and α_0 by taking THR $= 0$ in their derivations. Note that although the "detachment" removes the term of THR in the detached enzyme, it leaves the maximal velocity unchanged. The results obtained for the detached systems are given in Table 17.6 where a disturbance is applied via the ASPP entry point.

We note that in agreement with the results of [16], the detachment effect of AKI has the largest impact. We also note that detachments of the first two loops of the systems results in an elevated value of all the system norms.

The above "detachment" tests were readily repeated for a 90% initial concentration step disturbance at the entry of ASP. We obtain similar results to those obtained for the ASPP disturbance that are given in Table 17.6.

Table 17.6 Optimal analysis results for ASPP-disturbed detached systems. Shown are the minimal attenuation levels of γ for the feasibility interval of [0 1] for the H_∞, $L_2 - L_\infty$, and the $L_\infty - L_\infty$ system norm, and for the upper-bound on the H_2 norm

Case	Detached loop	H_2	H_∞	$L_2 - L_\infty$	$L_\infty - L_\infty$
0	Nominal	0.078	0.357	0.072	0.377
1	AKI	0.117	0.818	0.112	0.838
2	HDH	0.099	0.468	0.093	0.492
3	HK	0.078	0.357	0.072	0.377
4	AKI + HDH	0.196	1.83	0.193	1.86

17.7 The Threonine Synthesis—Controller Uncertainties

Following the state-space realization of the reaction system and the extraction of the controller gain matrix of (17.14), we apply the two most adequate optimality measures of Sect. 17.3. for further studies. Similarly to the previous section, in the H_∞ case, for example, we seek for a given interval for q and $R = I_3$, a minimal value for γ. We apply this measure, along with the peak to peak norm, to probe the closed-loop system sensitivity to uncertainties in the enzymes of the pathway—including several pairs. For each uncertain concentration of an enzyme, a set of two LMIs should be solved, each corresponds to edge point of the uncertainty interval. An uncertainty of $\pm 40\%$ in HDH, for example, is taken as an interval of $[0.6V_{HDH} \quad 1.4V_{HDH}]$ where V_{HDH} is the maximal velocity and where we note that this term is proportional to the concentration of HDH [16]. We thus obtain the system of (17.13) where the system matrices are given in (17.14). Applying a 90% step disturbance at the entry of ASPP, we obtain the following disturbance matrix:

$$B_1^T = \begin{bmatrix} -0.1122 \ 0 \ 0 \ 0 \ 0 \end{bmatrix}.$$

We begin by taking $Q = q I_5$, where q is a scalar in the feasibility interval $\in [0 \ 1]$ and a control effort weight of $R = I_3$. For the nominal system we obtain, for example, a minimum disturbance H_∞ attenuation level of $\gamma = 0.214$, where a 90% ASPP step disturbance is applied.

We concentrate next on the three enzymes that are the entry points of threonine, each of which constitutes a negative feedback loop. Here, we either keep the same nominal gain vector of $K^T = \begin{bmatrix} -0.1748 \ -0.0461 \ -0.0105 \end{bmatrix}$ or we introduce an uncertainty in the controller - resulting in two "terminal" controllers the K^- and the K^+. Thus, an uncertainty in AKI which causes 40% in the value α_2 results in the following two controllers $K^- = col\{0.6\alpha_2, \ \delta_3, \ \epsilon_2\}$ and $K^+ = col\{1.4\alpha_2, \ \delta_3, \ \epsilon_2\}$. In our study we introduce three uncertainty levels for all the tests. In each uncertain case we probe the two system norms. In Table 17.7 we bring the results of this test, where we apply 10, 50 and 90% uncertainty levels in the enzyme indicated in the first column of the table.

Table 17.7 Optimal analysis results for various uncertainties in the system enzyme concentrations, indicated by f—the uncertainty percentage, following a step disturbance in the entry point of ASPP. Shown are the minimal attenuation levels of γ for the feasibility interval of [0 1] for the H_∞ and the $L_\infty - L_\infty$, $(P2P)$ system norms, where three feedback gains were applied: K^+ stands for the upper gain controller, K^- stands for the lower gain one and .nom stands for the nominal controller. In all these tests the same initial steady- state values are kept constants

Enzyme	f (%)	H_∞ (nom.)	H_∞ (K^+)	H_∞ (K^-)
AKI	10	0.369	0.351	0.388
AKI	50	0.410	0.334	0.54
AKI	90	0.364	0.267	0.617
HDH	10	0.366	0.360	0.372
HDH	50	0.393	0.369	0.420
HDH	90	0.259	0.237	0.291
Enzyme	f (%)	$P2P$ (nom.)	$P2P$ (K^+)	$P2P$ (K^-)
AKI	10	0.488	0.475	0.503
AKI	50	0.530	0.475	0.627
AKI	90	0.566	0.483	0.818
HDH	10	0.490	0.485	0.495
HDH	50	0.529	0.509	0.552
HDH	90	0.557	0.525	0.607

Next, we apply the above test, indicated in Table 17.7, for the cases where specific pairs of enzymes concentrations are uncertain. We choose four pairs where in each case, only one enzyme of the pair is subject to a negative loop. In Tables 17.8 and 17.9, we bring the results of this test where we probe, as in Table 17.7, the H_∞ and the peak to peak norms. We thus note, in Tables 17.8 and 17.9, that for almost each row in the table, the attenuation level achieved by the K^- and by K^+ slightly deviate for the attenuation level of γ achieved by the "nominal" controller. We note that in the case of 90% uncertainty interval in the first enzyme AKI, the deviation is certainly more pronounced in the K^- case (rows 3 and 6 from the top in the H_∞ test [upper table] and in the peak to peak test [lower table]). Note also that the latter large deviation is rather moderate in the case of 90% uncertainty in the second enzyme that closes the loop—HDH. Similar results to the HDH case were obtained for the third enzyme that closes the loop—the HK enzyme [not shown in Tables 17.8 and 17.9]. Comparing the effect of the added uncertainty in an enzyme that does not affect the controller gain to the case where the uncertainty is only exerted on the enzyme that changes the controller gain, one can see that the added uncertainty has a very slight effect on the attenuation levels. Compare, for example, the results of the peak-to-peak case [the lower part of the table]. The attenuation levels for the AKI + AKIII and the AKI + ASD is very similar for all the uncertainty intervals. Similar results are obtained for the enzyme of the second loop centered around HDH. The pairs of HDH + AKIII and HDH + ASD show very similar attenuation levels in each level of the uncertainties indicated in Tables 17.8 and 17.9.

Table 17.8 Optimal analysis results for various uncertainties [indicated by f] in several pairs of the pathway enzymes, following a step disturbance in the entry point of ASPP. Shown are the minimal attenuation levels of γ for the feasibility interval of [0 1] for the H_∞ case, where three feedback gains were applied: K^+ stands for the upper gain controller, K^- stands for the lower gain one and .nom stands for the nominal controller. In all these tests the same initial steady- state values are kept constants

Enzyme	f (%)	H_∞ (n.)	H_∞ (K^+)	H_∞ (K^-)
AKI + AKIII	10	0.248	0.236	0.262
AKI + AKIII	50	0.336	0.269	0.456
AKI + AKIII	90	0.422	0.295	0.871
AKI + ASD	10	0.246	0.234	0.260
AKI + ASD	50	0.323	0.263	0.430
AKI + ASD	90	0.464	0.363	0.733
HDH + AKIII	10	0.236	0.232	0.240
HDH + AKIII	50	0.269	0.255	0.284
HDH + AKIII	90	0.296	0.279	0.359
HDH + ASD	10	0.234	0.230	0.238
HDH + ASD	50	0.259	0.243	0.278
HDH + ASD	90	0.317	0.299	0.347

Table 17.9 Optimal analysis results for various uncertainties in several pairs of the pathway enzymes for the P2P case [see the legend of Table 17.8 for details]

Enzyme	f (%)	$P2P$ (n.)	$P2P$ K^+	$P2P$ K^-
AKI + AKIII	10	0.331	0.322	0.341
AKI + AKIII	50	0.448	0.398	0.538
AKI + AKIII	90	0.567	0.481	0.925
AKI + ASD	10	0.330	0.321	0.340
AKI + ASD	50	0.440	0.395	0.520
AKI + ASD	90	0.571	0.504	0.783
HDH + AKIII	10	0.318	0.315	0.321
HDH + AKIII	50	0.380	0.368	0.394
HDH + AKIII	90	0.452	0.430	0.501
HDH + ASD	10	0.317	0.314	0.320
HDH + ASD	50	0.372	0.358	0.387
HDH + ASD	90	0.434	0.415	0.460

17.8 Conclusions

In this chapter, we have applied optimal control theory tools to asses the sensitivity of both the yeast glycolytic pathway (as presented via the minimal model of [13]) and the threonine synthesis pathway in the H_2, $L_2 - L_\infty$, $L_\infty - L_\infty$ and the

H_∞-norm senses, to various enzymatic kinetic rate constants uncertainties of the system, following a disturbance in the system input Glucose or ASPP, respectively. The H_2-norm test is also applied and was shown to largely agree with the $L_2 - L_\infty$ results in both pathways, rendering these two system norms as improper ones in the latter pathways.

Following a simple Taylor expansion of the system kinetics, we have obtained a linear model for the system from which we have derived the nominal controller gain matrix. We then apply the various optimality tests to various uncertainties in the system. The numerical results that were obtained clearly suggest that the nominal controlled system (i.e., with the nominal controller) is sensitive to the various enzyme kinetic constants uncertainties that were examined, judging via the H_∞ and the $L_\infty - L_\infty$ norms. It has also been shown that the sensitivity of the system to kinetic uncertainties is not similar for all the enzymatic kinetic constants that were tested.

The main finding of this chapter points at the issue of the adequate input–output relation in the system that better represents the sensitivity of the system to parameter uncertainties. This is stated in the following conjecture:

Conjecture 17.1 *The sensitivity of feedback-control-based biochemical pathways to various parameter uncertainties—judging form both the threonine synthesis and the glycolytic pathway—may be optimal in the H_∞ or the peak to peak optimal controlsense.*

It is interesting that both the H_∞ and the $L_\infty - L_\infty$ system norms have similar tendencies in the threonine synthesis and the glycolytic pathways. It may be that the system "seeks" a worst disturbance attenuation while "engaging" in the peak-to-peak input–output relation. Clearly, the results achieved in this study indicate the need for a further research of additional important issues such as the relative significance of the various negative loops that were missing in the minimal model of [13], in the glycolysis case.

References

1. Savageau, M.A.: Biochemical systems analysis. I. Some mathematical properties of the rate law for the component enzymatic reactions. J. Theor. Biol. **25**, 365–369 (1969)
2. Savageau, M.A.: Biochemical systems analysis. II. The steady-state solutions for an n-pool system using a power-law approximation. J. Theor. Biol. **25**, 370–379 (1969)
3. Savageau, M.A.: Biochemical systems analysis. 3. Dynamic solutions using a power-law approximation. J. Theor. Biol. **26**, 215–226 (1970)
4. Savageau, M.A.: Biochemical Systems Analysis: A Study of Function and Design in Molecular Biology. Addison Wesley, Reading (1976)
5. Alves, R., Savageau, M.A.: Effect of overall feedback inhibition in unbranched biosynthetic pathways. Biophys. J. **79**, 2290–2304 (2000)
6. Voit, E.O.: Biochemical systems theory: a review. ISRN Biomath. **2013**, Article ID 897658, 53 (2013). https://doi.org/10.1155/2013/897658
7. Fell, D.: Metabolic control analysis: a survey of it's theoretical and experimental development. Biochem. J. **286**, 313–330 (1992)

8. Fell, D.: Frontiers in Metabolism: Understanding the Control of Metabolism. Portland Press, London (1997)
9. Voit, E.O.: Computational Analysis of Biochemical Systems: A Practical Guide for Biochemists and Molecular Biologists. Cambridge University Press, Cambridge (2000)
10. Goel, G.: Reconstructing Biochemical Systems. Systems Modeling and Analysis Tools for Decoding Biological Designs. VDM Verlag Dr. Müller, Saarbrcken (2008)
11. Torres, N.V., Voit, E.O.: Pathway Analysis and Optimization in Metabolic Engineering. Cambridge University Press, Cambridge (2005)
12. Lehninger, A.L.: Principles of Biochemistry. Worth Publishers, New York (1982)
13. Wolf, J., Passarge, J., Somsen, O.J.G., Snoep, J.L., Heinrich, R., Westerho, H.V.: Transduction of intracellular and intercellular dynamics in yeast glycolytic oscillations. Biophys. J. **78**, 1145–1153 (2000)
14. Gershon, E., Yokev, O., Shaked, U.: H_∞ feedback-control of the threonine synthesis pathway via system uncertainty. In: Proceedings of the European Control Conference (ECC07), Kos, Greece (2007)
15. Gershon, E., Shaked, U.: H_∞ feedback-control theory in biochemical systems. Int. J. Robust Nonlinear Control **18**, 18–50 (2008)
16. Chassagnole, C., Rais, B., Quentin, E., Fell, D., Mazat, J.P.: An integrated study of threonine-pathway enzyme kinetics in Echerichia coli. Biochem. J. **356**, 415–423 (2001)
17. Chassagnole, C., Rais, B., Quentin, E., Fell, D., Mazat, J.P.: Threonine synthesis from aspartate in Echerichia coli cell-free extract: pathway dynamics. Biochem. J. **356**, 425–432 (2001)
18. Chassagnole, C., Rais, B., Quentin, E., Fell, D., Mazat, J.P.: Control of threonine-synthesis pathway in Echerichia coli: theoretical and experimental approach. Biochem. J. **356**, 433–444 (2001)
19. Gershon, E., Shaked, U.: Robust polytopic analysis of the feedback-control of glycolysis in yeasts via some system norms. In: Proceedings of the 20th Mediterranean Conference on Control and Automation (MED12), Barcelona, Spain (2012)
20. Gershon, E., Navon, M.: Robust feedback-control analysis of the threonine synthesis pathway via various system norms. In: Proceedings of the 22nd Mediterranean Conference on Control and Automation (MED14), Palermo, Sicily (2014)
21. Gershon, E., Navon, M., Shaked, U.: Robust peak to peak and H_∞ static output-feedback control of the threonine synthesis pathway. In: Proceedings of the European Control Conference (ECC15), Linz, Austria (2015)
22. Scherer, C., Weiland, S.: Linear matrix inequalities in control. http://www.imng.uni-stuttgart.de/simtech/Scherer/lmi/notes05.pdf
23. Segel, I.R.: Enzyme Kinetics - Behavior and Analysis of Rapid Equilibrium and Steady-State Enzyme Systems. Wiley, New York (1975)

Appendix A
Stochastic State-Multiplicative Noisy Systems

A.1 Introduction

In this Appendix, we bring some basic results concerning stochastic differential equations of the Ito type in which the state-multiplicative systems constitute a special case of. Stochastic differential equations received a comprehensive treatment in [1], mainly aimed at providing a rigorous framework for optimal state estimation of nonlinear stochastic processes. In the present Appendix, we provide only the main facts that are required to assimilate the main concepts and results which are useful in deriving optimal estimators and controllers for linear systems with state-multiplicative white noise. While the expert reader may skip this appendix, graduate students or practicing engineers may find it to be a useful summary of basic facts and concepts, before they read the text of [1]. Comprehensive treatment of stochastic differential equations in a form accessible to graduate students and practicing engineers is given also in [2] where the close connections between Ito-type stochastic differential equations and statistical physics are explored and where a few additional topics such as stochastic stability are covered. Also in [2], many communications oriented examples of nonlinear estimation theory can be found.

A.2 Stochastic Processes

Stochastic processes are a family of random variables parameterized by time $t \in \mathcal{T}$. Namely, at each instant t, $x(t)$ is a random variable. When t is continuous (namely $\mathcal{T} = \mathcal{R}$), we say that $x(t)$ is a continuous-time stochastic process, and if t is discrete (namely $\mathcal{T} = \{1, 2, \ldots\}$), we say that $x(t)$ is a discrete-time variable. For any finite set of $\{t_1, t_2, \ldots t_n\} \in \mathcal{T}$, we can define the joint distribution $F(x(t_1), x(t_2), \ldots, x(t_n))$ and the corresponding joint density $p(x(t_1), x(t_2), \ldots, x(t_n))$.

© Springer Nature Switzerland AG 2019
E. Gershon and U. Shaked, *Advances in H_∞ Control Theory*,
Lecture Notes in Control and Information Sciences 481,
https://doi.org/10.1007/978-3-030-16008-1

The first- and the second-order distribution functions, $p(x(t))$ and $p(x(t), x(\tau))$, respectively, play an important role in our discussion. Also, the mean $m_x(t) \overset{\Delta}{=} E\{x(t)\}$ and the autocorrelation $\gamma_x(t, \tau) \overset{\Delta}{=} E\{x(t)x(\tau)\}$ are useful characteristics of the stochastic process $x(t)$. When $x(t)$ is vector valued, the autocorrelation is generalized to be $\Gamma_x(t, \tau) = E\{x(t)x(\tau)^T\}$. The covariance matrix of a vector- valued stochastic process $x(t)$ is a measure of its perturbations with respect to its mean value and is defined by $P_x(t) \overset{\Delta}{=} E\{(x(t) - m_x(t))(x(t) - m_x(t))^T\}$.

A process $x(t)$ is said to be stationary if

$$p(x(t_1), x(t_2), \ldots, x(t_n)) = p(x(t_1 + \tau), x(t_2 + \tau), \ldots, x(t_n + \tau))$$

for all n and τ. If the latter is true only for $n = 1$, then the process $x(t)$ is said to be stationary of order 1 and then $p(x(t))$ does not depend on t. Consequently, the mean $m_x(t)$ is constant and $p(x(t), x(\tau))$ depends only on $t - \tau$. Also in such a case, the autocorrelation function of two time instants depends only on the time difference, namely $\gamma_x(t, t - \tau) = \gamma_x(\tau)$.

An important class of stochastic processes is one of Markov processes. A stochastic process $x(t)$ is called a Markov process if for any finite set of time instants $t_1 < t_2 < \cdots < t_{n-1} < t_n$ and for any real λ it satisfies

$$Pr\{x(t_n) < \lambda | x(t_1), x(t_2), \ldots, x(t_{n-1}), x(t_n)\} = Pr\{x(t_n) | x(t_{n-1})\}.$$

Stochastic processes convergence properties of a process $x(t)$ to a limit x can be analyzed using different definitions. The common definitions are almost sure or with probability 1 convergence (namely $x(t) \to x$ almost surely, meaning that this is satisfied except for an event with a zero probability), convergence in probability (namely for all $\epsilon > 0$, the probability of $|x(t) - x| \geq \epsilon$ goes to zero), and mean square convergence, where given that $E\{x(t)^2\}$ and $E\{x^2\}$ are both finite, $E\{(x(t) - x)^2\} \to 0$. In general, almost sure convergence neither implies nor it is implied by mean square convergence, but both imply convergence in probability. In the present book, we adopt the notion of mean square convergence and the corresponding measure of stability, namely mean square stability.

A.3 Mean Square Calculus

Dealing with continuous-time stochastic processes in terms of differentiation, integration, etc., is similar to the analysis of deterministic functions, but it requires some extra care in evaluation of limits. One of the most useful approaches to calculus of stochastic processes is the so-called mean square calculus where mean square convergence is used when evaluating limits.

The full scope of mean square calculus is covered in [1, 2] but we bring here only a few results that are useful to our discussion.

The notions of mean square continuity and differentiability are key issues in our discussion. A process $x(t)$ is said to be mean square continuous if $lim_{h \to 0} x(t + h) = x(t)$. It is easy to see that if $\gamma_x(t, \tau)$ is continuous at (t, t), then also $x(t)$ is mean square continuous. Since the converse is also true, then mean square continuity of $x(t)$ is equivalent to continuity of $\gamma(t, \tau)$ in (t, t). Defining mean square derivative by the mean square limit as $h \to 0$ of $(x(t + h) - x(t))/h$, then it is similarly obtained that $x(t)$ is mean square differentiable (i.e., its derivative exists in the mean square sense) if and only if $\gamma_x(t, \tau)$ is differentiable at (t, t). A stochastic process is said to be mean square integrable, whenever $\sum_{i=0}^{n-1} x(\tau_i)(t_{i+1} - t_i)$ is mean square convergent where $a = t_0 < t_1 < \cdots < t_n = b$, where $\tau_i \in [t_i, t_{i+1}]$ and where $|t_{i+1} - t_i| \to 0$. In such a case, the resulting limit is denoted by $\int_a^b x(t)dt$. It is important to know that $x(t)$ is mean square integrable on $[a, b]$ if and only if $\gamma_x(t, \tau)$ is integrable on $[a, b] \times [a, b]$. The fundamental theorem of mean square calculus then states that if $\dot{x}(t)$ is mean square integrable on $[a, b]$, then for any $t \in [a, b]$, we have

$$x(t) - x(a) = \int_a^t \dot{x}(\tau)d\tau.$$

The reader is referred to [1] for a more comprehensive coverage of mean square calculus.

A.4 Wiener Process

A process $\beta(t)$ is said to be a Wiener Process (also referred to as Wiener–Levy process or Brownian motion) if it has the initial value of $\beta(0) = 0$ with probability 1, has stationary independent increments and is normally distributed with zero mean for all $t \geq 0$. The Wiener process has then the following properties: $\beta(t) - \beta(\tau)$ is normally distributed with zero mean and variance $\sigma^2(t - \tau)$ for $t > \tau$ where σ^2 is an empirical positive constant. Consider now for $t > \tau$, the autocorrelation is

$$\gamma_\beta(t, \tau) = E\{\beta_t \beta_\tau)\} = E\{(\beta(t) - \beta(\tau) + \beta(\tau))\beta(\tau)\}$$
$$= E\{(\beta(t) - \beta(\tau))\beta(\tau)\} + E\{\beta^2(\tau)\}.$$

Since the first term is zero, due to the independent increments property of the Wiener process, it is readily obtained that $\gamma_\beta(t, \tau) = \sigma^2 \tau$. Since we have assumed that $t > \tau$ we have in fact that $\gamma_\beta(t, \tau) = \sigma^2 min(t, \tau)$. Since the latter is obviously continuous at (t, t), it follows that $\beta(t)$ is mean square continuous. However, a direct calculation (see [1]) of the second order derivative of $\gamma_\beta(t, \tau)$, with respect to t and τ at (t, t), shows that

$$\frac{min(t + h, t + h') - min(t, t)}{hh'} = 1/max(h, h'),$$

which is clearly unbounded as h and h' tend to zero. Therefore, $\gamma_\beta(t, \tau)$ is not differentiable at any (t, t) and consequently $\beta(t)$ is not mean square differentiable anywhere. It is, therefore, concluded that the Wiener process is continuous but not differentiable in the mean square sense. In fact, it can be shown that the latter conclusion holds also in the sense of almost sure convergence.

A.5 White Noise

We begin this section by considering discrete-time white-noise type stochastic processes. A discrete-time process is said to be white if it is a Markov process and if all $x(k)$ are mutually independent. Such a process is said to be a white Gaussian noise if, additionally, its samples are normally distributed. The mutual independence property leads, in the vector-valued case, to $E\{x(n)x^T(m)\} = Q_n\delta_{n,m}$ where $\delta_{n,m}$ is the Kronecker delta function (1 for equal arguments and zero otherwise) and where $Q_n \geq 0$. The discrete-time white noise is a useful approximation of measurement noise in many practical cases. Its continuous-time analog also appears to be useful. Consider a stationary process $x(t)$ whose samples are mutually independent, taken at large enough intervals. Namely,

$$\gamma(\tau) = E\{x(t + \tau)x(t)\} = \sigma^2 \frac{\rho}{2} e^{-\rho|\tau|},$$

where $\rho >> 1$. As ρ tends to infinity $\gamma(\tau)$ rapidly decays as a function of τ, and therefore the samples of $x(t)$ become virtually independent and the process becomes white. Noting that for ρ that tends to infinity, $\frac{\rho}{2} e^{-\rho|\tau|} \to \delta(\tau)$ where δ is the Dirac delta function [1], a vector valued white process $x(t)$ is formally considered to have the autocorrelation of $\gamma(\tau) = Q(t)\delta(\tau)$ where $Q(t) \geq 0$. Namely, $E\{x(t)x(\tau)\} = Q(t)\delta(t - \tau)$. Defining the spectral density of $x(t)$ by the Fourier transform of its autocorrelation, namely by

$$f(\omega) = \int_{-\infty}^{\infty} e^{-i\tau\omega}\sigma^2 \frac{\rho}{2} e^{-\rho|\tau|}d\tau = \frac{\sigma^2}{1 + \omega^2/\rho^2},$$

we see that this spectral density is constant and has the value of σ^2 up to about the frequency ρ where it starts dropping to zero. Namely, the spectrum of $x(t)$ is nearly flat independently of the frequency, which is the source of the name "white" noise, in analogy to white light including all frequencies or wavelengths. We note that for finite $\rho >> 1$, $x(t)$ is said to be a wide-band noise (where 1 may represent the measured process bandwidth and ρ the measurement noise bandwidth). In such a case, modeling $x(t)$ as a white noise is a reasonable approximation. We note, however, that constant $f(\omega)$ or white spectrum corresponds to infinite energy by Parseval's theorem . Alternatively, looking at the autocorrelation at $\tau = 0$, we see that

$$\gamma(0) = E\{x^2(t)\} = \frac{1}{2\pi} \int_{-\infty}^{\infty} f(\omega)d\omega \to \infty.$$

Therefore, white noise is not physically realizable but is an approximation to wide-band noise. To allow mathematical manipulations of white noise, we relate it to Wiener processes which are well defined. To this end, we recall that the autocorrelation of a Wiener process $\beta(t)$ is given by

$$\gamma(t, \tau) = E\{x(t)x(\tau)\} = \sigma^2 min(t, \tau).$$

Since expectation and derivatives can be interchanged, namely

$$E\left\{\frac{d\beta(t)}{dt}\frac{d\beta(\tau)}{d\tau}\right\} = \frac{d^2}{dtd\tau}E\{\beta_t\beta_\tau\},$$

it follows that the autocorrelation of $\dot{\beta}(t)$ is given by $\sigma^2\frac{d}{d\tau}[\frac{d}{dt}min(t, \tau)]$. However, $min(t, \tau)$ is τ for $\tau < t$ and t otherwise; therefore, its partial derivative with respect to t is a step function of τ rising from 0 to 1 at $\tau = t$. Consequently, the partial derivative of this step function is just $\sigma^2\delta(t - \tau)$. The autocorrelation of $\dot{\beta}(t)$ is thus $\sigma^2\delta(t - \tau)$, just as the autocorrelation of white noise, and we may, therefore, formally conclude that white noise is the derivative, with respect to time, of a Wiener process.

A.6 Ito Lemma

Ito lemma is a key lemma which is widely used in the present monograph to evaluate differentials of nonlinear scalar valued functions $\varphi(x(t))$ of solutions $x(t)$ of Ito type stochastic differential equations. Consider a scalar process $x(t)$ which satisfies

$$\frac{dx}{dt} = f(x(t), t) + g(x(t), t)\dot{\beta}(t).$$

Then, using Taylor expansion, we have

$$d\varphi = \varphi_t dt + \varphi_x dx + \frac{1}{2}\varphi_{xx}dx^2 + \frac{1}{3}\varphi_{xxx}dx^3 + \cdots.$$

Discarding terms of the order $o(dt)$, recalling that $d\beta^2(t)$ is of the order of dt, and substituting for dx in the above Taylor expansion, it is found that [1]

$$d\varphi = \varphi_t dt + \varphi_x dx + \frac{1}{2}\varphi_{xx}g^2 d\beta^2(t).$$

Substituting $\sigma^2 dt$ for $d\beta^2(t)$ we obtain

$$d\varphi = \varphi_t dt + \varphi_x dx + \frac{\sigma^2}{2}\varphi_{xx}g^2 dt.$$

For vector valued $x(t)$, where $Qdt = E\{d\beta d\beta^T\}$, the latter result reads:

$$d\varphi = \varphi_t dt + \varphi_x dx + \frac{1}{2}Tr\{gQg^T\varphi_{xx}\}dt,$$

where φ_{xx} is the Hessian of φ with respect to x.

A.7 Application of Ito Lemma

Ito lemma is useful in evaluating the covariance of state multiplicative processes considered in the present monograph. Consider

$$dx = Axdt + Dxd\xi + Bd\beta,$$

where ξ is a scalar-valued standard Wiener process and where β is also a scalar Wiener process independent of ξ so that $E\{d\beta d\beta^T\} = Qdt$. We define $w = col\{\beta, \ \xi\}$. The intensity of w is given by $\tilde{Q} \triangleq \begin{bmatrix} Q & 0 \\ 0 & 1 \end{bmatrix}$. We also define $\tilde{G} = [\ B \ Dx\ ]$.

Defining $\varphi(x(t)) = x_i(t)x_j(t)$, where x_i is the ith component of x, we get (see [1]), using Ito lemma, that

$$d\varphi = x_i(t)dx_j(t) + x_j(t)dx_i(t) + \frac{1}{2}Tr\{\tilde{G}\tilde{Q}\tilde{G}^T\Sigma\},$$

where the only nonzero entries in Σ are at locations i, j and j, i. Consequently, we have that

$$d(xx^T) = (xdx^T + dxx^T + \tilde{G}\tilde{Q}\tilde{G}^T)dt.$$

Taking the expectation of both sides in the latter, and defining $P(t) = E\{x(t)x^T(t)\}$, the following result is obtained.

$$\dot{P} = AP + PA^T + BQB^T + DPD^T.$$

References

1. Jazwinsky, A.H.: Stochastic Processes and Filtering Theory. Academic Press (1970)
2. Schuss, Z.: Theory and Applications of Stochastic Differential Equations. Wiley, New York (1980)

Appendix B
The Input–Output Approach to Retarded Systems

B.1 Introduction

In Chap. 12, we apply the input–output approach to linear time invariant delayed systems. This approach transforms a given delayed system to a norm-bounded uncertain system that can be treated, in the stochastic context, by the various solutions methods that can be found in [1]. The major advantage of the input–output approach lies in its simplicity of use such that the resulting inequalities that emerge are relatively tractable and simple, for both: delay dependent and delay-independent solutions. However, this technique entails some degree of conservatism that can be compensated by a clever choice of the Lyapunov function that is involved in the solution method. In the following two subsections we introduce the input–output approach for continuous-time and discrete-time stochastic systems.

B.1.1 Continuous-Time Case

We consider the following system:

$$
\begin{aligned}
dx(t) = {} & [A_0 x(t) + A_1 x(t - \tau(t))]dt + H x(t - \tau(t))d\zeta(t) \\
& + G x(t)d\beta(t), \quad x(\theta) = 0, \theta \le 0,
\end{aligned}
\tag{B.1}
$$

where $x(t) \in R^n$ is the state vector and A_0, A_1 and G, H are time invariant matrices and where $\beta(t)$, $\zeta(t)$ are zero-mean real scalar Wiener processes satisfying:

$$
\mathcal{E}\{\beta(t)\beta(s)\} = min(t, s), \quad \mathcal{E}\{\zeta(t)\zeta(s)\} = min(t, s),
$$

$$
\mathcal{E}\{\beta(t)\zeta(s)\} = \bar{\alpha} \cdot min(t, s), \quad |\bar{\alpha}| \le 1.
$$

© Springer Nature Switzerland AG 2019
E. Gershon and U. Shaked, *Advances in H∞ Control Theory*,
Lecture Notes in Control and Information Sciences 481,
https://doi.org/10.1007/978-3-030-16008-1

In (B.1), $\tau(t)$ is an unknown time delay which satisfies:

$$0 \le \tau(t) \le h, \quad \dot{\tau}(t) \le d < 1. \tag{B.2}$$

In the input–output approach, we use the following operators:

$$(\Delta_1 g)(t) \overset{\Delta}{=} g(t - \tau(t)), \quad (\Delta_2 g)(t) \overset{\Delta}{=} \int_{t-\tau(t)}^{t} g(s)ds. \tag{B.3}$$

In what follows we use the fact that the induced L_2-norm of Δ_1 is bounded by $\frac{1}{\sqrt{1-d}}$, and similarly to [2], the fact that the induced L_2-norm of Δ_2 is bounded by h. Using the above operator notations, the system (B.1) becomes a special case of the following system:

$$\begin{aligned}
dx(t) &= [A_0 + m]x(t)dt + (A_1 - m)w_1(t)dt - mw_2(t)dt + Gx(t)d\beta(t) \\
&\quad + Hw_1(t)d\zeta(t) - \Gamma_\beta dt - \Gamma_\zeta dt, \\
\bar{y}(t) &= [A_0 + m]x(t) + (A_1 - m)w_1(t) - mw_2(t) - \Gamma_\beta - \Gamma_\zeta,
\end{aligned} \tag{B.4}$$

where

$$\Gamma_\beta = m\int_{t-\tau}^{t} Gx(s)d\beta(s), \text{ and } \Gamma_\zeta = m\int_{t-\tau}^{t} Hw_1(s)d\zeta(s), \tag{B.5}$$

and where

$$w_1(t) = (\Delta_1 x)(t), \text{ and } w_2(t) = (\Delta_2 \bar{y})(t). \tag{B.6}$$

Remark B.1 The dynamics of (B.1) is a special case of that of (B.4) as follows: Noting (B.6) and applying the operators of (B.3), Equation (B.4a) can be written as

$$dx(t) = [A_0 + m]x(t)dt + (A_1 - m)w_1(t)dt - m\{\int_{t-\tau}^{t} \bar{y}(t^\cdot)dt^\cdot\}dt$$

$$+ Gx(t)d\beta(t) + Hw_1(t)d\zeta - \Gamma_\beta dt - \Gamma_\zeta dt, \quad w_1(t) = x(t - \tau(t)).$$

Now, recalling $\bar{y}$ of (B.4b) one can write

$$dx(t) = \bar{y}(t)dt + Gx(t)d\beta(t) + Hw_1(t)d\zeta(t)$$

and therefore

$$\bar{y}(t^\cdot)dt^\cdot = dx(t^\cdot) - Gx(t^\cdot)d\beta(t^\cdot) - Hw_1(t^\cdot)d\zeta(t^\cdot).$$

Hence,

$$-mw_2(t) \stackrel{\Delta}{=} -m \int_{t-\tau}^{t} \bar{y}(t')dt' = -m \int_{t-\tau}^{t} \{dx(t') - Gx(t')d\beta(t') - Hw_1(t')d\zeta(t')\}$$

$$= -mx(t) + mx(t-\tau) + \Gamma_\beta + \Gamma_\zeta = -mx(t) + mw_1(t) + \Gamma_\beta + \Gamma_\zeta,$$

where Γ_β and Γ_ζ are defined in (B.5a, b) respectively. Replacing the right-hand side of the latter equation for $-mw_2(t)$ in (B.4a), the dynamics of (B.1a) is recovered.

We note that the matrix m is a $n \times n$ unknown constant matrix to be determined. This matrix is introduced into the dynamics of (B.4) in order to achieve additional degree of freedom in the design of the various controllers in all the linear continuous time-delayed systems considered in this book. Using the fact that $||\Delta_1||_\infty \leq \frac{1}{\sqrt{1-d}}$ and $||\Delta_2||_\infty \leq h$, (B.4) may be cast into what is entitled: the norm-bounded uncertain model, by introducing into (B.1) the above new variables of (B.6) where $||\Delta_1||_\infty \leq \frac{1}{\sqrt{1-d}}$ and $||\Delta_2||_\infty \leq h$ are diagonal operators having identical scalar operators on the main diagonal.

B.1.2 Discrete-Time Case

We consider the following linear retarded system:

$$\begin{aligned} x_{k+1} &= (A_0 + Dv_k)x_k + (A_1 + F\mu_k)x_{k-\tau(k)}, \\ x_l &= 0, \ l \leq 0, \end{aligned} \tag{B.7}$$

where $x_k \in \mathcal{R}^n$ is the system state vector and where the time delay is denoted by the integer τ_k and it is assumed that $0 \leq \tau_k \leq h$, $\forall k$. The variables $\{\mu_k\}$ and $\{v_k\}$ are zero-mean real scalar white-noise sequences that satisfy:

$$E\{v_k v_j\} = \delta_{kj}, \ E\{\mu_k \mu_j\} = \delta_{kj},$$

$$E\{\mu_k v_j\} = 0, \ \forall k, j \ \geq 0.$$

The matrices in (B.7) are constant matrices of appropriate dimensions.

In order to tackle the stability (and hence the BRL) of the retarded discrete-time system, we introduce the following scalar operators which are needed, in the sequel, for transforming the delayed system to an equivalent norm-bounded nominal system:

$$\Delta_1(g_k) = g_{k-h}, \quad \Delta_2(g_k) = \sum_{j=k-h}^{k-1} g_j. \tag{B.8}$$

Denoting

$$\bar{y}_k = x_{k+1} - x_k$$

and using the fact that

$$\Delta_2(\bar{y}_k) = x_k - x_{k-h},$$

the following state-space description of the system is obtained:

$$x_{k+1} = (A_0 + Dv_k + M)x_k + (A_1 - M + F\mu_k)\Delta_1(x_k) - M\Delta_2(\bar{y}_k)$$

$$x_l = 0, \ l \le 0,$$

where the matrix M is a free decision variable to be determined. Similarly to the continuous-time case, this matrix is introduced into the dynamics of (B.7) in order to achieve additional degree of freedom in the design of the state-feedback controller or estimators for retarded stochastic discrete-time systems.

References

1. Gershon, E., Shaked, U.: Advanced Topics in Control and Estimation of State-Multiplicative Noisy Systems. LANCIS - Lecture Notes in Control and Information Sciences, vol. 439. Springer, Berlin, (2013)
2. Kao, C.Y., Lincoln, B.: Simple stability criteria for systems with time-varying delays. Automatica **40**, 1429–1434 (2004)

Index

© Springer Nature Switzerland AG 2019
E. Gershon and U. Shaked, *Advances in H∞ Control Theory*,
Lecture Notes in Control and Information Sciences 481,
https://doi.org/10.1007/978-3-030-16008-1

Printed in the United States
By Bookmasters